Landscapes and Labscapes

One explorer of the lab-field frontier, William Ganong, 1901. New Brunswick Museum, St. John, New Brunswick (William Francis Ganong Collection).

Landscapes & Labscapes

Exploring the Lab-Field Border in Biology

Robert E. Kohler

The University of Chicago Press
Chicago and London

Robert E. Kohler is professor of the history and sociology of science at the University of Pennsylvania and author of *Lords of the Fly: Drosophila Genetics and the Experimental Life,* published by the University of Chicago Press.

The University of Chicago Press, Chicago 60637
The University of Chicago Press, Ltd., London

Printed in the United States of America

11 10 09 08 07 06 05 04 03 02   1 2 3 4 5
ISBN: 0-226-45009-0 (cloth)
ISBN: 0-226-45010-4 (paper)

Library of Congress Cataloging-in-Publication Data

Kohler, Robert E.
Landscapes and labscapes : exploring the lab-field border in biology / Robert E. Kohler.
p. cm.
Includes bibliographical references (p.).
ISBN 0-226-45009-0 (cloth : alk. paper) — ISBN 0-226-45010-4 (pbk. : alk. paper)
1. Biology—Field work. I. Title.

QH318.5 .K58 2002
570'.7'2—dc21

2002023331

⊗ The paper used in this publication meets the minimum requirements of the American National Standard for Information Sciences—Permanence of Paper for Printed Library Materials, ANSI Z39.48-1992.

To Frances
la miglior fabbro

CONTENTS

List of Illustrations *xi*
Preface *xiii*

Chapter 1 Borders and History ***1***
Field and Lab in Biology *1*
Lab and Field as Place *6*
Borders and Frontiers *11*
An Overview, with Caveats *19*

Chapter 2 A New Natural History ***23***
The New Natural History *25*
New Naturalists and Educational Reform *35*
Labscapes: Marine Stations and Biological Farms *41*
Labscapes: Vivaria and Field Stations *48*
Conclusion *55*

Chapter 3 Border Crossings ***60***
The Rise and Decline of Biometry *63*
Ecology: Physiology or Natural History? *74*
Up "Brush Creek" and Back Again *78*
Guardians of the Faith: Genetics and Physiology *88*
Conclusion *95*

Chapter 4 Taking Nature's Measure ***97***
Trust in Numbers: Quadrats *100*
Taking Nature's Measure: Instruments *108*
Plant-Machine: Atmometers and Phytometers *118*

Instrumental Eye: The Camera *124*
Making the Place Right: Forrest Shreve *127*
Conclusion *132*

Chapter 5 Experiments in Nature *135*
Experimental Evolution *138*
Shifting Ground: Field and Lab *145*
Experiments in Nature: Ecology *154*
Experimental Taxonomy: Harvey M. Hall and Jens Clausen *163*
Conclusion *172*

Chapter 6 Troubled Lives *175*
Midlife Crises: Ecology *177*
Ends and Means: Experimental Evolution *190*
Identity *194*
Genetics, True and False *199*
Physiologists and the Field *205*
Conclusion *210*

Chapter 7 Nature's Experiments *212*
Nature's Experiments *214*
Places in Process *218*
Reading Places *224*
Physiographic Ecology *230*
Panoramas *238*
Evolution *241*
Conclusion *248*

Chapter 8 Border Practices *252*
Geographical Speciation: Ernst Mayr *257*
Hybrid Introgression: Edgar Anderson *265*
Ecosystem Ecology: Raymond Lindeman *274*
Gradients and Continua: Robert Whittaker *283*
Conclusion *290*

Chapter 9 Border Biology: A Transect *293*

The Border Zone: A Bird's-Eye View *295*

Conclusion *306*

Abbreviations *309*

Bibliography *311*

Index *319*

ILLUSTRATIONS

William F. Ganong *frontispiece*
2.1 Charles O. Whitman *31*
2.2 Charles C. Adams *34*
2.3 Collecting expedition, Woods Hole *44*
2.4 Carnegie Station for Experimental Evolution *47*
2.5 University of Pennsylvania vivarium *50*
2.6 Field station, Gull Lake *52*
2.7 Carnegie Institution Desert Laboratory *53*
3.1 Charles B. Davenport *64*
3.2 Carl H. Eigenmann at Donaldson's Cave *66*
3.3 Student fieldwork, Syracuse Lake *67*
3.4 Henry F. Nachtrieb and "Megalops" *68*
3.5 "Driftwood Lake" (Sister Lake) *79*
3.6 Henry C. Cowles and Homer L. Shantz *81*
4.1 Counting a quadrat, Mount Garfield *101*
4.2 A chart quadrat *104*
4.3 Eggbeater psychrometer *112*
4.4 Edward A. Birge and Chancey Juday, Lake Mendota *113*
4.5 Atmometers *121*
4.6 Phytometer garden, Pike's Peak *123*
4.7 Edith and Frederic E. Clements *125*
4.8 Stations in the Santa Catalina Mountains *132*
5.1 Francis B. Sumner, Woods Hole *141*
5.2 Francis B. Sumner with "Perodipus" *144*
5.3 Mexico, William Tower's field sites *147*
5.4 Arthur Banta, Cold Spring Harbor *150*
5.5 Tahiti, topographic map *152*
5.6 Victor E. Shelford, University of Chicago *159*
5.7 Field stations, Sierra Nevada *166*
5.8 Harvey Hall's field garden *169*
5.9 Taxonomy of a species group *171*

5.10 Jens Clausen's field gardens *173*
6.1 Conway MacMillan *180*
6.2 William F. Ganong *182*
6.3 Henry C. Cowles and students *184*
6.4 Henry A. Gleason *188*
6.5 Edith and Frederic E. Clements, Pike's Peak *196*
6.6 Francis B. Sumner *206*
6.7 Forrest Shreve *208*
7.1 Glacier Bay, Alaska *222*
7.2 Salton Sea *223*
7.3 Lodgepole pine forest, Estes Park *229*
7.4 Concentric glacial bog *231*
7.5 Glacial physiography, Lake Michigan *233*
7.6 Indiana dunes *234*
7.7 Fossil beaches, Lake Michigan *235*
7.8 Lakeshore stream series *237*
7.9 San Francisco Mountains *244*
7.10 Humboldt Bay research site *245*
7.11 Santa Rosa Island *247*
8.1 Biogeography of subspecies ring *259*
8.2 Double invasion of an island *260*
8.3 Edgar Anderson *269*
8.4 Cajun farms, Mississippi Delta *272*
8.5 Distribution of hybrids *273*
8.6 Seasonal fauna, Cedar Creek Bog *278*
8.7 Cedar Creek Bog, Minnesota *280*
8.8 Transect, Cedar Creek Bog *281*
8.9 Vegetation map, Great Smoky Mountains *287*
9.1 Silver Springs, Florida *301*
9.2 Silver Springs sampling sites *302*
9.3 Sonoran desert expeditions *305*

PREFACE

The germ of this largish book was a small episode that I came across in my research for a previous book on the history of genetics. In his oral history the geneticist Theodosius Dobzhansky describes how he came to invent population genetics, a practice that took him into the field and caused him to abandon his beloved *Drosophila melanogaster,* the standard fly, for a wild cousin, *D. pseudoobscura.* Some of his colleagues, Dobzhansky recounts, were incredulous: how could he give up a wonderful and proven instrument, they expostulated. What could the field possibly offer that could compare with laboratory work?[1] Their reaction took Dobzhansky aback. Having begun his career as a field entomologist, he moved freely between field and lab, untroubled by the invisible boundary that, for other biologists, separated distinct and unequal areas of science.

Dobzhansky's story made me curious about this cultural glass wall, the existence of which Dobzhansky revealed by bumping his head on it. My curiosity was further piqued when I learned in an archive that ecologists with whom Dobzhansky worked in the Sierra Nevada thought him still in the "milk-

1. Robert E. Kohler, *Lords of the Fly:* Drosophila *Genetics and the Experimental Life* (Chicago: University of Chicago Press, 1994), p. 262.

bottle stage" of genetics and so inept in the field that he needed constantly watching lest he get lost or injure himself.[2] To his lab friends he had gone native; to field workers he was still a lab man. It was not a simple line that separated lab and field, I realized, but a cultural zone with its own complex topography of practices and distinctions. What is this cultural frontier, and how does it exercise such powerful effects?

To answer these questions meant leaving laboratory biology behind and venturing into the unfamiliar terrain of field biology—ecology, evolution, systematics, areas of which I knew little or nothing, having spent my own career first working in labs, then studying what others do in them. Fortunately, historians of science were beginning to take an interest in field science, including my own colleagues and graduate students, so I was encouraged to think that total ignorance of the terrain was no disqualification for making a voyage of exploration.

Finding my way about that cultural frontier proved more formidable than I had imagined. It was interesting but boundless, with few familiar landmarks, and easy to get lost in. Finding a coherent narrative structure also proved irksome. More than once along the way I had to entertain seriously the possibility that the whole project was misconceived and would miscarry. That had not happened to me in my studies of laboratory science, and it was an unpleasant feeling. Colleagues opined that it was a sign that the project was worth doing, but I sometimes wished I had stayed on the lab side of the border, where issues were more clear-cut and easier to control. But an ungainly embryonic narrative proved viable after all and has grown into something now able to take its chances in the world.

My subject is the lab-field border, and I determined at the outset to stay well on the field side of it, to break old habits of seeing the world from a laboratory window, and to make retreat more difficult. I had to remind myself regularly that I was not writing a history of field biology but just exploring the dynamics of lab-field relations: what it is like to do field biology in a world of labs and experiments. I have not addressed, much less answered, the question of how labs and lab culture came to dominate the world of science. That is perhaps the germ from which other books may one day grow.

I hope that field biologists as well as historians of science will be interested in this narrative of my excursions in the field. Also general readers, whose firsthand knowledge of natural places and their goings-on may have led them to wonder how scientists ever produce true knowledge of such complex things. How do they do it? With difficulty, is part of the answer, and

2. Ibid., p. 282.

one thing I try to show is that behind scientists' most acclaimed achievements lie frustrations and failures. It is how science works, especially field science, and knowing that will, I hope, enable readers to see that science is not some arcane mystery or magic act but a peculiar form of the practical arts by which we all understand and make use of the world we live in.

Acknowledgements

Research for this book was supported in part by a grant from the National Science Foundation (SBR-9700645).

CHAPTER 1

Borders and History

The border between laboratory and field science is one of the most important in the cultural geography of modern science. It cuts across a range of biological and physical sciences and demarcates differences of standing and credibility, physical location, and modes of scientific practice. It marks different connections to the world of middle-class work and leisure. And it is well patrolled. Field biologists are sometimes made to feel that they do not live in the best neighborhood, so to speak. They keep having to explain and justify their work to a world that takes for granted that experiments are the better, or even the only, way of knowing nature. But if the lab-field border is a familiar feature of our intellectual landscape, few have inquired how it came to be and how it works. My aim in writing this book is to begin to understand its cultural geography.

Field and Lab in Biology

As Dorinda Outram has observed, the differences between indoor and outdoor science have been a subject of debate among biologists since the late eighteenth century or even earlier. Then, however, it was not a difference between laboratory and field, but between "closet" or armchair science and expeditionary science. Thus a museum man like Georges Cuvier could grant that fieldwork had the advantage of direct and vivid impressions, but assert that for breadth of comparison and objective analysis the closet naturalist had the advantage.[1] This was an important difference, yet closet naturalists and naturalist voyagers asked essentially the same questions and dealt with the same material specimens, the only difference being that one dealt with it alive and in nature and the other, dead and indoors. Closet science and fieldwork were two ways of doing natural history, not two distinctly different kinds of science.

In modern science the difference between laboratory and field is also one between indoor and outdoor, sessile and mobile work; but it is more than that. The objects of laboratory work are not at all like those of herbarium or museum collections; they may be acquired in the field but are not specimens and are often so transformed for the purpose of experiment as to be quite unlike anything in nature. Not only the objects but the objectives of laboratory work are often quite different from those of field biology. Experimenters analyze and reveal causes and effects; field biologists more often describe, compare, name, classify, map. In fieldwork spatial and locational ways of knowing have equal standing with causal reasoning. In labs there is one best way.

The sociologist Karen Knorr Cetina has pointed to some salient differences between laboratory and field objects. Those who work in labs need not work with whole plants and animals but can substitute partial or less literal versions, such as organ preparations. Nor must they put up with creatures' natural environments, or their irregular or cyclic appearance in nature, all of which greatly complicate analysis. These simplifications are enormously advantageous. In laboratories, objects can be called forth on demand and studied continuously in part or in whole on the experimenter's own terms, not nature's, and that ensures credibility. "The power of the laboratory (but of course also its restrictions) resides precisely in its enculturation of natural objects," Knorr Cetina writes. "The laboratory subjects natural conditions to a social overhaul and derives epistemic effects from the new situation."[2] We are well beyond the advantages of leisurely reflection

1. Dorinda Outram, "New spaces in natural history," in *Cultures of Natural History,* ed. Nicholas Jardine, James A. Secord, E. C. Spary (Cambridge: Cambridge University Press, 1996), 249–65.

2. Karin Knorr Cetina, "The couch, the cathedral, and the laboratory: on the relationship between experiment and laboratory in science," in *Science as Practice and Culture,* ed. Andrew Pickering (Chicago: University of Chicago Press, 1992), pp. 113–38, on pp. 116–17. Knorr Cetina, *Epistemic Cultures: How the Sciences Make Knowledge* (Cambridge: Harvard University Press, 1999), chap. 2.

and comparison that kept Cuvier happily at home, indoors. Modern laboratories reshape and transform natural objects to suit the needs of experiment. Nature is socialized.

How this all came about we have hardly begun to understand, but one thing at least seems clear: the lab-field border in biology is of recent origin, probably no older than the mid-nineteenth century, when laboratories outgrew museums and herbaria as the premier places of modern science. Indeed, we could say that our concept of the "field" was a by-product of the laboratory revolution of the 1840s to the 1870s. The categories of field and laboratory were coinvented and are mutually (and changeably) defining: like matter and antimatter, town and country, good and fallen angels. Observation and comparison, once performed by closet and voyager naturalists alike, became second-best practices in a landscape dominated by labs. And whole organisms and organisms in situ became less real than their disassembled parts. The laboratory revolution made the cultural landscape in which field biologists now live. It created the lab-field border.

Exactly how this coevolution worked we do not know, because its history has not been written. Did laboratory builders deploy invidious distinctions with fieldwork to capture public resources from museum men? I'd bet they did: important cultural categories are often the result of some such public contest. And there are the odd bits of evidence. For example, Julius Sachs, the founder of plant physiology and a zealous advocate of labs and experiment, dismissed Charles Darwin's experiments on plant-root growth by simply declaring that true knowledge was made only in laboratories, not the country houses and backyards of untrained naturalists. That ended the argument (though it turned out that Darwin got it right).[3]

Victorian laboratory-builders like T. H. Huxley also appropriated the idea of nature's authenticity for laboratory instruments and practices. The chemist Henry Roscoe likened chemical experimentation to a walk in the country: "[L]aboratory work brings the student into direct contact with Nature. In the lecture room the student forms an idea, as in a panorama, of the general appearance of the country; but it is in the laboratory, as in a walk through a given district, that he first learns what the land he is travelling through is really like." Worries about the artificiality of microscopic observations and the risks of visual artifacts were solved, Graeme Gooday argues,

3. Soraya De Chadarevian, "Laboratory science versus country-house experiments: The controversy between Julius Sachs and Charles Darwin," *British Journal for the History of Science* 29 (1996): 17–41. See also Sophie Forgan, "The architecture of display: Museums, universities and objects in nineteenth-century Britain," *History of Science* 32 (1994): 139–62, on pp. 140–42, 153–56.

by bringing nature indoors. By a kind of cultural body snatching, labs were made more natural than nature itself.[4]

But in the absence of a full history of the lab-field border we can only take its existence as given and pick up the story in 1890, when it was a recognized but still novel feature of biology's cultural topography, and memories of how it came to be were still fresh and raw. The field biologists who came of age in the early 1900s were the first who could not operate exclusively by their own rules on their own cultural ground. They lived in a world of laboratories, in which they felt bound to use lab methods and understood that their own practices and achievements would be judged by lab standards. Few field biologists could ignore these cultural-geographic realities, not even those who would never do an experiment. All lived to some degree in the shadow of laboratory science, and their successors still do.

The lab-field border was marked in the 1890s by ritualized exchanges of jokes and caricatures, often good humored, always a little barbed, and occasionally poisoned when important things like jobs and grants were at stake—like ethnic jokes and stereotypes. The cytologist Edmund B. Wilson recalled the "customary exchange of compliments—bug hunters vs section cutters or worm slicers . . . or egg shakers."[5] The ecologist Charles C. Adams likened the tensions between lab and field sciences to the conflict between social classes:

> [I]ndividuals of each class were inclined to feel that their group only was concerned with "fundamental" problems, and that the others perhaps might be "all right" personally, but unfortunately were on the wrong trail! The closet [laboratory] student often felt that he only was doing the careful "permanent" (is there such?) scientific work, while the field worker was superficial and untrained, and therefore his results were of little value. The field worker was also often inclined to look upon his closet friend as one who devoted his time to trifles.[6]

The entomologist William Morton Wheeler was sorry to see the older field naturalists dying off, but mused that "from one point of view there is no

4. Graeme Gooday, "'Nature in the laboratory': Domestication and discipline with the microscope in Victorian life science," *British Journal for the History of Science* 24 (1991): 307–41, on p. 313.

5. Edmund B. Wilson, "Aim and methods of study in natural history," *Science* 13 (1901): 14–23, on p. 19.

6. Charles C. Adams, "The new natural history—ecology," *American Museum Journal* 17 (1917): 491–94, on p. 491.

harm in labeling a dead entomologist 'Natural History specimen. Dried. Handle with care,' as that is all he really is, I suppose, after he leaves off dealing with the numerous dried insects." (That was an in-joke: Wheeler was an avid field man and friend of amateur naturalists.)[7]

The lab-field border was intangible but real; people were affected by it, talked about it, took it seriously. (What is more serious than jokes?) But what kind of thing was this border, exactly? Was it—is it—a physical place, a part of the geography of our natural and built environments? Laboratories are physical places, after all, and so are the places where field naturalists do their work. Are there places where we would say to ourselves, "Oh look, here we are on the lab-field border," as we know when we stand on a forest edge or continental divide? Or are we dealing with a cultural place or "space" (the more usual term for cultural terrain)? Is the geography of the lab-field border a geography merely of the mind and imagination, nowhere in particular?

The quick answer is both, but it is more complicated than that. The domains of laboratory and field are cultural domains first and foremost, where different languages, customs, material and moral economies, and ways of life prevail. But these cultural differences are experienced in actual places, and the experience differs from place to place. One may sense the power of laboratory culture strongly in some places in nature, which resemble labs or are used in lablike ways, and hardly at all in others. In nature, other conventions of understanding landscape may prevail over those of the laboratory: for example, concepts of "wilderness."[8] There are also differences of cultural location. Ecologists feel the tension between laboratory and field values strongly, because they aspire to be experimental; biogeographers and taxonomists may resent the imperiousness of lab culture but are immune to its appeal.

In this book I treat the lab-field border as an intangible cultural space. But that space or terrain is embodied and experienced in particular tangible places, and much of the evidence for its qualities is evidence of place—physical, geographical, nameable, pickable, kickable places. I use the word

7. William M. Wheeler to R. P. Dow, 16 Jan. 1916, WMW, box 9.

8. John R. Stilgoe, "A New England coastal wilderness," *Geographical Review* 71 (1981): 33–50. William Cronon, "The trouble with wilderness; or, getting back to the wrong nature," in *Uncommon Ground: Toward Reinventing Nature,* ed. Cronon (New York: Norton, 1995), 69–90. John F. Sears, *Sacred Places: American Tourist Attractions in the Nineteenth Century* (New York: Oxford University Press, 1989). Cultural anthropologists, landscape and environmental historians, and historical geographers have produced large literatures on cultural "space" and how we read it.

"place," not "space" to underline the physicality and reality of these places where border culture is experienced and lived.

I shall begin somewhat abstractly with reflections on some of the pertinent cultural differences between field and laboratory: not any lab in any historical period but the modern lab of the late nineteenth century, the generic place that appears in our mind's eye when we think of labs. Actual laboratories vary a good deal; but it is useful for my purpose to start with a Weberian ideal type, and to begin by characterizing the field as place in terms of how it is unlike a lab. For that is how the field came to be defined after the laboratory revolution: as not-lab. If the definition is not exactly true, it is at least an actors' category.

Lab and Field as Place

Some differences between field and our generic modern laboratory are fairly self-evident. First, natural places are not just neutral stages for measuring and experiment, as laboratories are, but are themselves the *objects* of study. Plants and animals are elements of natural environments, along with topography, habitat, and weather: they are not mere passive guests as they are in labs, but actively alter their environments. Thus place must figure quite differently in lab and field practices. Laboratory workers eliminate the element of place from their experiments. Field biologists use places actively in their work as tools; they do not just work *in* a place, as lab biologists do, but *on* it. Places are as much the object of their work as the creatures that live in them.

Another obvious difference between lab and field is that natural places are *particular* and *variable* places, none quite like another, each the result of a unique local history, never the same from one moment to the next, unpredictable, unrepeatable, beyond human control. These qualities are precisely what draw biologists to study life in nature, but they make it difficult to apply methods devised for use in the invariant and generic environment of a lab. Field biologists spend much time and energy inventing methods for dealing with particularity and variability and turning these problems to analytic advantage.

Finally, laboratory and field differ socially in the variety and kinds of people who inhabit and use them. Natural places, unlike labs, have *multiple* uses: outdoor recreation, foraging, religious retreat and restoration, travel and trade, finding oneself or losing oneself, and so on. Naturalists share the field with hunters, fishers, poachers, trappers, surveyors, tramps, madmen, shamans, loggers, prospectors, bird watchers, bandits, vacationers, herbalists, cowboys, students, con men, true and false prophets, and green terrorists. Moreover, field biologists' activities and paraphernalia are sometimes not so different from those of tourists or other sorts and can cause confusions of identity. Laboratories, in contrast, are socially homogeneous; access is restricted to those who are qualified and have legitimate business there. Socially as well as physically and biologically, the field is a more ambiguous and unstable place than any lab. Labs are separate, a world apart from the world; nature connects field biologists to other social worlds.

These features of physical and cultural topography go a long way toward explaining why laboratory science seems always to be granted a higher standing than field science, and why observation and comparison are taken to be less credible ways of knowing than experiment. It is precisely the stripped-down simplicity and invariability of labs—their placelessness—that gives them their credibility. Generic places sustain the illusion that their inhabitants' beliefs and practices are everyone's beliefs and practices. We credit knowledge and practices that are universal and mistrust what is merely local and particular, and laboratories are meant to seem universal, the same everywhere. The variability and unexpected occurrences of nature have no place in labs. Such things would only undermine the reason why we trust experiments, which is that they turn out the same wherever they are performed.

As Steven Shapin has argued, laboratories help solve the great problem of trust in science: making knowledge that is produced locally—as all knowledge is—acceptable anywhere.[9] We credit science done in labs because we know, or think we know, what kind of place they are, and that the same rules of procedure and evidence apply in all. Since no one can personally visit every lab, their placelessness become the symbolic guarantee that science done there is everyone's, not just someone's in particular. The social homogeneity of laboratories is another powerful resource of trust. We know that access to laboratories is restricted to qualified people. (We know them by their white coats and badges.) We can trust work we have not seen performed because we know that a trustworthy sort of person must have done the work.[10] The place where knowledge is made is not supposed to matter,

9. Steven Shapin, "Pump and circumstance: Robert Boyle's literary technology," *Social Studies in Science* 14 (1984): 481–520.

10. The logic here follows the argument of Steven Shapin, *A Social History of Truth: Civility and Science in Seventeenth-Century England* (Chicago: University of Chicago Press, 1994); and

and it is an ominous sign when scientists ask about it, because when scientists suggest that a phenomenon happens only in a particular place, they mean that it does not happen at all. It was the end for cold fusion, for example, when people decided that it only happened in Salt Lake City.

Placelessness was not always a guarantee of the credibility of science done in laboratories. Before the laboratory revolution of the nineteenth century, labs were often located in various borrowed places, to capture the social credibility that was automatically granted to their inhabitants. Robert Boyle set up shop in gentlemen's (or gentlewomen's) houses, thus ensuring that his reports of pneumatic experiments would be credited, as the word of a gentleman was automatically credited in seventeenth-century England. (Being of independent means and beholden to no man, gents had no reason not to tell the truth—that was the social logic.) Likewise with knowledge produced in noblemen's castles, master craftsmen's chambers (but ambiguously), civic humanists' townhouses, and Renaissance scholars' villas; and, more recently, with museums, monastic workshops, pubs, and the country estates of the industrial bourgeoisie.[11] The credibility of experiments was guaranteed by the particular places where the work was done, and by the social qualities of the people known to live there. Shapin puts it proverbially: for Boyle's contemporaries it was "gentlemen in, genuine knowledge out."[12]

But since the industrial and social revolutions of the nineteenth century, we have lived in a culture that values universality over locality, and in such a culture placelessness is a reason to trust. For us moderns, it is "expert in, genuine knowledge out." The culture of the West (and now the world) can

---

Shapin, "The house of experiment," *Isis* 79 (1988): 373–404. See also Shapin, "Trust, honesty, and the authority of science," in *Society's Choices: Social and Ethical Decision Making in Biomedicine,* ed. Ruth E. Bulger, Elizabeth M. Bobby, and Harvey V. Fineberg (Washington: National Academy Press, 1995), pp. 388–408.

11. Shapin, "House of experiment" (cit. n. 10). Steven Shapin, "Who was Robert Hooke?" in *Robert Hooke: New Studies,* ed. Michael Hunter and Simon Schaffer (Woodbridge, U.K.: Boydell Press, 1989), pp. 253–86. Owen Hannaway, "Laboratory design and the aim of science: Andreas Libavius versus Tycho Brahe," *Isis* 77 (1986): 585–610. Paula Findlen, *Possessing Nature: Museums, Collecting, and Scientific Culture in Early Modern Italy* (Berkeley: University of California Press, 1994), chap. 3. Forgan, "Architecture of display" (cit. n. 3). Anne Secord, "Science in the pub: Artisan botanists in early-nineteenth-century Lancashire," *History of Science* 32 (1994): 269–315. Myles Jackson, "Illuminating the opacity of achromatic lens production: Joseph Fraunhofer's use of monastic architecture and space as a laboratory," in *Architecture and Science,* ed. Peter Galison and Emily Thompson (Cambridge: MIT Press, 1999), pp. 423–55. Simon Schaffer, "Physics laboratories and the Victorian country house," in *Making Space for Science,* ed. Crosbie Smith and Jon Agar (London: Macmillan, 1998), pp. 149–80. Sophie Forgan, "Bricks and bones: Architecture and science in Victorian Britain," in *Architecture and Science,* ed. Galison and Thompson, pp. 181–208.

12. Shapin, "House of experiment" (cit. n. 10), on p. 397.

be thought of as a weedy culture. We credit beliefs and practices that travel fast and assimilate anywhere because they seem not local but universal, what everyone does and thinks. Practices and ways of life that stay local must have something wrong with them, we think. That is just what weeds would think, who build and inhabit airports and autobahns, chain stores and malls, suburbs, and cyberspace—the placeless places of a weedy, global capitalism. How we have come to judge the value of knowledge by its power to disperse is a problem that historians have yet to solve, but probably it is a deeply ingrained habit of our weedy species, now amplified by technologies of rapid long-distance travel and communication.

In any case, laboratories and science are very much part of this weedy modern landscape and no doubt one of its causes. Science is the paradigmatic knowledge of modern culture, and experiment is the paradigmatic scientific practice. Scientific practices and rationales are essential to the operation of state and corporate bureaucracies.[13] The placelessness of modern labs, like corporate parks and capital cities, advertise the universality and authority of the culture that builds and inhabits them.

Obviously field biologists cannot avail themselves of the useful resources of this logic of placelessness, which lab workers take for granted. In nature's variable and particular places the logic of place is inverted. When place affects laboratory experiments we know that something went wrong. Field biologists, however, know that something is wrong if place does *not* affect the behavior of plants and animals; it indicates that human observers have been indiscreet and intrusive in the lives of their subjects and disturbed the natural relations of creatures and their habitats. In laboratories, experiments that turn out different on repetition are suspect. In nature, experiments that turn out the same every time and in every place may be suspect, because life in nature is not so uniform. The logic of placelessness does not operate in the field, and that puts field biologists at a real disadvantage in a weedy world of labs, where placelessness is the accepted guarantee of credibility.

The ambiguous social identity of field naturalists also deprives them of the credibility that automatically attaches to the inhabitants of labs. As Steven Shapin argues, we judge the worth of knowledge claims by the social character of the claimants, and we know that from the places they inhabit and the company they keep. Thus the exclusivity of labs inspires trust. However, being known to work in nature confers no such authority: quite the

13. James C. Scott, *Seeing Like a State: How Certain Schemes to Improve the Human Condition Have Failed* (New Haven: Yale University Press, 1998). Bruno Latour, "Visualization and cognition: Thinking with eyes and hands," *Knowledge and Society* 6 (1986): 1–40. Latour, *Science in Action* (Cambridge: Harvard University Press, 1987).

contrary. That guy in a checked shirt with a gun, net, or pack could just as well be having a day off, making a buck, or looking for trouble. In the field there is no equivalent to "White coat in, genuine knowledge out." With field-workers, appearances may deceive.

In the nineteenth century field scientists could still bolster their credibility by borrowing the social characteristics of nonscientific groups. The early-nineteenth-century "gentlemen geologists" described by Roy Porter are a good example. Field geology acquired authenticity when it was assimilated to a gentry culture of riding and outdoor recreation. Martin Rudwick has suggested that field geologists experience expeditions as analogous to religious pilgrimages: as pilgrims go forth expecting to perceive the world in new and deeper ways, so geologists who go afield are more open to new theoretical insights than those who stay at home. Similarly, glaciologists and other field scientists identified themselves with heroic explorers. We trust explorer-heroes (or used to) because we read their displays of endurance, fortitude, and self-sacrifice as guarantees of their credibility as witnesses. We know from their actions that heroes are men of physical courage, whom we can trust to tell the truth courageously come what may. Assimilating the social role of explorer, field scientists acquire a similar authenticity. "Hero in, genuine knowledge out." The logic was especially compelling for field scientists in the late Victorian period, when explorers of tropical and polar regions were lionized public figures. (A similar logic operated occasionally in labs, for example, in the once-honored custom of experimenting on yourself.)[14]

Such borrowing of social identities lingered into the mid-twentieth century in corners of field science like cultural anthropology, which required extended residence in isolated and sometimes dangerous places. But for most field biologists it ceased to be an option. The cultural power of experimental science is so strong that borrowing from not-science no longer solves the problem of credibility. The logic of placelessness is too dominant in modern

14. Roy S. Porter, "Gentlemen and geology: The emergence of a scientific career, 1660–1920," *Historical Journal* 21 (1978): 809–36. Martin Rudwick, "Geological travel and theoretical innovation: The role of 'liminal' experience," *Social Studies of Science* 26 (1996): 143–59. Bruce Hevly, "The heroic science of glacier motion," *Osiris* 11 (1996): 66–86. Naomi Oreskes, "Objectivity or heroism? On the invisibility of women in science," *Osiris* 11 (1996): 87–116. Henrika Kuklick, "After Ishmael: The fieldwork tradition and its future," in *Anthropological Locations: Boundaries and Grounds of a Field Science,* ed. Akhil Gupta and James Ferguson (Berkeley: University of California Press, 1997), pp. 47–65. George W. Stocking, Jr., "The ethnographer's magic: Fieldwork in British anthropology from Tylor to Malinowski," in *Observers Observed: Essays on Ethnographic Fieldwork* (Madison: University of Wisconsin Press, 1983), pp. 70–120. Beau Riffenburgh, *The Myth of the Explorer,* ed. George W. Stocking, Jr. (New York: Oxford University Press, 1994).

science and modern culture to be circumvented; it can only be lived with. The most obvious strategy for field biologists to achieve credibility—but difficult, because of the places where they work and the things they work on—is to become more like laboratory scientists. The trick is to assimilate elements of laboratory practice and make them appropriate to field conditions, to seek out natural places that resemble laboratories in some way. Another strategy is to do what lab workers cannot: namely, use the very particularity of nature to create knowledge that is true of nature generally.

Field biologists have used both these strategies to make themselves credible in a world of laboratories, inventing many ingenious ways of dealing with particular, variable, diversely occupied places. They have designed instruments and protocols that stand up to field conditions, studied creatures where they live without intruding, deciphered the incomplete records of "Nature's experiments," and skillfully exploited the places in which simple comparison reveals cause and effect relations without experimental manipulation of variables—a field analogue of laboratory experiment. As experimental biologists make the most of the placelessness of labs to gain credibility, so field biologists turn the particularities of nature to their advantage.

Borders and Frontiers

I have so far emphasized the differences between laboratory and field, but they are by no means totally different or forever separate worlds. They meet in a zone of active interaction and exchange. Field biologists look across the border to what the people in labs think of what they are doing. There are places with qualities of both lab and field, and practices that are as much of one as the other. Field and lab biologists read each others' works and even engage in cooperative projects. They know when they have crossed a cultural boundary, but they cross quite freely.

The cultural geography of field and lab is like a biogeographic ecotone, an area where two biota intermingle and where neither has a clear advantage, and where we expect to find the odd hybrid. It is for this reason that I use the term "border," which implies a permeable zone, rather than "boundary," which reminds one of political or property maps, where sharp definition matters (orange flags in the woods, Checkpoints Charley).

Laboratory and field are best seen as two parts of one thing, as the environmental historian William Cronon sees city and countryside: linked opposites, mutually supporting and defining, neither one viable or even conceivable without the other. Cities depend economically on their hinterlands, and vice versa; so do lab and field. Laboratory and field biologists have more in common with each other than either has with distant disciplines. (French cities and countryside are different, but both are French, not Italian or Canadian.) City and country are experienced not as separate places but as movement from one to the other: from Green Lake, Wisconsin, to the yellow cloud of industrial Chicago and back.[15] So, too, border biologists experience lab and field as movement and border crossing. Physically, fieldwork is movement between indoor places (museum, laboratory) and outdoor ones. Culturally, too, field practices take biologists from one realm to another, from observation to experiment and back. Border biologists inhabit a complex cultural ecotone in which elements of lab and field mingle and sometimes combine in new ways.

Historians and sociologists of science have invented several useful ways of thinking about the cultural topography of borders—most notably the sociologist Thomas Gieryn. Gieryn urges us to think topographically and to create cultural cartographies of science. He points to boundaries as crucial features of cultural terrain and argues that they are created and maintained by what he calls "boundary work," the efforts that scientists make to defend or expand boundaries between science and overlapping or competing cultural activities, such as religion or technology. It is primarily rhetorical and ideological work, in Gieryn's view: constructing public images of science that justify scientists' possession of valuable pieces of cultural territory and persuade the public that competing groups are less worthy occupants.[16]

In science, boundary workers were especially active in the mid-nineteenth century, when much of the infrastructure of modern science was constructed. On the border between physical science and engineering, for example, that indefatigable boundary worker John Tyndall projected an image of science as fundamental, theory driven, and financially disinterested, in contrast to cut-and-try, profit-driven technology. The idea that engineering is "applied science" was a most effective bit of boundary work: except by historians and sociologists of science, it is still widely credited. A different assemblage of traits has been equally effective in defending the boundary between science and religion. Boundary work has also been deployed to banish deviants or suspected frauds across disciplinary borders, making them someone else's credibility problem. (For example, psychologists relabeled Cyril Burt, when

15. William Cronon, *Nature's Metropolis: Chicago and the Great West* (New York: Norton, 1991). See also John R. Stilgoe, *Borderland: Origins of the American Suburb, 1820–1939* (New Haven: Yale University Press, 1988).

16. Thomas F. Gieryn, *Cultural Boundaries of Science: Credibility on the Line* (Chicago: University of Chicago Press, 1999), pp. 1–25.

he was suspected of faking data, as an "applied" psychologist—i.e., not a "real" one.)[17]

Gieryn mainly concerns himself with the contested boundaries between science and not-science, but the idea applies equally well to the internal borders of science, including the lab-field border. For example, the barbed ethnoscientific jokes quoted earlier are a kind of boundary work: a way of mending fences, keeping standards and cultures pure and ownership secure.

"Boundary object" is another border concept, one invented by Susan Leigh Star and James Griesemer to explain how actors with different values and purposes cooperate in common enterprises. Boundary objects are things whose meanings are pliable enough that people from different subcultures can view them as embodying a shared cause. They have both local and global identities and can serve as vehicles of cultural exchange. In Star and Griesemer's case of a natural history museum, the essential actors are wealthy patrons, hunter-collectors, taxidermists, and biologists, and examples of boundary objects include animal specimens, field notes, maps, and the museum itself. Even the state of California qualifies, being "an object which lives in multiple social worlds and which has different identities in each." It seems useful, though one wonders if putting an insect on a pin in the same category with the state of California might not deprive the idea of any power to make analytically revealing distinctions between objects. Bruno Latour's concept of "immutable mobile," which applies more specifically to compressed and standardized objects like maps, files, and tables, is a similar and perhaps more effective analytic tool for understanding the special power of some objects to travel and enlist others in their projects. (Are boundary objects *mutable* mobiles?)[18]

A third way of thinking about the cultural geography of borders is the idea of "trading zones," which the historian Peter Galison uses to explain how sciences that speak different mathematical languages may in certain places combine in novel forms. In Galison's words, a trading zone is "an arena in which radically different activities [can] . . . be *locally,* but not globally, coordinated." The wartime atomic laboratory at Los Alamos was such a place, where physicists and computer programmers developed the statis-

17. Ibid. Thomas F. Gieryn and Anne E. Figert, "Scientists protect their cognitive authority: The status degradation ceremony of Sir Cyril Burt," *Sociology of the Sciences Yearbook* 10 (1986): 67–86.

18. Susan Leigh Star and James R. Griesemer, "Institutional ecology, 'translations' and boundary objects: Amateurs and professionals in Berkeley's Museum of Vertebrate Zoology, 1907–39," *Social Studies of Science* 19 (1989): 387–420, quote on p. 409. Latour, "Visualization and cognition" (cit. n. 13).

tical method of "Monte Carlo" simulation to model nuclear fusion reactions. A potentially subversive practice (because it substituted mathematical "experiments" for proofs, replaced physical experiments by modeling, and empowered computer technicians), Monte Carlo would have been unlikely to have evolved in, say, academic departments of physics or mathematics. But it could evolve at Los Alamos, where scientists of diverse sorts, corralled together in extreme isolation, set aside territorial claims to labor toward a vital practical end. Galison compares the formation of trading zones with the evolution of new languages, which begin with a pidgin, a rudimentary hybrid that allows basic communication, and develop into a distinct language of a new mixed community with its own practices and culture—a creole.[19]

These boundary concepts serve the particular purposes for which they were invented, but they have their limits. The analogy with languages works well for mathematics, which is a kind of language, and for isolated locales, but it is hard to see how it might be used to analyze the interactions that occur within such complex terrains as molecular genetics, neuroscience, and the disciplines of the lab-field border. Likewise, defining boundary work as rhetoric and ideologizing applies nicely to the public activities that define boundaries between science and its publics (committee reports, political lobbying, press conferences, polemics). But does it apply as tidily to the private and semipublic borrowing (or avoiding) of practices that have shaped borders within science? And the concept of "boundary objects" seems to miss entirely the vital elements of place and people. In general, the concepts of science studies tend to be a little too abstract and timeless, too purely cultural to deal well with natural places and large-scale features of the scientific landscape.

For areas like the lab-field border, a more complex and capacious approach serves best, and for that we do well to look outside science studies to historians or historical geographers, whose stock in trade is cultural interaction and change on the grand scale. Concepts of science studies are simple and travel well: like laboratory science, science studies is a weedy culture. But for the historical geography of field science, concepts more in the natural-history style fit better.

19. Peter Galison, "Computer simulations and the trading zone," in *The Disunity of Science: Boundaries, Contexts, and Power,* ed. Galison and David J. Strump (Stanford: Stanford University Press, 1996), pp. 118–57, quote on p. 119. See also Galison, *Image and Logic: A Material Culture of Microphysics* (Chicago: University of Chicago Press, 1997), pp. 803–44. It is not always clear if Galison means the "trading zone" to be the simulation itself or the place where it evolved, or both; the actual place seems to me a more appropriate object for a spatial metaphor.

There is an extremely rich and varied literature on the social geography of borders and frontiers: of migrations and minglings, of political boundaries thrown across cultures or suddenly removed. The European peninsula, with its kaleidoscope of cultures, affords many such cases, in the Pyrenees, the northern river deltas, the Balkans, and the two Germanys. The historical geography of European colonial expansion is another rich source. Richard White's study of the "middle ground" of French and Indian fur trading in the Great Lakes region suggests how ritualized cultural misunderstandings facilitated interaction. James Merrill's study of the role of negotiators and translators on the Appalachian frontier is another analysis of cultural dynamics that historians of science can use.[20]

But the cases that I have found most fruitful are Owen Lattimore's classic and still vital study of the inner Asian frontiers of imperial China, and Charles Whittaker's work on the Rhine frontiers of the Roman Empire.[21] The Roman frontier is the case closer to my own, because in Europe new border cultures emerged along what was initially a defensive frontier, while in China two cultures became more distinct and different as a result of contact. However, Lattimore advances general ideas of the cultural dynamics on frontier zones that invite extension to other kinds of cultural boundaries, including those of science.

One general principle of frontier cultural dynamics is that boundaries are not lines but broad zones, in which one society and culture gradually gives way to another and where the balance point shifts over time. Frontiers are zones of movement and change, where two sets of social rules apply and where power is uncertain. The linear boundaries of modern maps are a quite recent invention, useful legally and politically, but hardly an apt representation of the complexity of cultural geographic borders. The Romans never drew such lines but distinguished only between organized provinces

20. Edward W. Fox, *History in Geographic Perspective: The Other France* (New York: Norton, 1971). Peter Sahlins, *Boundaries: The Making of France and Spain in the Pyrenees* (Berkeley: University of California Press, 1989). Daphne Berdahl, *Where the World Ended: Re-unification and Identity in the German Borderland* (Berkeley: University of California Press, 1999). Mark Mazower, *The Balkans: A Short History* (New York: Modern Library, 2000). Richard White, *The Middle Ground: Indians, Empires, and Republics in the Great Lakes Region, 1650–1815* (New York: Cambridge University Press, 1991). Stephen Aron, *How the West Was Lost: The Transformation of Kentucky from Daniel Boone to Henry Clay* (Baltimore: Johns Hopkins University Press, 1996). James H. Merrell, *Into the American Woods: Negotiators on the Pennsylvania Frontier* (New York: Norton, 1999).

21. Owen Lattimore, *Inner Asian Frontiers of China* (Washington: American Geographical Society, 1940; reprinted New York: Oxford University Press, 1988); citations are to the reprint edition. Charles R. Whittaker, *Frontiers of the Roman Empire: A Social and Economic Study* (Baltimore: Johns Hopkins University Press, 1994).

and outer areas that were controlled but not formally administered. The Great Wall between China and Mongolia is actually a set of walls, like tidemarks on the sand of the ebb and flood of peoples.

The great systems of frontier walls—the Great Wall, Hadrian's Wall between England and the Scottish uplands, and Offa's Dyke between England and Wales—were built not to keep barbarians out of civilized areas but to control the movement across frontiers of people on both sides. Frontier boundaries are not places where movement ceases but where it is most intense and defining. Movement of traders, raiders, transhumants, outlaws or exiles, would-be settlers, armies, refugees, parvenu provincials, and imperial officials is the characteristic experience of frontier zones, not stasis.[22] These zones are not the end of the world but connections between worlds.

A second feature of the cultural geography of frontiers is that they are not just political, but also ecological and economic zones, where one way of making a living from the land gives way to another. Lattimore was the first to insist on this point, and it is fundamental. Often frontiers mark a change from intensive to extensive economies. China's frontier with Mongolia is the ecological or environmental frontier between irrigated rice agriculture, one of the most intensive uses of land ever devised, and nomadic pastoralism, one of the most extensive. Irrigated rice farming only works in areas where there are moderately good soils, adequate rainfall, streams or rivers to divert, and a very dense population. Nomadic pastoralism only works in large grassland areas of marginal soils and rainfall and thinly scattered populations. The Great Wall marks roughly where the tributary streams of the Yellow River drainage gradually become too small and irregular to sustain irrigation. Likewise, the Roman frontiers in Germany and Britain came to an uneasy rest in the ecological zones where clay and loess meet sand, and where plough agriculture of cereals gradually gave way to pastoral economies of hay and livestock.

Frontiers represent the ecological outer limits of expanding empires, where the economic and political benefits of further expansion no longer outweigh the costs. The Chinese empire, for example, grew from its ancient center in the middle Yellow River valley by converting areas of mixed agriculture (dry farming or makeshift irrigation, plus husbandry and foraging) to areas of full irrigation. Where soils and environment favored large-scale irrigation, as in the south, expansion was virtually unlimited. The agricul-

22. Lattimore, *Inner Asian Frontiers* (cit. n. 21), on pp. 238–50, 434. Whittaker, *Frontiers of the Roman Empire* (cit. n. 21), on pp. 17, 60–62, 68–73, 82–83, 121. Lucien Febvre, *A Geographical Introduction to History* (London: Kegan Paul, Trench, Truber, 1932), pp. 296–315.

tural surpluses and augmented workforce made the effort worthwhile. In the north, however, topography and climate gradually became less suitable, and smaller surpluses of grain and labor made further expansion more trouble than it was worth. It was the same at the fringes of Roman expansion in Germany, where declining agricultural surpluses and thinning population meant a costly military occupation with no economic or political return.[23]

A third significant feature of frontiers is that they are culturally constructed. "Barbarians" are the invention of expanding imperial societies—non-Chinese, not-yet-Roman. Ethnic boundaries rarely define frontiers, just as topographical features seldom do. The Rhine River had the same people on both banks when the Romans arrived; the trans-Rhine "Germans" were the Romans' invention. The Mongol nomads were an invention of the Chinese in a more literal sense. Lattimore argues that it was the arrival of irrigators, with their intolerance of mixed agriculture, that drove some inhabitants of the zone of mixed economy to abandon settled farming for a fully nomadic way of life that required refined techniques of animal husbandry; this shift occurred even in areas ecologically suited to mixed farming.[24] The constructive process is easiest to see from the barbarians' point of view.

Finally, there is the tendency of new hybrid societies to arise in frontier zones, a process that was well marked in the European frontiers of the Roman Empire. Such societies evolve when inhabitants of the frontier zone, imperial provincials and barbarians alike, begin to develop customs and identities that resemble each other's more than they do those of either the imperial or the barbarian core areas. These societies develop gradually at the grass roots, through trade and military service. Provincials and barbarians who benefit from commerce, for example, develop interests that conflict with imperial policy of suppressing activities that may strengthen competitors. It was Roman policy, also, to co-opt local tribesmen into ethnic units of the imperial army, and to delegate authority for frontier administration to local German elites, who gradually adopted Roman ways (villas, baths, trade goods, markets) without at first entirely abandoning their tribal ways. Thus there arose in mixed border zones societies whose interests and customs were distinct from and even antagonistic to those on both sides.[25]

23. Lattimore, *Inner Asian Frontiers* (cit. n. 21), on pp. 38–39, 56–61, 240–47, 327–32. Whittaker, *Frontiers of the Roman Empire* (cit. n. 21), on pp. 85–97.

24. Lattimore, *Inner Asian Frontiers* (cit. n. 21), on pp. 276–78, 410–12. Whittaker, *Frontiers of the Roman Empire* (cit. n. 21), on p. 73.

25. Lattimore, *Inner Asian Frontiers* (cit. n. 21), on pp. 249–50. Whittaker, *Frontiers of the Roman Empire* (cit. n. 21), on pp. 130–33, 158–61, 219–23, 231–42.

We cannot expect the cultural geography of the lab-field border in science to resemble exactly those of the Chinese or Roman imperial frontiers; there are too many differences for any simple correspondence. But some general similarities suggest that the underlying principles may be sufficiently alike to make the comparison, if nothing more, a healthy stimulus to the imagination.

First, the boundary between lab and field cannot be demarcated by a line, as political boundaries are drawn; rather, it is a zone of mixed practices and ambiguous identities (as are also boundaries between disciplines). Boundaries are political fictions, the creations of boundary workers; borders are indefinite zones, in which ways of life are not sharply differentiated. The analogy with cultural frontiers is a useful reminder not to take the ideological representations of imperial sciences too seriously and to see matters, if possible, from the field (or barbarian) point of view. There is also an analogy between the cultural creation of "the other" and what happens in science. The social category of the field came into existence as the result of the expansion of laboratory science, I argue, as the nomadic barbarians were the result of the expansion of the irrigating Han, or the Germans of the wheat-eating Romans. Before the laboratory became the dominant institution in science, much the same kind of people occupied what became, as a result of that domination, two sides of a border zone. Field is defined in opposition to lab, as nomad is in opposition to farmer, and barbarian to cosmopolite.

So, too, is the lab-field border zone fundamentally rooted in ecology and economy. Laboratory and field biology are distinct modes of knowledge production and have distinct political economies. To compare them to intensive and extensive or mixed economies may seem a stretch, but consider it. Experiment is an intensive mode of knowledge production: capital intensive and requiring close control to realize the benefits of a more complex and expensive organization. Experiment as a mode of production works only in an artificial and ecologically simplified environment constructed for the purpose, as do plough or paddy agriculture. Some forms of fieldwork might be compared to nomadic pastoralism (expeditions, for example, are highly specialized forms of extensive land use), but much of field biology seems more like mixed farming and foraging: sedentary and mobile in alternation; dependent on natural environments and at times irregular in output; flexible and opportunistic. It is science before the arrival of the lords of land and water: a bit fanciful, perhaps, but not a bad way to think about field science.

Does the lab-field boundary circa 1890 represent the limits of expansion of an imperial form of science? Had experimental laboratory culture expanded into the domain of natural history to a point where the costs of further expansion—expensive travel, loss of control, compromised standards, inefficiencies—were greater than the potential rewards? Perhaps this is stretching the analogy beyond its tensile strength, but we won't know until the history of the laboratory movement is written.

Motion is as characteristic a feature of the lab-field border as it is of political-ecological frontiers. Like transhumant pastoralists, field biologists spend summers on lakes or in alpine meadows or forests, winters in more protected domestic spaces. Expeditions begin and end in museums or laboratories. Border crossers are an important social type in field science, importers of laboratory culture and translators between cultures with different languages and practices. The careers of border biologists move between lab and field.

Finally, and most important, the lab-field border zone is a place of mixed cultures, where biologists on either side adopt each others' practices and develop approaches that are neither pure lab nor pure field. The various branches of ecology mix experimental and field practices in ways that serve their practitioners well without satisfying purists on either side. Evolutionary biology may likewise be seen as a blend of experimental and taxonomic practices. New scientific disciplines often begin as border provinces, then take shape through the accumulated experience of border traffic and exchange and so become cultures as distinct as those of Franks, Germans, and Burgundians.

Readers should not take the analogy with ecopolitical frontiers too literally; the differences between them and the cultural boundaries of science are plain enough. The value of the comparison lies in the family resemblances of border phenomena. I read border histories less for specifics than to put myself in a cultural-geographic frame of mind. Nor would I recommend trying to refine or "theorize" spatial concepts like place and border; it would only weaken their heuristic power. The analogy with cultural frontiers will have served its purpose if it keeps us thinking spatially or cartographically, as Tom Gieryn puts it, and attentive to the crucial role of place in the history of field science.

An Overview, with Caveats

Readers may pick up this book expecting a survey of field biology. It is not. Its subject is the subset of field disciplines that lie closest to the laboratory side of the lab-field border and that have been most strongly influenced by laboratory culture—ecology and evolution, mainly. Other important field disciplines, such as paleontology, paleoecology, and biogeography, are not covered because they are more exclusively of the field and so less apt to illustrate my theme of cultural interaction. Likewise systematic botany and zoology, which were initially part of this book but proved so difficult to fit in (too big, too purely of the field) that I left them for another book. All these sciences of the field lie largely outside the border zone, or at least did in the period I cover. They are important and understudied historical subjects, just not my subject here.

I have likewise steered clear of laboratory sciences that deal in some way with field material, such as animal behavior or population genetics. Including the lab side would have made the book intolerably long, and I wanted to position myself firmly and safely on the field side. (As my previous training and interests were in the lab sciences, the risk of backsliding into the usual one-sided laboratory perspective was real, and I wanted to steer clear of it.) I deal with both landscapes and "labscapes," but more the land than the lab.

The subject of this book, as its title indicates, is the border zone itself and the field disciplines that lie most squarely within it. I aim less at mapping and surveying the whole terrain than at analyzing its spatial and cultural dynamics: what it was like to do fieldwork in a world that took laboratory methods as universally the best for all sciences. It is about the places and practices of the field biologists who occupied the border zone, who sometimes moved between laboratory and field but were primarily field-workers; and who, even when they were in the field, were attracted to (or repelled by) laboratory values and culture.

I also cover a fairly limited domain of place and time—North America between 1890 and about 1950. The time frame is easily justified: it is the period when field biologists first came to grips with the fact that they operated in a world of labs, and when they gradually discovered how to do that without losing their field souls. The generations of 1890–1950 are the crucial ones for border-zone history and worth particular study. The geographical limits will doubtless seem less justifiable in our globalizing age, and were this book a general history of field biology, it would certainly have had to include European developments, because European ideas and practices crossed the Atlantic and influenced deeply (if spottily) what was done on this side.

However, this book is not such a history; and given the importance of place in my approach, it seemed wise and proper to limit myself to one region of the world and one that I know as historian and inhabitant. Adding examples of European places and practices would no doubt enrich my account, but I doubt it would fundamentally alter its analysis of places and practices in general.

Finally, this book is about the dynamics of historical change. It is not a snapshot or map of an unchanging cultural topography, but a history of a scientific terrain in a period of undramatic but fundamental transformation. The sciences of the lab-field border region in 1950 were nothing like what they were in the early 1890s. At the beginning of my story the lab-field border was a new and relatively shallow terrain, thinly inhabited and marked by a sense of difference rather than of possibilities of interaction. By midcentury it was a deep zone, well settled and with a range of distinctive cultural practices. It was like the Rhine frontier in the early Middle Ages: no longer a line between Roman and barbarian but the territories of the Franks, Germans, and so on. How this transformation happened on the lab-field border is my subject: how distinctive border practices and cultures gradually evolved as a result of the daily activities of a generation or two of border crossers.

A border zone was imagined before it actually existed, I will argue, when academic biologists, alarmed by the intellectual and social failures of laboratory morphology, sought to broaden its scope and public appeal by reinjecting into it elements of the older natural history. Flying the banner of a "new natural history," these lab-based biologists imagined a mixed border culture and opened possibilities for mixed practice; chapter 2 examines this process.

How this ill-defined cultural space began to take specific form is the subject of chapter 3. Much depended, I argue, on what sorts of people first occupied the border zone. In ecology a little gold rush occurred in the early 1900s, as naturalists who often were young or ill trained pursued traditional natural history, calling it ecology. Their ineptness led to traditional field methods being discredited, and leading ecologists reacted by embracing laboratory ideals and practices, in part to exclude practitioners of dubious professional standing. Border practitioners were thus saddled with a rather scientistic and unrealistic ideal of what scientific field practice must be.

Chapters 4 and 5 focus on the practices and experiences of early border dwellers as they attempted to import laboratory material culture and procedures into their field practice. I follow instruments and techniques of counting and experimentation into the border zone to make visible the cultural dynamic of border practice. Where did instruments come from, and how were they transformed by use in the field? Did they serve as vehicles of laboratory values?

Generally field biologists accepted the superiority of laboratory science and borrowed from it in ways that were virtually certain to disappoint. A generation of border crossers experienced frustrated ambitions, with effects that are examined in chapter 6—midcareer burnout, fitful movement between lab and field, and departures for less risky work in traditional disciplines.

At the same time, field biologists also pursued practices that were more of the field, homegrown practices that were appropriate to complex natural objects and places—practices of place, I call them. It was around such practices of place, I argue, that stable border cultures eventually coalesced. One such practice, common in the 1910s and 1920s, used techniques for reading natural places and phenomena as "Nature's experiments." Though "Nature" does not experiment, the layout of certain special places could be read as a record of "experiments" in succession and speciation. These practices are the subject of chapter 7.

Another kind of mixed practice appeared in the 1930s and 1940s: traditional natural-history methods amplified by simple thoroughness and quantitative techniques to the point where they were as good as laboratory experiments. In chapter 8 I describe some examples of these border practices, among them Ernst Mayr's evolutionary biogeography and Raymond Lindeman's ecosystem ecology. These border practices did not depend on laboratory models or laboratory paraphernalia but constituted cultures that transcended the old, simple distinction between laboratory and field.

The border disciplines of the later twentieth century were the result of these decades of active exchange and border traffic, and they form a broad cultural-geographic zone where there was once a shallow and uneasy frontier. This historical transformation is the principal subject of this book.

CHAPTER 2

A New Natural History

The biological world of 1890 was contested terrain. Rapid expansion of laboratories and experimental disciplines since the 1850s had created a zone of tension between an expansive, confident scientific culture and an older culture of field science that was cast as old fashioned (though in fact it was not) and pushed to the periphery of the new scientific world. The situation resembled the frontiers of expanding political empires—it was not a defensive line, but neither was it the region of cultural interaction that it would eventually become. Laboratory and field biologists were not warring camps, despite their exchanges of jibes and jokes. Their relations were more like the rivalry between generations, or among the inhabitants of rival neighborhoods.

The tension was asserted even as it was denied. Thus Herbert Spencer Jennings was taken aback when the biologist-philosopher William Ritter depicted biologists as divided into separate worlds of holists and reductionist experimenters. "I am a laboratory man," Jennings wrote his combative friend, "not because I have any theory that that is the only scientific method of work, but because, it being necessary for me to work on something specific if I am to get anywhere, this seems here and now the most direct method of getting light on [my] . . . particular problems." Most experimentalists, Jennings thought, would admit the value of Ritter's natural-history approach "if approached in such a way as not to arouse an instinctive self-protective mechanism."[1] But in the late-nineteenth century, laboratory and field biologists often did approach each other defensively. The aggressive expansion of laboratory disciplines was a living memory, and few biologists had experience in blending lab and field.

The dynamics of the lab-field border began to change around the turn of the century, and in the following decades an uneasy frontier gradually evolved into a zone of interaction and mixed border practices. In the United States this change was marked initially by a spate of public pronouncements by leading biologists that the laboratory movement had gone too far and that it would be to all biologists' advantage to combine the older natural history with the more recent experimental tradition. A "new" or "scientific" natural history—that is, a natural history that borrowed laboratory techniques and values—was conjured up. Already a few biologists had departed laboratories and herbaria to work in the field, among them the botanists Conway MacMillan and Volney Spalding, who became known as pioneering ecologists. Some enterprising zoologists were beginning to think of experiments on the mechanism of evolution.

Calls for a "new" natural history were thus partly a symptom of grassroots changes in practice. But only partly; they were also an ideological projection of a future that did not resemble the present. The new natural history was an imagined future, not a lived present. It was a stimulus and cause of changes in practice. It takes ideological work to justify new ways of doing science and to encourage individuals to choose new and risky ways over the tried and true. The spokesmen for laboratories and experiments had done just such work a generation earlier, prising power from an older generation of naturalists who controlled biological institutions. The "new naturalists" of the turn of the century, facing the unanticipated and undesired consequences of the laboratory revolution in biology, now did the same. It was the first step in the evolution of a mixed border zone. So we begin with this ideological movement, its social contexts and cultural meanings, and its consequences for biologists of the field.

To anticipate briefly, the new natural history was in part a movement of lab-trained biologists hoping to remedy the shortcomings of microscopic morphology. It began as an internal agitation for reform of a laboratory discipline that had become too narrowly focused on a few instruments and

1. Herbert S. Jennings to William E. Ritter, 23 March 1922, WER box 12.

practices and too reckless in its speculative claims. An infusion of the older natural history was for reformers one way to broaden their practice and regain credibility. The new natural history was also in part a response to developments in educational institutions that were threatening biologists' supply of talented recruits. The excesses of microscopic morphology were discouraging students from studying zoology and botany and from choosing biology for careers, laboratory biologists feared, and they hoped that a revitalized natural history would have wider appeal. Reinstating traditional natural history was not what these reformers wanted; for them natural history was a means to the larger ends of disciplinary and educational reform.

The new natural history was an idea, but it had real effects. Most concretely, it was partly responsible for new kinds of laboratories—"vivaria" and "biological farms"—that enjoyed a brief popularity in the early 1900s, and for the numerous marine and inland field stations that dotted the continent from coast to coast. We are dealing not just with an infrastructure of the imagination, but a built environment designed to encourage mixed lab-field practices. New scientific field disciplines like ecology were another solid consequence of the new naturalists' ideological work, an altered cultural geography. The agitation for a scientific natural history stimulated a generation of field biologists to devise novel practices that combined elements of lab and field. It helped create new places, in which new kinds of border practice could arise and flourish.

The New Natural History

In December 1900 in Baltimore, Edmund B. Wilson stood before an audience of his fellow zoologists and commended their revived interest in traditional natural history:

> We see the signs of this revival in many directions, in precise and far-reaching inquiries into the habits and instincts of insects and birds, and the life of animal communities; in renewed and more accurate studies on plants and animals of almost every group, in the examination of the plankton of inland waters and the sea, in the rapid development of exact statistical methods in the study of variation.[2]

2. Edmund B. Wilson, "Aims and methods of study in natural history," *Science* 13 (1901): 14–23, on pp. 18–19.

Wilson's observations encapsulate the essential features of the changes that were then percolating in biology. One was the scientific breadth of the movement: studies of wild creatures foreshadowed a discipline of animal behavior, and plankton surveys gave a quantitative shape to food relations. Biometric measurement of variation in natural populations promised a revitalized systematics and a refreshingly empirical approach to evolution. Ecologists were moving beyond descriptive floristics and devising ways of studying the dynamics of plant communities. Lab and field were beginning to interact along a broad front circa 1900.

Second, the new natural history was concerned less with structure than with function. New naturalists studied living animals and plants in their natural habitats; their activities and relations with their environments and with each other. And new naturalists studied nature not just with field methods but also with the more powerful methods of experimental science. This was for Wilson the third essential feature of the new movement: it combined in equal measure the old naturalists' breadth of vision and sympathy for living things with the experimentalists' control and precision—the best of both worlds.

Other leading American biologists echoed Wilson's appeal for a rejuvenated, scientific natural history. The zoologist Jacob Reighard heralded the rise of a "rational natural history" that studied living animals in their natural environments and reunited laboratory zoologists with amateur naturalists. Charles Otis Whitman thought a "modern natural history" or "experimental natural history," best expressed the desire of academic biologists for an appealing general biology. For Thomas Harrison Montgomery "natural history" and "biology" were synonymous. Charles C. Adams identified "new natural history" as his own discipline of ecology. Edwin G. Conklin welcomed the fashion for studying living animals at home: their development, food and foraging, enemies and friends, parasites and messmates, crossing and inbreeding, daily and nightly activities, habits and instincts. (However, he also thought that fieldwork must consist "largely of collection, classification, and scattered observations," and that "more serious work must be transferred to the laboratory.")[3]

3. Jacob E. Reighard, "The biological sciences and the people," *Science* 11 (1900): 966–72, on p. 972. Charles O. Whitman, "Natural history work at the Marine Biological Laboratory, Wood's Holl," *Science* 13 (1901): 538–40, on p. 539. Whitman, "Some of the functions and features of a biological station," *Biological Lectures, Marine Biological Laboratory* 5 (1896–97): 231–42, on p. 240. Thomas H. Montgomery, Jr., to William M. Wheeler, 26 Sept. 1908, WMW box 26; see also Whitman to Wheeler, 9 Jan. 1903, WMW box 36. Charles C. Adams, "The new natural history—ecology," *American Museum Journal* 17 (1917): 491–94. Edwin G. Conklin, "Advances in methods of teaching: Zoology," *Science* 9 (1899): 81–84, quotes on pp. 83 and 84.

Botanists were less inclined than zoologists to use the terminology of natural history but shared the vision. John Merle Coulter was the likely author of an editorial heralding "the strong movement in botany which is impelling [botanists] to emerge from herbaria and laboratories and to come in contact with the larger problems of plant-life." William Ganong, one of the best of the early ecologists, also noted the movement away from an overrefined indoor botany: "[T]his conception of the plant as a living, working, struggling, plastic being is . . . the one that excites the greatest human sympathy and interest, and therefore, is . . . the best 'method' the science has to offer." The botanist Franklin S. Earle thought amateur naturalists saw nature most truly: "The busy man, whose love of nature compels him to spend his Sundays and holidays in the woods and fields, often gains that intimate knowledge of plants as they really are, and of the relationship to their environment that is sometimes sadly lacking in the professional botanists whose horizon is bounded by his laboratory walls."[4]

The movement for a scientific natural history was not the only program of reform in biology. Many laboratory biologists simply looked to experimental physiology, and high priests of experimentalism like Jacques Loeb were at the peak of their influence in the early 1900s. If there was a mainstream position it was for a general biology that included lab and field without being dominated by either. "Biology" was the generally preferred term, rather than natural history, with its somewhat provocative implication of a rejection of modernity. Younger biologists shied away from both extremes. Jennings was typical of the new generalists, an experimentalist who also knew systematics and natural history. He came out of an interview with Loeb bemused at Loeb's "lack of acquaintance with the *animals,*—their structure etc.—just as some zoologists suffer from lack of chemistry and physics."[5] Others in the middle of the road were the biostatistician Raymond Pearl and the entomologist William Morton Wheeler. The movement for a revitalized natural history was one of a variety of reform programs within the experimental tradition, by no means dominant but widely influential.

The new natural history was usually defined less in terms of what it was than of what it was not. It was a corrective to the excesses of microscopic morphology; on that there was general agreement. The ideologues of a new natural history were themselves morphologists, and they defined the new

4. Editorial [by John M. Coulter?], *Botanical Gazette* 22 (1896): 57. William F. Ganong, "Advances in methods of teaching: Botany," *Science* 9 (1899): 96–100, on p. 98. Franklin S. Earle, "New species," *Botanical Gazette* 24 (1897): 58–59.

5. Herbert S. Jennings to Charles B. Davenport, 5 Jan. 1900, CBD-Gen; see also Jennings to William E. Ritter, 23 March 1922, WER box 12.

natural history as a solution to problems within their own discipline. The evolutionary morphology and embryology that was high fashion in the 1880s was an easy target. Of all the varieties of morphology it was the most narrowly reductive and cultish. Relying on the so-called biogenetic rule (ontogeny recapitulates phylogeny), evolutionary morphologists believed that a minute study of anatomy and embryology could reveal the long-lost history of vertebrate evolution. Attention focused narrowly on a few anatomical features that were thought to hold clues to phylogenetic origins, and on marine invertebrates, which were thought to be primitive relics of earlier stages of life. And it was done exclusively at the bench, indoors, with microtome and microscope.

Evolutionary morphology was unlike the evolutionary science of such earlier naturalists as Charles Darwin, whose working methods ranged from traditional taxonomy to backyard experiment. What gave evolutionary morphology its enormous popularity was its simplicity of method and grandiose promises of important results. To ambitious young biologists it looked like a shortcut to discoveries that would otherwise require long and uncertain labors in field and lab. As John Merle Coulter quipped, evolutionary morphology was "phylogeny made easy."[6] This promise faded rapidly in the 1890s, however, as the practical failings of the method became more evident. Mountains of descriptive data piled up, and wildly divergent reconstructions of phylogenetic history made all look like mere speculations. As the risks became clear, devotees bailed out.[7] Advocates of a scientific natural history capitalized on biologists' desire to distance themselves as quickly as possible from an embarrassing episode in their recent past. Returning to the older, broader approach to evolution meant no shortcuts, but the work was virtuous and safe.

More broadly, the new naturalists reacted against excessive reliance on microscopy in biology. The most common refrain of reformers was that microscopic morphology was too much an indoor game, too artificial and detached from nature. Edmund Wilson, who as a young man had thrown over natural history for microscopic cytology, recalled how a movement that had be-

6. John M. Coulter, "Development of morphological conceptions," *Science* 20 (1904): 617–24, on p. 620.

7. Lynn K. Nyhart, *Biology Takes Form: Animal Morphology and the German Universities, 1800–1900* (Chicago: University of Chicago Press, 1995). Garland E. Allen, *Life Science in the Twentieth Century* (New York: Wiley and Sons, 1975), chaps. 2–3. Allen, "The transformation of a science: T. H. Morgan and the emergence of a new American biology," in *The Organization of Knowledge in Modern America, 1860–1920,* ed. Alexandra Oleson and John Voss (Baltimore: Johns Hopkins University Press, 1979), pp. 173–210. Stephen J. Gould, *Ontogeny and Phylogeny* (Cambridge: Harvard University Press, 1977), pp. 167–69, 186–206.

gun as a healthy reaction against the dullness of rote systematics had become narrow, isolated, and dull. The naturalist Stephen Forbes, who had witnessed the whole cycle of hope and disappointment without taking part, took a similar view: "[T]he mere book-worm," he wrote, "is hardly narrower and more mechanical than the mere laboratory grub." William Ritter quipped that "the idea of learning biology proper in a laboratory or a museum is as preposterous as the idea of learning navigation from a toy ship on a mill pond." Alpheus Packard put zealous histologists and morphologists at the same intellectual level as the "species describers" they disdained: "Expert in the use of the microtome and of reagents, they appear to have but little more general scientific or literary culture than high-class mechanics."[8]

Laboratory instruments became objects of derision, especially the microscope, the symbol and emblem of morphologists' identity, their stethoscope or slide rule. New naturalists ridiculed the naïveté of young morphologists who treated the microscope as a kind of magic wand for quick and easy discovery. "The sooner we can get rid of the notion that a microscope is a magic instrument, which when touched transforms one into an original investigator, the better," one editor inveighed. "It is just as magical as a hand-saw, of which tool one may possess a chestful, and yet not be a carpenter." Physiological instruments were likened to instruments of torture, and polemics were waged against the unnatural practices of tormenting plants with instruments and preservatives. Volney Spalding, a botanist with ecological inclinations, ridiculed over-zealous experimentalists who seemed to feel that "a plant will never tell its story correctly until it is brought to the rack." Hermon Bumpus warned of the disease of "microtome-mania" and urged educators to prevent its spread among the young and vulnerable, lest it become an epidemic.[9]

Microscopists' ignorance of systematic zoology and botany also drew fire. Willis Blatchley, for example, poked fun at the

> evoluted histological and embryological specialist with a B.S. after his name, and a summer or two's experience at some seaside laboratory to

8. Wilson, "Aims and methods of study" (cit. n. 2), on p. 19. Stephen A. Forbes, "The Biological Station," *University of Illinois Trustees Report* 17 (1893–94): 311–18, on pp. 312–13. William E. Ritter, "The marine biological station at San Diego: Its history, present conditions, achievements, and aims," *University of California Publications in Zoology* 9 (1912): 137–248, on p. 215. Alpheus S. Packard, "A half-century of evolution, with special reference to the effects of geological changes on animal life," *American Naturalist* 32 (1898): 623–74, on p. 628.

9. Editorial, *Botanical Gazette* 16 (1891): 86–87. Volney M. Spalding, "The rise and progress of ecology," *Science* 17 (1903): 201–10, on p. 204. Hermon Bumpus, "Laboratory teaching of large classes—zoology," *Science* 1 (1905): 260–63, on p. 262.

> give him added prestige. He is an expert in the use of the microscope and microtome. He knows every detail concerning the embryology of the sea-squid and the development of the amphioxus but he don't know a jumping mouse from a long-tailed shrew, an oriole from a cat bird, nor a Hessian fly from a chinch bug. The only field of nature which he has ever explored . . . is the field beneath the lenses of his microscope.

Botanists likewise lampooned a generation of young botanists who knew everything about plant cells but nothing of their local floras.[10]

Generational differences sometimes gave the quarrel between morphologists and new naturalists a bitter edge. Microscopic and evolutionary morphology were the fast-track careers of 1880s biology, promising quick and easy successes. Their techniques were relatively easy to learn and could be practiced without the deep experience that was required for first-rate systematic zoology and botany. Celebrated for their bold quest for evolutionary origins and their sensational "discoveries," morphologists adopted the cultural style of an elite priesthood, secure in the knowledge that they had discovered the short way to true knowledge, leaving older folk plodding along in the customary ways. Evolutionary morphology appealed to ambitious young men on the make and devalued older naturalists' hard-won knowledge. It caused great resentment when youngsters fresh out of college disdained experienced systematists as outdated relics who knew nothing of real science.[11] (A similar cultural dynamic operated in genetics in the 1910s and in molecular biology in the 1950s and 1960s.)

Biologists born in the 1840s, when systematics was what the best did, watched the fashion for morphology from the sidelines—sidelined. As "new naturalists" they were most likely to lament the decline of systematics and call for its reinstatement. This generation included Stephen Forbes (born 1844), Volney Spalding (1849), Alpheus Packard (1839), Charles V. Riley (1843), William H. Dall (1845), and William E. Ritter (1856). A younger cohort was less concerned with restoring honor to systematics than tempering the excesses of morphology. Born in the 1860s, they came of age when morphology was what ambitious young men did, and they experienced it as converts from natural history, eager to make their mark in the vanguard mode. Mem-

10. Willis S. Blatchely, "The relations of the high schools of Indiana to the proposed biological survey," *Proceedings of the Indiana Academy of Sciences* 9 (1893): 199–204, on pp. 200–201. Blatchley was an entomologist and the Indiana state geologist. Coulter, "Development of morphological conceptions" (cit. n. 6).

11. See, e.g., editorial [Robert P. Bigelow?], "A plea for systematic work," *American Naturalist* 32 (1898): 350–51.

2.1 Charles O. Whitman in his columbarium 1908. Photograph by Kenji Toda. University of Chicago Archives (photographic files, Whitman, informal no. 3). Courtesy of the Department of Special Collections, The University of Chicago.

bers of this morphological cohort included Edmund B. Wilson (born 1856), Henry Nachtrieb (1857), Jacob Reighard (1861), Edwin G. Conklin (1863), Henry B. Ward (1865), William Morton Wheeler (1865), Charles B. Davenport (1866), Thomas Harrison Montgomery (1873), and Charles C. Adams (1873). Charles Otis Whitman (born 1842) was biologically older but a slow bloomer and culturally more like the generation of the 1860s. Not having been declassed, they were comfortable with microscopy and lab culture and sought to improve and reform them, not restore the *status quo ante bellum.*

The crisis of faith in morphology enabled these generations of biologists to make common cause. Charles V. Riley could welcome "the murmurings of the reactionary wave which will carry us back toward the more comprehensive methods of the older school of naturalists." And Charles Davenport, an energetic modernist, could defend the Society of Naturalists and its journal, the *American Naturalist,* as valuable vehicles for a reformed biology.[12]

The agitation for a revitalized natural history was not a reactionary or

12. Charles V. Riley, "Problems of zoology," *Science* 22 (1893): 133–34, on p. 134. Charles B. Davenport to James McKeen Cattell, 1 Feb. 1906, CBD-Gen.

fringe movement. Its keenest promoters were not field naturalists but mainstream figures. Wilson, Conklin, and Montgomery were leading cytologists; Whitman was a pioneer in cell-lineage work. They were the founders of departments and graduate programs and builders of laboratories at Chicago, Columbia, Pennsylvania, Princeton, Michigan, and Texas. They had an ecumenical vision and aimed to broaden microscopic morphology and extend experimental methods to traditional lines of zoology and botany. The idea of a scientific natural history was in their view not a retreat to the past but an advance to a better future through revival of earlier traditions.

Going forward by seeming to go backward is a tricky maneuver, common in religious revivals but rare in science, and for good reason. In a cultural activity like Western science, which takes as given that the future is always better than the present and disdains the past as an inferior state, a movement that appeals to the virtues of past practices risks being dismissed as reactionary. This was especially so in the 1890s, when many biologists could recall how hard the founders of experimental disciplines had labored to distinguish themselves from descriptive taxonomists and amateurs.

The trick was to borrow selectively from natural history: for example, older naturalists' breadth of interest in plant and animal activities—their "living interest"—was held up as a cure for microscopists' myopia. Reformers also held up naturalists' feeling for living creatures and natural places and their alliance with amateur nature-lovers. Edmund Wilson decried lab zealots' condescension to the sentimental side of natural history. Making great discoveries required imagination and love of nature: "[T]he scientific naturalist should welcome every movement toward the cultivation of general interest in natural history," he wrote, and welcomed schools' growing interest in teaching fieldwork and natural history as "a happy omen for the future of our science." Charles C. Adams celebrated the "spontaneous naturalists" who were "largely self-trained, . . . introduced to animals for the first time not in schools or in museums but by direct contact with them as they tramped the fields and woods or while hunting and fishing." Jacob Reighard thought it a fine thing that the retreat from morphology reopened biology to amateur naturalists, whose work did not require elaborate equipment. "Add to the unnatural product of our school and university labs the naturalness and love of the old-time naturalists," Henry Nachtrieb urged, "so that trained intellect will be balanced by a heart full of love and sympathy for the animals that may come under observation."[13]

13. Wilson, "Aims and methods of study" (cit. n. 2), on p. 22. Adams, "New natural history" (cit. n. 3), on p. 491. Reighard, "Biological sciences and the people" (cit. n. 3), on p. 72. Henry

Another way of making the old new is to honor saints and heroes who embody customary virtues, and the old naturalist most commonly held up as an ideal was, of course, "our great master, Charles Darwin."[14] It was not just Darwin's fame and iconic status that made him a useful model. Darwin had in fact been practicing a "scientific" natural history a whole generation before the new naturalists. It was less the Darwin of *The Origin of Species* or of barnacle taxonomy that appealed, than the Darwin of orchids and climbing plants, insect pollinators and mimics, floating seeds, and earthworms. The hero of the new naturalists was the Darwin who observed the habits and adaptations of plants and animals in their natural homes in the lawns and gardens of Down House; the Darwin who showed how backyard phenomena could reveal evidence of natural selection at work in protective coloring, mimicry, symbiosis, and parasitism.

Darwin was the exemplary scientific naturalist, who worked out of doors and observed so well that observation was as good as experiment; whose experimental methods were simple, home-made and accessible to anyone; who was both analytical scientist and empathic, intuitive nature lover. "[H]e was a great exponent of ecology . . . before it had a name," wrote Volney Spalding. "He, more than any other man before or since, worked in such sympathy with living things—not dried in the herbarium, nor tortured on the klinostat, nor pickled in formalin, but living, living in their own way—that they unfolded to him secrets they would tell no other." But for Darwin, the plant physiologist Joseph C. Arthur claimed, ecology would have "lain largely uncultivated and uninteresting." Darwin's unequalled achievements and modest demeanor were a rebuke to fast-track morphologists and cultists of the microscope. Darwin was the ideal ancestor for new naturalists.[15]

No one proposed throwing microscopes out of windows or putting a moratorium on experiment, of course. The keenest field naturalist could see that the achievements of the laboratory revolution were too important to give up. As Whitman put it, "The biology of today . . . has not too much laboratory, but too little of living nature." Stephen Forbes and his younger colleague, Charles Kofoid, foresaw a synthesis of the best of old and new: a "happy combination of the sympathetic observation of the old-time naturalist, the technical skill and searching logic of the morphologist, and the pa-

F. Nachtrieb, "Zoological stations, what they are and where they are," MS sent to Jens K. Grondahl, 28 Sept. 1899, HFN box 4 f. 47.

14. M. E. Hendriksen, "A biological station in Greenland," *American Naturalist* 39 (1905): 505–506. Whitman, "Some of the functions and features" (cit. n. 3), on p. 240.

15. Spalding, "Rise and progress of ecology" (cit. n. 9), on p. 204. Joseph C. Arthur, "The development of vegetable physiology," *Botanical Gazette* 20 (1895): 381–402, on pp. 393, 386–87.

2.2 An ecologist in the field, probably Charles C. Adams, circa 1915. Charles C. Adams, "An ecological study of prairie and forest invertebrates," *Bulletin of the Illinois Natural History Survey* 11:2 (1915): 33–276, plate 8.

tient zeal and ingenuity of the experimental physiologists." Or as Charles Adams put it, "The new natural history . . . takes the laboratory problems into the field and brings the field problems into the laboratory as never before."[16] Fieldwork alone produced undisciplined and "sentimentally colored" generalizations, while laboratory work alone was as unreal as "cloister theology."[17] The ideal of the new naturalists was a synthesis of sympathy and intellect, observation and experiment, spontaneity and control, laboratory and field. It was the blueprint of a border culture.

To be sure, it was easier to imagine a hybrid border science than to devise methods that really achieved its ideals, as we will see in later chapters. But the agitation for a scientific natural history created the potential for a larger

16. Charles O. Whitman, "A biological farm for the experimental investigation of heredity, variation and evolution and for the study of life-histories, habits, instincts and intelligence," *Biological Bulletin* 3 (1902): 214–24, quote on p. 216. Charles A. Kofoid, "The fresh-water biological stations of America," *American Naturalist* 32 (1898): 391–406, quote on p. 406; see also Stephen A. Forbes, "Special report of the Biological Experiment Station," *University of Illinois Trustees Report* 18 (1895–96): 302–26, on pp. 308–10. Adams, "New natural history" (cit. n. 3), quote on p. 492.

17. Robert M. Yerkes, "The Harvard laboratory of animal psychology and the Franklin Field Station," *Journal of Animal Behavior* 4 (1914): 176–84, on p. 181; and Ritter, "Marine biological station" (cit. n. 8), on p. 215.

cultural space in which lab and field could mingle: that is the essential point. It made a mixed culture imaginable in a way that it had not been before. The ideals of the new naturalists were a blueprint for creating a border zone of practice, in which field naturalists and laboratory experimentalists could explore, poach, borrow, share, collaborate. In the early 1900s this border region was not a reality but an idea, a kind of "imagined community," to use a concept invented by Benedict Anderson to understand quite different kinds of things (nation states, which Anderson believes can never be actual communities).[18] The lab-field border existed as an idea, an imagined place, before it was an actual one. Scientific cultures need not arise in this way, but they did in the lab-field border of biology. The idea of a hybrid culture encouraged biologists on both sides to make it a reality.

New Naturalists and Educational Reform

But the movement for a revived natural history was more than an internal reaction to microscopic morphology. There were more immediate and practical dangers to all biologists than catching a case of "microtome mania" or becoming the butt of ethnoscientific jokes. Most worrisome was that microscopists' excesses seemed to be turning off the one thing that no biologists could do without: a steady supply of new talent. It was widely felt in the late 1890s that high school and college students were losing interest in studying zoology and botany and in making these sciences their career. And the reason they had lost interest, the new naturalists believed, was that they were being fed a steady diet of dry microscopic morphology. The larger purpose of a revitalized natural history was to head off a demographic crisis.

Viewed in a wider social context, then, the new natural history was part of a broad movement for educational reform, and the intellectual crisis among morphologists became an institutional concern shared by all biologists. For most biologists the new natural history was less an end in itself than a means to a larger end, and in the long-term view the creation of lab-field border cultures in biology was a by-product of an educational movement driven by practical imperatives.

Microscopic morphologists had been able to dominate college and high school biology curricula for a number of reasons. One was an outstanding textbook and laboratory manual that made comparative anatomy and mi-

18. Benedict Anderson, *Imagined Communities: Reflections on the Origin and Spread of Nationalism* (London: Verso, 1983).

croscopy accessible to the uninitiated: namely, Thomas Henry Huxley's famous and much imitated text book (1875). Another was an educational reform movement that emphasized hands-on work over passive book learning, and a wave of laboratory construction emulating the earlier wave in German universities. Finally, there was the oversupply of young Ph.D. biologists turned out by the new graduate schools, especially Johns Hopkins and Harvard. These acolytes regarded the diffusion of the new laboratory culture as a sacred mission and themselves as a kind of scientific priesthood with a calling to spread the word and the work in non-elite institutions. In addition, microscopic anatomy was cheaper and easier to teach en masse than experimental physiology, and in the favorable social and demographic environment of the 1880s and early 1890s it became the centerpiece of modern education in biology.[19]

In the mid-1890s, however, biologists noticed an alarming decline of student interest in zoology and botany, which some blamed on the recent "improvements" in curricula. Laboratory manuals were austere, rigorous books of rules and etiquette for aspirants to an elite. Their stern discipline appealed to zealous young morphologists, but others were put off by it, especially those whose interest in biology came from direct experience of nature. As Charles C. Adams put it: "[M]any persons who already had developed an interest in live animals were repelled, and even driven from this field of activity . . . by the narrowness of their leaders and instructors. Almost every teacher interested in the natural history side of zoology can recall such examples." The paleontologist William H. Dall agreed, and C. Hart Merriam remarked caustically that the few young talents who did manage to become field naturalists were those whom poverty had prevented from going to college and falling into the hands of laboratory zealots.[20]

Contemporary testimony supports this diagnosis. the zoologist Barton Evermann, who had taught in high and normal schools in Indiana in the late 1880s, had observed that students trained only in morphology rarely took courses in biology when they got to normal school, and not one in a hundred chose to become a teacher of biology. The students who did continue were

19. Philip J. Pauly, "The appearance of academic biology in late nineteenth-century America," *Journal of the History of Biology* 17 (1984): 369–97. John P. Campbell, *Biological Teaching in the Colleges of the United States,* U.S. Bureau of Education Circular of Information no. 9 (1891), pp. 128–31. Robert E. Kohler, "The Ph.D. machine: Building on the collegiate base," *Isis* 81 (1990): 638–62. Larry Owens, "Pure and sound government: Laboratories, gymnasia, and playing fields in nineteenth-century America," *Isis* 76 (1985): 182–94.

20. Adams, "New natural history" (cit. n. 3), on pp. 491–92. William H. Dall to Carl E. Eigenmann, 24 July 1906, CIW f. Eigenmann. C. Hart Merriam, "Biology in our colleges: A plea for a broader and more liberal biology," *Science* 21 (1893): 342–55, on p. 343.

those who had acquired "the spirit of the naturalist" through experience in the field. These students were the backbone of biology classes and natural history clubs, already skilled observers and ready for a tramp in the woods. Laboratory work, Evermann thought, killed this living interest. John Merle Coulter also thought that the field was a better road than the lab to careers in botany. "What naturalist has not begun with the fever for collecting?" he asked. "And to what more natural impulse in the young can appeal be made?" Why were there no manuals of field botany as there were for microscopic anatomy, when fieldwork was "the chief means of exciting a living interest in . . . botany"? The Dartmouth zoologist Winterton C. Curtis reported that his keenest students were those with outdoor experience, and that those who took field courses were the ones who actively proselytized other students—a crucial grass-roots mechanism of recruitment. The zoologist Henry Ward thought biology was actually better taught in 1900 than in 1880, when he was a student, but a less effective filter for recruiting new generations of zoologists.[21]

Reformers also blamed morphologists' neglect of their vital connections with amateur naturalists, an important source of support and recruits. As Jacob Rieghard observed, teachers in local colleges and normal schools were the people who led natural history societies, generated local interest in science, and served as role models for aspiring young naturalists. Yet zealous young morphologists disdained knowledge of local flora and fauna and were, so to speak, lost in their own backyards. C. Hart Merriam exaggerated when he conjured up an "unfathomable abyss" separating specialists and intelligent citizens, but he was probably right to worry that biologists were alienating an important constituency.[22] If nothing else, local amateurs were often the best or only source of small gifts for field trips and collecting expeditions.

The new naturalists were not disinterested observers, obviously, and we should not take everything they said too literally. It is not clear, for example, that enrollments in biology courses really were declining. Unfortunately there are no data for colleges and next to none for high schools. What we do know is that in public high schools in 1900–01, 16.9 percent of students took botany and 10.5 percent took zoology, or about the same as took physics and

21. Barton W. Evermann, "The teaching of biology in the public schools," *Plant World* 1 (1897): 119–24. Editorials [John M. Coulter?], *Botanical Gazette* 12 (1887): 87–88, 140–41. Winterton C. Curtis to William M. Wheeler, 1 Jan. 1904, WMW box 6 f. C. Henry B. Ward, "The fresh-water biological stations of the world," *Science* 9 (1899): 497–508, on p. 507.

22. Reighard, "Biological sciences and the people" (cit. n. 3), pp. 970–71. Merriam, "Biology in our colleges" (cit. n. 20), p. 354.

chemistry (18.9 and 9.3 percent) and less than took "physiology"—a hodgepodge of medical biology, temperance, sex education, and a bit of physiology—at 22.4 percent. Since this was the only year such data were reported for botany and zoology, we do not know how they were changing. The percentage of students taking physics and chemistry was declining in between 1890 and 1906, however, and it is likely that the percentage taking the life sciences was as well.[23] On the other hand, total enrollments in both high schools and colleges were rising steadily, so it is likely that biologists' perception of a demographic crisis reflected a relative decline, as compared to more popular general education courses. But that would have been reason enough to worry.

Statistics on the employment of college graduates are also suggestive of demographic changes affecting recruitment to the life sciences. The fastest growing occupations for graduates in the 1890s were education, commerce, and "miscellaneous," that is, the diverse white-collar occupations opened up by the large institutions of an industrializing society. This underlying change in occupational structure fueled the movement for general education, and any specialized curricula, especially one like morphology, would lose out to more general curricula for nonspecialists.[24] We could see the new natural history as an attempt by zoologists and botanists to create a gen-ed package to compete with the more medically oriented "physiology."

I do not mean to suggest that natural history ever again became a staple of college curricula. It did not, for various reasons: field courses were expensive and hard to organize, and students' annual cycle of activity was just the opposite of nature's. Courses in field biology were commonly taught in summer schools and served mainly hard-core naturalists. Winters were a time for laboratory work. For these and other reasons the experimental life sciences retained their dominant position in collegiate curricula. But we can see why field biologists could for a time have seen a revived natural history as the solution to their threatened position.

For the same reasons new naturalists were also active in the reform of high school curricula. High schools were the critical level for channeling youngsters interested in nature into academic work, and they were expanding rapidly in the 1890s and early 1900s, replacing grade schools as the min-

23. *Report of the U.S. Commissioner of Education* (1900–1901): 1907; ibid. (1907): 1050–52.

24. Bailey B. Burritt, "Professional distribution of college and university graduates," *United States Bureau of Education Bulletin* 19 (1912): 10–147, on pp. 74–79. Laurence R. Veysey, *The Emergence of the American University* (Chicago: University of Chicago Press, 1965). Olivier Zunz, *Making America Corporate 1870–1920* (Chicago: University of Chicago Press, 1990).

imum requirement for college entrance. Reformed high schools were an important source of recruits and a significant job market for college graduates in biology, and professors of biology took an active role in reshaping high schools to make sure that they would be organized in a way that would serve their institutional purposes.[25] Charles Davenport and his wife, Gertrude, coauthored a high school textbook in which natural history virtually replaced microscopic anatomy. As one reviewer observed, it was "an attempt to restore the old natural history in a newer garb . . . [and] will be of value to any person who, while on his holiday trip, wishes to learn about the animals he may come across." Charles C. Adams actively promoted the teaching of ecology and fieldwork in high schools, as did John Coulter and William Ganong.[26] Midwestern botanists and zoologists were especially active, because land-grant universities were legally charged with rationalizing state systems of education. Henry Ward was expected to restructure the zoology curricula of Nebraska's high schools. In Illinois, Stephen Forbes and Charles Kofoid were responsible for seeing that high schools had accreditable courses in zoology, and they used the opportunity to inject the ideas of the new natural history into collegiate as well as high school curricula.[27]

The rival claims of laboratory and fieldwork were as much a bone of contention in fights over high school curricula as at universities. "It is scarcely necessary to emphasize the importance of the laboratory method," wrote one zoologist. "It is the only possible way; and if it cannot be adopted the boys had better be turned out in the woods to study nature first hand there." Field trips were better at least than books and pickled specimens. For another writer, however, the school laboratory was merely "a retreat for rainy days; a substitute for out of doors; a clearing house for ideas brought in from

25. Edward A. Krug, *The Shaping of the American High School 1880–1920* (Madison: University of Wisconsin Press, 1965). *Report of the U.S. Commissioner of Education* (1907): 1050–51. Philip J. Pauly, "The development of high school biology: New York City, 1900–1925," *Isis* 82 (1991): 662–88. A good contemporary source is Marion R. Brown, "The history of zoology teaching in the secondary schools of the United States," *School Science and Mathematics* 2 (1902–3): 201–9, 256–64.

26. Charles B. Davenport and Gertrude C. Davenport, *Introduction to Zoology* (New York: Macmillan, 1900). "Zoology" (review), *Popular Science Monthly* 57 (1900): 440. Charles C. Adams to Stephen A. Forbes, 29 Apr. 1904, INHS box 9. John M. Coulter, *Plant Studies: An Elementary Botany* (New York: Appleton, 1900). Coulter, "Botany in secondary schools," *Journal of Applied Microscopy* 2 (1899): 489–90. Ganong, "Advances in methods of teaching," (cit. n. 4).

27. Henry B. Ward to Charles B. Davenport, 14 July, 25 Aug. 1897, 1, 9 June 1898, 2 March 1899, CBD-Gen. Charles A. Kofoid to William E. Ritter, 26 Oct. 1900, WER box 13. Edmund James to Stephen A. Forbes, 5 Oct. 1898; Charles C. Adams to Forbes, 29 Apr. 1904; Newell D. Gilbert to Forbes, 26 March, 22 Apr. 1901, all in INHS box 9.

the outside." Any course "confined within four walls" was, he thought, "in some measure a failure."[28] Charles Kofoid wrote a colleague in 1900 that the question of biology and physiology in high schools was "a live one" and going their way "toward the natural history idea rather than the microscope and dissection in the laboratory." High school reform in California was a few years behind Illinois but progressing in the same direction, Kofoid thought.[29]

Most observers agreed that microscopy was far too advanced and specialized for high school students, but few wished to go back to systematics. As one observer of Indiana's high schools put it, the "fossils" (aging taxonomists) were as bad as the "carpetbaggers" (morphologists fresh out of college); all-round "biologists" were the preferred type.[30] But exactly what was the right balance between lab and field? Experimental biology was seen as the best, but impracticable for large classes. Natural history lacked the rigor and prestige of lab work but was appropriate general education for life in rural or suburban towns, and good training for students who were college bound. It built on childhood experience of nature and was scientific but not forbiddingly arcane. And it could make free use of the open countryside that still surrounded many towns at the time. In 1899, for example, Henry Linville found abundant animal life for his high school classes in the Fort Lee Woods, just across the Hudson River from Manhattan, and in the wilds of Staten Island. (We forget how close city and country were before the age of the automobile.) Coulter's high school botany textbook of 1899 advised frequent trips to the woods, and was praised by one reviewer as a book that could be read "by [any] boy or girl who knows out-of-door life" and so could "do no harm . . . for its touch with nature is so close." That is, it would not crush youthful interest in plants with a load of professional erudition.[31] On the other hand, field courses were hard to do in a properly scientific way. To William Holt "real laboratory work in the field" meant careful field notes, herbarium work, and training in ecology, lest it degenerate into outdoor entertainment.[32]

Natural history never did gain the position in high school and college cur-

28. Clarence M. Weed, "Zoology in the high school," *American Naturalist* 28 (1894): 1003–10, on p. 1006. Herbert E. Walter, "An ideal course in biology for the high school," *School Science and Mathematics* 9 (1909): 717–24, 840–47, on p. 719.

29. Charles A. Kofoid to William E. Ritter, 26 Oct. 1900, WER box 13.

30. Blatchley, "Relations of the high schools of Indiana" (cit. n. 10).

31. Henry R. Linville to Charles B. Davenport, 5 Apr., 4 June 1899, CBD-Gen. John M. Coulter, *Plant Relations: A First Book of Botany* (New York: Appleton, 1899).

32. William P. Holt, "The value of field and herbarium work," *School Science and Mathematics* 4 (1904): 121–28. See also E. D. Congdon, "Fieldwork in high school botany and zoology," ibid. 5 (1905): 291–96. This journal is a rich source on curriculum issues in the early 1900s.

ricula that some would have liked. But the movements for educational reform gave an immediate, practical significance to ideas of a revived scientific natural history and promoted vigorous public discussions of the pros and cons. It made what might have remained an inconspicuous reform trend within morphology into a public issue for biologists in general, and the public. And as a by-product of these discussions, I argue, biologists' eyes were opened to the possibility of new kinds of practices that combined laboratory and field methods. For about ten years professional journals were full of open debates on the comparative merits of experiment and observation, field and laboratory practice. A generation of students was taught that natural history was a proper road to professional careers in biology. Fieldwork acquired academic standing without losing its association with outdoor recreation. By advocating a revived natural history, educators opened up the cultural geography of biology, focused attention on the lab-field border and its potential, and helped imagine a future for biology in which lab and field were not at odds but formed a common culture.

Labscapes: Marine Stations and Biological Farms

The border zone of mixed practices that the new naturalists imagined was not purely imaginary. Their ideals were inscribed not only in words and projects, but in their workplaces. Intellectual and social agendas can create places appropriate to their unfolding, and imagined futures become lived practices in places designed to make them happen. "Built places materialize identities for the people, organizations, and practices they house," as Thomas Gieryn observes. "Through their very existence, outward appearances, and internal arrangements of space, research buildings give meanings to science, scientists, disciplines, and universities." If one knows how to read the evidence, such places are often the best evidence we have of tacit cultural values and imaginings. Historians of laboratory science have used the evidence of place to very good effect, to illuminate cultural systems of science from the early modern period to high-energy physics and technoscience.[33] Built and natural places likewise afford some of the clearest evidence of the new naturalists' imagined future, especially the new kinds of

33. Thomas F. Gieryn, "Two faces on science: Building identities for molecular biology and biotechnology," in *The Architecture of Science,* ed. Peter Galison and Emily Thompson (Cambridge: MIT Press, 1999), 423–55, quote on p. 423. See also Sharon Traweek, *Beamtimes and Lifetimes: The World of High Energy Physicists* (Cambridge: Harvard University Press, 1988).

laboratories that began to appear in the landscape—or "labscape"—of biology around the turn of the century.

In marine and field laboratories, and in the briefly popular "vivaria" and "biological farms," the ideals and imaginings of the new natural history are physically embodied. These places express in their arrangement and natural surroundings the idea that laboratory objects and practices had become too enclosed and needed to be reconnected with the world of nature. Even the most modest steps in that direction required some minimal alteration to the outfit and routines of standard labs: going afield to collect living or fresh "material" and keeping it alive or fresh indoors. But the new naturalists were inclined to more radical alterations to the physical spaces of laboratories and their sites. Studying the whole life of animals indoors meant making laboratories more natural, more like home to their involuntary guests. Combining laboratory work and fieldwork ideally meant relocating laboratories from urban or suburban campuses to places in the field, where natural surroundings invite a regular traffic between lab and field. These infrastructural solutions to the shortcomings of microscopic morphology afford visible and compelling evidence of the new naturalists' imagined future, because it was in these places that they meant their future to unfurl.

Marine laboratories are the oldest of these infrastructural experiments, appearing first in the 1870s in Europe and not long after in the United States. They were the places of the then-new morphology, bringing morphologists closer to the source of the marine organisms that were their favored material. Marine stations were located near deep water, where these creatures spawned, and wherever possible in attractive seaside towns, to make summer courses attractive to students and summer research a working vacation. Physically and culturally these were domesticated border places. As Philip Pauly has observed, the marine lab was "the marine faunal equivalent of a hothouse, a structure designed specifically to recreate environments that would sustain certain organisms during the period of study."[34]

Station biologists looked from their laboratory windows upon a world of

34. Philip J. Pauly, "Summer resort and scientific discipline: Woods Hole and the structure of American biology, 1882–1925," in *The American Development of Biology,* ed. Ronald Rainger, Keith R. Benson, and Jane Maienschein (Philadelphia: University of Pennsylvania Press, 1988), 121–50, quote on p. 135. See also the following. Frank R. Lillie, *The Woods Hole Marine Biological Laboratory* (Chicago: University of Chicago Press, 1944). Various authors, "The Naples Zoological Station and the Marine Biological Laboratory: One hundred years of biology," *Biological Bulletin* 168, supplement (1985). Keith R. Benson, "Laboratories on the New England shore: The 'somewhat different direction' of American marine biology," *New England Quarterly* 61 (1988): 55–78. Charles A. Kofoid, *The Biological Stations of Europe,* United States Bureau of Education Bulletin 440 (1910).

nature: "What a happy hunting ground!" one rhapsodized. "What a paradise for the naturalist! The sandy shores, the rocky points, the muddy bays, tide pools and bottoms . . . are all astir with life which he may study at first hand."[35] Yet marine stations had elaborate customs for keeping laboratory and fieldwork separate, and the more prestigious the lab the more fieldwork was subtly discouraged. At the most famous of them all, the Naples Zoological Station, biologists arrived at work in the morning to find the organisms they required waiting for them in aquaria at their benches, brought in earlier by local fishermen. Boats were available for those who wished to collect for themselves, but for most, collecting in the field was all too reminiscent of the bad old days before stations existed and morphologists had no alternative but to collect their own material, spending time that could have been more productively spent doing real work at their microscopes. It was the custom at most stations to leave collecting in the field to local fishermen, who knew where to find things. The division of labor was modern and efficient, and it was institutionalized in marine stations' networks of local collectors and supply departments.[36]

Thus despite their natural surroundings, marine stations did relatively little to unite lab and field practice. It was customary for beginning students at the Marine Biological Laboratory at Woods Hole (MBL) to be taken on field trips, to give them some sense of the natural environments in which their organisms lived, as well as a pleasant memory of station life to take home. Likewise, nature-study courses with field trips were popular among the school teachers who flocked to seaside labs to acquire needed extra credentials.[37] But these were low-status activities; the real work of marine stations was done indoors at the bench with microtome and microscope, not in the field. A few senior biologists did collect in the fields and ponds surrounding Woods Hole, William Morton Wheeler and Thomas Harrison Montgomery, for example, but they were unusual. When you came upon an overturned rock, Edwin Conklin recalled, you knew Montgomery had been there: obviously there were not many it could have been. Thomas Hunt Morgan advised Conklin to avoid Woods Hole if he wanted to do outdoor

35. Dallas L. Sharp, "The marine biological laboratory," *Science* 22 (1893): 127–28, quote on p. 128.

36. Kofoid, *Biological Stations of Europe* (cit. n. 34), pp. 11, 18–19. Bashford Dean, "Notes on the marine biological laboratories of Europe," *American Naturalist* 27 (1893): 697–707, on p. 702. Edmund B. Wilson to Charles O. Whitman, 27 Dec. 1891, CBD-Gen f. Whitman, Woods Hole, Blanchard. Lillie, *Woods Hole Marine Biological Laboratory* (cit. n. 34), pp. 87–88.

37. Edwin G. Conklin, "The Marine Biological Laboratory," *Science* 11 (1900) 333–43, on pp. 338–39. Whitman, "Natural history work" (cit. n. 3).

2.3 Collecting excursion, Marine Biological Laboratory, Woods Hole. Edwin G. Conklin, "The Marine Biological Laboratory," *Science* 11 (1900) 333–43, on p. 339.

work. Other woods were a safer bet: "[G]o to the mountains," he urged his friend, "and not to Woods Hole where the temptation to work indoors is hard to resist."[38]

Marine stations, despite their seaside location, were essentially extensions of campus labs, bound tightly by the web of teaching and supply to laboratory culture. In marine labs it was not the natural surroundings but cultural habits and customs that shaped practices most powerfully.[39] Morphologists' desire for fresh material was a harbinger of the ideal of a new natural history, but it was just a small step across the laboratory threshold. Microscopic morphology was a laboratory practice wherever it was performed, and its cultural geography is visible in the siting and spatial customs of marine labs.

The new naturalists envisioned more radical alterations to traditional labs. In fact, the most concrete images we have of a scientific natural history come from proposals for building projects. "Biological farms," for example, were facilities designed specifically for experiments on evolution. European evolutionists were the first to take up the cause in the 1890s. G. J. Romanes thought such a laboratory "the one great desideratum of modern bi-

38. Thomas H. Montgomery to William M. Wheeler, 31 July 1902, 11 July 1909, WMW box 23 f. M. Formicidae, and box 26. Edwin G. Conklin, "Professor Thomas Harrison Montgomery, Jr.," *Science* 38 (1913): 207–13, on p. 211. Thomas H. Morgan to Conklin, 30 May 1911, PU-Bio box 1 f. Biological plans and projects.

39. In the stations of the Pacific Northwest coast, with its rich littoral fauna, biologists did more ecological fieldwork. Keith Benson, personal communication.

ology." Henry de Varigny conjured up an imagined "institut transformiste," a laboratory, but not one on a city street: "No, a laboratory in the country, not encumbered with delicate and expensive apparatus but equipped modestly, its essential element an experimental garden. . . . [O]n an ample site, a little farm just for experiment, and some ordinary greenhouses stocked with essential plants." That was the essential ground plan of all such proposals: a rural complex with facilities for large-scale breeding of animals, a farm for experimental evolution.[40] Francis Galton and Herbert Spencer tried to get Andrew Carnegie interested, but that "castle—or rather menagerie—in the air . . . did not descend to solid ground."[41] Nor did any such project on that side of the Atlantic.

American biologists took a broader view of the possibilities of biological farms, which seemed to them just the place to foster a new, more scientific natural history. Charles Otis Whitman, for example, envisioned a place for experiments on every aspect of animal life, from physiology and development to life history and ecological relations. "The laboratory is too narrow," he wrote, "and the world too wide for the continuous study of living organisms, under conditions that can be definitely known and controlled." He envisioned a place that blended elements of lab and field, where animals could be studied in nature but experimentally—a "biological farm."[42] Charles Davenport's ideal facility was "a farm or an extensive zoological reserve with hothouses, breeding ponds, insectaries and vivaria of various sorts." In such a place a comprehensive program of research could combine experimental evolution and heredity with field studies of distribution and variation in nature.[43] Edwin Conklin predicted in 1899 that experimental farms would soon become as essential to a modern university as laboratories already were. A biological farm, the psychologist James R. Angell thought, was the only place where animals could be studied "with scientific rigor, and amid surroundings which, in distinction from those of the ordinary laboratory, would . . . preserv[e] . . . the genuine conditions of nature." As Daven-

40. E. Ray Lankester, E. B. Poulton, and G. J. Romanes, "An Institut Transformiste," n.d. [c. 1891], CBD-CHS-1 f. Romanes. Henry C. de Varigny, "Le transformisme expérimental," *Revue Scientifique* 47 (1891): 769–77, quote on p. 776.

41. Francis Galton to Charles B. Davenport, 29 July 1897, 16 Oct. 1902, CBD-Gen. Theodore Cockerell, review of books by J. C. Ewart, *Science* 13 (1901): 423–24 (quote).

42. Charles O. Whitman, "A biological farm," n.d. [1906–7?], on p. 4, CBD-Gen f. Whitman biological farm. Whitman, "Biological farm" 1902 (cit. n. 16), quotes on pp. 219–20, 222.

43. Charles B. Davenport, "The biological problems of today—morphogenesis," *Science* 7 (1898): 158–61, quote on p. 161. See also Davenport, "Biological experiment station for studying evolution," *Carnegie Institution of Washington Year Book* 1 (1902): 280–81. Davenport to Carnegie Institution, 5 May 1902, CIW f. Genetics director.

port put it, evolution should be studied in the places where it had occurred, but with laboratory methods and to laboratory standards.[44]

Strictly speaking no "biological farm" was ever built, but from the plans that were laid for specific sites we can get a concrete idea of its physical and cultural geography. Whitman, for example, had his eye briefly on the hinterland of Woods Hole, where he hoped to build a biological farm to complement the Marine Biological Laboratory. Where the MBL looked seaward, Whitman looked to the ecologically rich and scientifically neglected landscape of Cape Cod: to the "extensive shore fauna and flora, . . . numerous accessible islands rich in forms of peculiar interest, and in many perfectly isolated fresh-water ponds, brackish ponds, and salt-water ponds of easy control." It was in such places that evolution must have occurred, and it was there that it should be studied. It was a grand but impracticable scheme and was abandoned when the Carnegie Institution declined to underwrite it.[45]

Charles Davenport, meanwhile, was pressing a similar project on officials of the University of Chicago. (When he left in 1904, Whitman picked it up.) The site that inspired Davenport was a tract in southwestern Indiana between Lake Michigan and the Calumet River. Comprising 880 acres of lakeshore dune country and an adjacent farm of 270 acres, the place was ecologically diverse: lakeshore, dunes, swamps, peat bogs, river bottom and seasonally flooded forest, upland fields with tillable ground and pasture, and a lake and small brook (then filled with rubbish) and natural springs, all just twenty-five miles from the university and one mile from the nearest railroad stop. It seemed a perfect site for a biological farm. As Davenport and Whitman envisioned the place it was a scientific Noah's ark: six laboratories—including one for physiology with an animal house, an aviary for Whitman's pigeon work, a vivarium for work with insects and mollusks, and a lab for animal psychology—plus facilities for animal breeding and three experimental ponds. Whitman envisioned a series of artificial ponds for fish and amphibia, and several acres for experiments on the changes in habits and in-

44. Edwin G. Conklin, "Advances in methods of teaching" (cit. n. 3), quote on p. 84. James R. Angell to M. H. MacLean, 14 Oct. 1907, in McLean, "Report on the scientific and financial requirements of a proposed Biological Farm for the University of Chicago," 29 Oct. 1907, UC-Pres box 18 f. 7. Davenport, "A summer of progress in experimental evolution," 5 March 1903, CBD-Gen f. CSH beginnings.

45. Whitman, "Biological farm", 1902 (cit. n. 16), pp. 219 (quote), 222–23. On the plan's demise see the following. Whitman, "Report to the trustees of the MBL to the trustees of the Carnegie Institution," 25 Aug. 1902, part 6, CIW f. MBL. Whitman, "A marine biological observatory," *Popular Science Monthly* 42 (1893): 459–71, on pp. 460–61. Lillie, *Woods Hole* (cit. n. 34), pp. 47–62.

2.4 Carnegie Station for Experimental Evolution, Cold Spring Harbor, New York, 1906. The main laboratory and vivarium (under construction) are on the far right. Poultry runs are in the foreground, and experimental gardens behind them. *Carnegie Institution of Washington Year Book* 5 (1906), on p. 100.

stincts that occurred as wild animals were domesticated.[46] Poised on the ecological boundary between city and country, and with its mix of built, altered, and natural environments, this place—at once real and vividly imagined—makes visible the ideals of the new natural history.

Whitman never realized his dream of a biological farm: the experimental biologists at the university were not really interested (they wanted state-of-the-art labs on campus), and the project always ended up at the bottom of donor wish lists.[47] The nearest thing to a biological farm that was ever built was the Carnegie Institution's Station for Experimental Evolution at Cold Spring Harbor—Davenport's creation. It included experimental labs

46. Charles B. Davenport to William R. Harper, 16 Apr., 3 Dec. 1903, CBD-Gen. Davenport memo, 15 Apr. 1903; Grey and Heckman report on biological farm, 19 May 1903; both in Trustees Minutes, vol. 4, pp. 246, 258, UC-Trus. M. H. McLean, "Report on the scientific and financial requirements of a proposed Biological Farm for the University of Chicago," 29 Oct. 1907; Charles O. Whitman to Harry P. Judson, 11 July 1907, both in UC-Pres. box 18 f. 7. Whitman, "A biological farm," n.d. [1906–7?]; Whitman et al. to Acting President, 14 Jan. 1907; Whitman to Lloyd Morgan, n.d., all in CBD-Gen f. Whitman biological farm. Whitman et al., memo to President and Board, 11 May 1906, UC-Pres box 15 f. 12.

47. Charles O. Whitman to Lloyd Morgan, n.d. [1900–2?], CBD-Gen f. Whitman biological farm. Harry P. Judson to Wallace Heckman, 17 March 1914, Trustees Minutes vol. 4, UC-Trus.

and gardens and facilities for large-scale breeding of diverse domesticated animals, plus bird and animal rooms and terraria, even (a few years later) an artificial basement "cave" for experiments on evolution of cave forms. It was not a marine station, though it was located on salt water. Davenport's gaze was turned inland to the varied landscapes of Long Island: the sand-spits and inlets of varying ecology, salt, brackish, and freshwater ponds, open fields and woodlands, salt and freshwater marshes. The hundreds of springs pouring forth pure water of constant temperature year-round would supply experimental aquaria—a natural medium circulating through the labs. Staff were encouraged to collect a variety of wild creatures and bring them indoors for study, at least in the station's early years.[48] It was not a biological farm, exactly, but more biodiverse and integrated with its natural surroundings than any laboratory, then or now.

Labscapes: Vivaria and Field Stations

A vivarium was a more modest and affordable variant of a biological farm: a campus laboratory fitted out with facilities for keeping living animals in natural conditions, and with some experimental grounds. Three such labs were built. Two were the work of Edwin Conklin: the first at the University of Pennsylvania (1900), and another like it at Princeton University (1910). The third, at the University of Illinois, was built for the ecologist Victor Shelford in 1913. (The University of Wisconsin may have had a small vivarium, and sporadic attempts were made to have one built at the University of Chicago.) Conklin's vivarium most nearly embodied the idea of a new natural history, to which Conklin was devoted. The ideal vivarium, as Conklin imagined it, was equipped with freshwater and saltwater aquaria; terraria; hives for bees and insectaries; special rooms for amphibia, reptiles, birds, and small mammals; hatcheries for eggs of vertebrate and invertebrate animals; appliances for studying living creatures; and a synoptic collection of species. It was in Conklin's view "the most essential part of any laboratory of zoology." Standard labs, with their lecture rooms, laboratories,

48. Charles B. Davenport, reports of director, *Carnegie Institution of Washington Year Book* 1904–10. Davenport, "Some of the lines of work of a Station for the Experimental Study of Evolution," 5 March 1903, CBD-Gen f. CSH beginnings. Davenport to Carnegie Institution, 15 Feb. 1904, CIW f. Genetics director. Davenport to Willet M. Hays, 20 Feb. 1904; Davenport to Bradley M. Davis, 20 June 1904; Davenport to Raymond Pearl, 19 Feb. 1904, all in CBD-Gen. Davenport to Robert S. Woodward, 21 Oct. 1909, 1 Oct. 1912, CIW f. Genetics, buildings, quarters, grounds.

and museums, existed to support the vivarium—strong words. A separate building was best; at a minimum, greenhouses and outdoor ponds attached to a standard campus lab.[49]

The idea of a vivarium was a radical departure from normal laboratory design. In the 1890s facilities for keeping live animals were rudimentary: typically, a few aquaria and terraria relegated to "some dark corner of the cellar" near the coal and boiler rooms. (The one at Harvard consisted of a few small cages for mice, snakes, etc., in a corner of a basement lumber room.) As Conklin shrewdly observed, this disposition of space revealed all too clearly how little microscopists valued the study of living animals in natural conditions.[50] Vivaria, in contrast, were intended to embody in their spatial layout and fittings a practice that addressed all the activities of animals in their natural habitats.

Penn's vivarium came closest to Conklin's ideal: a utilitarian structure fitted out with four salt and twelve fresh-water aquaria, terraria, hives, formicaries, and insectaries, with a special room for reptiles and small mammals and another planned for birds and bats. Environments were made as natural as possible and were stocked with as many species of animals as could be induced to live there. Surrounding the vivarium building were experimental gardens and a small botanical garden with a fernery, experimental greenhouses, and an aquatic garden stocked with native and exotic species for experimental work. So attractive were the large aquaria (especially the exotic marine aquaria) and gardens that they became public resorts and tourist attractions—"a picturesque spot in the heart of the city."[51] It was a place designed for the practice of scientific natural history: at once controlled and natural, lablike and a patch of nature in an urban campus.

The Princeton vivarium was built on the same model but in a more elaborate natural setting. The building itself was a modest appendage to a large biology laboratory, but Conklin planned to turn the university's extensive grounds to scientific uses. There were large gardens on the campus, some for public use and some for research that were hidden from public view; also a

49. Conklin, "Advances in methods of teaching" (cit. n. 3), p. 84. See also Charles Zeleny, "The need of a vivarium," 23 Jan. 1911, CZ box 6 f. Vivarium early plans.

50. Edward L. Mark to Charles B. Davenport, 23 Dec. 1900, CBD-Gen. Harvard College, *Report of the President* (1888–89): 182; (1892–93): 210; (1902–03): 266. Conklin, "Advances in methods of teaching" (cit. n. 3), p. 84.

51. John M. Macfarlane, "Botany Department," *Old Penn Weekly Review* 6 (4 Jan. 1908): 1 (quote). Edwin G. Conklin, "Report of the director of the University Vivarium," University of Pennsylvania *Provost's Report* (1898–99): 153–55; (1900–1901): 163–65. Philip P. Calvert, "The University Vivarium," University of Pennsylvania *Alumni Register* 4 (1900): 13–14. Macfarlane, "The Vivarium," *Old Penn Weekly Review* 1 (14 March 1903): 6.

2.5 University of Pennsylvania vivarium and experimental garden, circa 1905. Collections of the University of Pennsylvania Archives.

small zoological park, and a "biological pond." Along a brook and in an adjacent wood, cages and retreats were arranged in which native wild animals could be kept and studied in natural conditions and used for teaching and public exhibition. A botanical garden and arboretum were planned, and Lake Carnegie and the university's extensive fields and woodlands afforded opportunities for collecting and observing.[52] For a time, campus landscaping (a favorite cause of wealthy alumni) went hand in hand with outdoor research; lab and landscape fruitfully intermingled in a natural and cultural border zone.

In contrast, the Illinois vivarium (1913) was more building than grounds. A substantial, no-nonsense laboratory with two climate-controlled greenhouse wings, it had the usual saltwater and freshwater aquaria (no pictur-

52. "Princeton University Department of Biology," n.d. [c. 1916]; "The Biological Laboratories," n.d. [c. 1916]; Edwin G. Conklin to Mr. McLapin, 6 Dec. 1910; Henry B. Thompson to Conklin, 28 June 1911; Robert A. Harper to Conklin, 5 June 1911, all in PU-Bio f. Biological plans and projects. Conklin, "Memorandum of the urgent needs of the department of biology," 9 March 1909, EGC box 43 f. Woodrow Wilson.

esque grottoes, though, at taxpayer expense), and just two shallow ponds flanking the front door to keep live fish for experiments. (Big enough to breed mosquitoes but not big enough for skating in the winter, one disgruntled local observed.) It may seem odd that an ecologist would design a vivarium so lablike, but Victor Shelford believed that the future of ecology lay in precise laboratory work, as we shall see.[53]

The fashion for vivaria was short-lived, and the three that were built never fully realized their builders' dreams. But they vividly and concretely embody the cultural geography of the new natural history. Vivaria softened the distinctions between lab and field culture that had been hardening for half a century. They brought elements of nature indoors and extended the working space of labs into their natural surroundings. In this expanded border region, lab practices could become more natural and field practices more controlled and lablike, and animals could be studied in natural conditions and by prevailing laboratory standards.

If numbers indicate success, the most successful border institutions were the field stations that dotted the North American interior. Concentrated in the lake districts of the glaciated regions and the alpine parks of western mountains, these stations were more numerous even than the coastal marine stations—some forty-eight were built between the 1890s and 1940.[54] Here is the "labscape" of modern field science, the active sites of a cultural borderland, where laboratory and field practices can meet and mingle.

In many ways field stations are inland versions of marine stations, though not always with the refinements of a seaside resort town. Some were on the fringes of the North Woods timbering frontier, like the University of Minnesota's station at Gull Lake. Others were situated in established tourist districts, and all afforded visiting scientists and students a mix of work and recreation. Like coastal labs, field stations mainly served as summer schools for colleges and universities, especially the land-grant universities of the upper Middle West. The hinterlands of inland stations were ecologically more diverse and interesting than marine shores, but culturally field labs were no less extensions of the domesticated world of campus labs. In the late 1930s

53. Charles Zeleny, "The need of a vivarium," 23 Jan. 1911; Zeleny to Henry B. Ward, 14 June 1913; Victor E. Shelford, "Research facilities in connection with the Vivarium," 1912; Shelford to Ward, 30 June 1914, enclosing "Specification for vivarium from V. E. Shelford," n.d., all in CZ box 6 f. Vivarium early plans. "The Vivarium," University of Illinois *Alumni Quarterly* 1 no. 14 (1916): 294–95.

54. Homer A. Jack, "The Biological Field Stations of the World: A Comparative and Descriptive Study," Ph.D. diss., Cornell University, 1940, vol. 2, pp. 334–59. Jack, "Biological field stations of the world," *Chronica Botanica* 9 (1945): 5–73 is a short summary. Robert E. Kohler, "Labscapes: The nature and culture of biology laboratories," unpublished MS 2001.

2.6 University of Minnesota field station at Gull Lake, Minnesota, 1893. Seated in the front, left to right, are Henry Nachtrieb, Conway Macmillan, and Josephine Tilden. The desperados surrounding them are visiting Swedish professors and graduate students. University of Minnesota Archives (photographic files, Botany).

only one-quarter were devoted exclusively to research; a third were exclusively teaching institutions, and the rest were mixed. No other country was so well supplied with field stations, and in none was education such a dominant concern.[55] It was less a demand for field research that created this vast inland labscape, than the desire of every university and college, however small, to have its own place for summer teaching.[56]

Field stations occupy an environmental middle ground between town and country. Because researchers and students need relatively undisturbed

55. Charles A. Kofoid, "Fresh-water biological stations of America" (cit. n. 16), pp. 391–92. Morton J. Elrod, "The University of Montana biological station and its work," *Science* 20 (1904): 205–12. Henry Nachtrieb to Cyrus Northrop, 5 Apr. 1900, UM, Comptroller's Papers f. 318. Charles E. Bessey, "Summer-school of botany in the mountains," *American Naturalist* 29 (1895): 845–47.

56. Jack, "Biological Field Stations of the World" (cit. n. 54), vol. 2, table 4, p. 45. Three-fifths of European field stations were exclusively for research, the rest mixed.

areas for collecting and observing, stations could not be too close to the expanding fringes of towns and suburbs. Yet they could not be in the deep outback because of the practical logistics of operating a laboratory and getting crowds of students in and out safely and cheaply. The distribution of inland stations thus maps a complex cultural ecotone: North America's scientific inner frontier, one might say, where zones of expanding settlement pressed in upon regions that remained lightly settled or had been plundered and abandoned.

Stations also map a region defined by the logistic and cultural requirements of both lab and field practice. Scouting Arizona for a site for the Carnegie Institution's proposed desert laboratory, Daniel MacDougal and Frederick Coville were most attracted to the mountain ranges rising up from the desert floor: "charming situations . . . remote from civilization, rich and remarkable in their flora . . . and altogether delightful in their surroundings." Unfortunately, what were "treasure spots for the camping naturalist" were quite impracticable for a laboratory with a permanent staff and a constant shuttle of visitors. The scouts settled on a site in the high desert on the outskirts of Tucson. The University of Michigan's station at Douglas Lake was similarly poised between civilization and the great North Woods: "distinctly an out-of-door station," Henry Gleason wrote, "out in the woods [and] . . . away from all the distracting influences of urban or sum-

2.7 The Carnegie Institution's Desert Laboratory at Tucson, Arizona, in the Sonoran desert. Carnegie Institution of Washington Archives.

mer resort life." But not too far away: just six miles from the railroad station and with daily mail and telephones, the place "embodies all the desirabilities of camp life and avoids all its discomforts." Morton Elrod's pitch for the University of Montana's station on Flathead Lake struck a similar balance: lots of nature and scenery, but also comfortable accommodations, well-tended trails, and an auto road.[57]

Station building began in earnest in the early 1900s, slightly later than marine stations, and so overlapped almost exactly the revival of natural history. Early station builders meant lake and river stations to be places in which ecology and life history could be investigated. As marine stations were the paradigmatic sites for a reformed morphology, field stations were the exemplary places for a scientific natural history. "It is the biological station," wrote Stephen Forbes, "which is to restore to us what was best in the naturalist of the old school united to what is best in the laboratory study of the new." The sentiment was echoed by Charles Kofoid, Forbes's chief lieutenant at the University of Illinois station on the Illinois River: field stations "bring the student and the investigator into closer connection with nature, with living things in their native environment," he exclaimed. "They encourage in this day of microtome morphology . . . the old natural history or, in modern terms, oecology."[58] In field laboratories, lab and field practices could be combined in a way that was not possible either in campus laboratories or on wilderness expeditions.

At least, that was the ideal; in practice, balancing lab and field practice was a perpetual tug-of-war. On the one hand, stations' natural hinterlands afforded opportunities for ecological and natural-history work, for those who wished to take advantage of them, and many clearly did wish. On the other hand, the routines of summer teaching kept field stations firmly in the force-field of laboratory culture. Homer Jack, who surveyed field stations firsthand in the 1930s, was struck by their preference for standard laboratory courses in cytology, histology, and morphology rather than courses in ecology or natural history that might exploit their natural advantages. Charles C. Adams also saw many cases where "the *field* school merely repeated the city class work, only using fresher material than in the city, and

57. Frederick V. Coville and Daniel T. MacDougal, *Desert Botanical Laboratory,* Carnegie Institution Publication no. 6 (1903): 1–58, quote on pp. 12–13. Henry A. Gleason, "The biological station of the University of Michigan," *School Science and Mathematics* 13 (1913): 411–15, quote on p. 411. Elrod, "University of Montana biological station" (cit n. 52).

58. Forbes, "Special report" (cit. n. 16), pp. 302–3, 308 (quote). Kofoid, "Fresh-water biological stations of America" (cit. n. 16), quotes on pp. 391 and 406. See also Kofoid to William E. Ritter, 7 Nov. 1912, WER box 13.

without the slightest idea that this was not a sane procedure."[59] In field stations, as in marine stations, the infrastructure of laboratory culture proved stronger than the open-ended opportunities of nature.

The relentless sprawl of towns and suburbs into the inner frontiers, fueled by automobility, made it increasingly difficult for field stations to realize the ideals of the new natural history. Adams complained in 1929 that no field station possessed a really wild area where animals could be studied in natural environments. A decade later Victor Shelford saw no place with both a laboratory and the access to undisturbed nature that he needed to test his ecological ideas.[60]

But if the new naturalists' high hopes for vivaria and biological farms were not fully realized, their plans and the institutions they did build are vivid evidence of their imagined future. And if field stations were never used exactly as the new naturalists intended, they remain the visible landmarks of a cultural terrain in which laboratory and field culture can and often do mingle.

Conclusion

Historians of biology have not been accustomed to seeing the turn of the century as a time of revived interest in natural history. On the contrary, most have depicted it as the time when biologists rejected natural history as outmoded and second-rate science. In this view biologists' "revolt from morphology" took them in one clean step from descriptive taxonomy and speculative evolutionary morphology to experimental physiology and genetics.[61] That historians could lump such different subjects as natural history and evolutionary morphology together to banish them from history is distinctly odd, but perhaps not too surprising, since that is what experimental biologists did themselves in the 1910s.

The new naturalists became invisible because experimental biologists

59. Jack, "Biological field stations of the world" (1945) (cit. n. 54), p. 27, quoting Charles C. Adams to Jack, 25 March 1940.

60. Charles C. Adams to William E. Ritter, 31 Dec. 1929, WER box 5. Victor E. Shelford, "Grassland as a site for basic research on terrestrial animals," *Science* 90 (1939): 564–65, on p. 564.

61. Garland E. Allen, "Naturalists and experimentalists: The genotype and the phenotype," *Studies in History of Biology* 3 (1979): 179–209. Jane Maienschein, Ronald Rainger, Keith R. Benson, "Introduction: Were American morphologists in revolt?" *Journal of the History of Biology* 14 (1981): 83–87; also papers by various authors, ibid., pp. 88–176. A more radical critique is adumbrated by David Magnus, "In Defense of Natural History: David Starr Jordan and the Role of Isolation in Evolution," Ph.D. diss., Stanford University, 1993, pp. 20–22.

became the dominant, expansive culture and were thus the ones who told the stories of how it happened, stories that simplified the actual history. Viewed from the laboratory, modern experimental sciences like genetics, physiology, and animal behavior had a pure, unambiguous ancestry in earlier experimental traditions. It was easy to forget that for a time many biologists expected that the road from a morphological past to an experimental future would take them through a border region of scientific natural history, believing that fieldwork would help them cast off a discredited laboratory practice. Following their subjects, historians of experimental biology likewise simplified the past. But can the new natural history perhaps be seen as merely a false start, an imagined future that turned out to be only imagined? I think not; and in any case the history of modern field biology is impossible to understand without the movement for a new natural history.

As a cultural movement the new natural history was short-lived, lasting just ten, perhaps fifteen years. By 1910 calls for a revitalized natural history had all but ceased, and the very term had been largely replaced by "experimental biology" or just plain "biology," which embodied the ideal of a general, ecumenical discipline. The ideals of a scientific natural history (if not the term) survived in border disciplines like ecology, but they were not a public issue for all life scientists. Field biologists were far from resolving the problems of operating in the field by laboratory standards, but it was not a matter for public agitation.

Why did the agitation for a new natural history fade so rapidly? Partly, I think, because laboratory biologists no longer needed it. The extraordinary achievements of the 1910s in experimental genetics, general physiology, and animal behavior gave these disciplines such a solid reputation that memories of the embarrassments of morphology were quite forgotten by an awed scientific public. Nor did experimental biologists need natural history to assure a source of supply of recruits and jobs, once a nationwide reorganization of medical education transformed American universities between 1904 or so and the early 1920s. Pushed by a powerful professional lobby and fueled by massive private patronage, medical schools and biomedical sciences became a dominating presence in most universities. Experimental biologists were one of the chief beneficiaries, acquiring an educational mission and public legitimacy that remains in place a century later. As medical schools began to require college degrees for entrance instead of high school diplomas, undergraduate biology courses became premedical requirements, greatly increasing the demand for the laboratory branches—anatomy, embryology, genetics, physiology. For the same reason, high school biology courses became preparation for laboratory biology at the college level. Laboratory biologists thus acquired a captive audience, a large pool of potential recruits, and an expansive job market.[62]

The institutional rationale for a scientific natural history thus disappeared. Natural history had always been a means to an end for laboratory biologists, and when the problems vanished that gave rise to their anxieties, their interest in the field side ceased; indeed, it became if anything a mild embarrassment, as any failed strategy will. Institutionally the field branches became pursuits for devoted but marginalized specialists. Zoologists at Cornell saw the handwriting on the wall as early as 1898, when the authorities made it clear that the new medical school and premedical courses in anatomy and physiology were going to absorb any funds that might have gone to zoology.[63] It happened sooner or later in just about every university.

Nowhere in the educational system did natural history ever achieve the central position that it seemed it might in the early 1900s. Systematics was crowded out of college curricula and became largely a museum practice. Field excursions never became a regular part of high school courses, and textbooks never featured natural history; "general" biology was a little bit of this and a little bit of that.[64] Automobility and urban sprawl were the final nails in the coffin of regular field instruction in high schools. Natural history ended up at the bottom of the educational heap in the lower grades, as nature study—kid stuff. And even there it was replaced in the 1910s by courses in preparation for life that were deemed more appropriate to an urban, industrialized society.[65]

Yet the fading of the movement for a new natural history was at bottom a sign that its essential aims had been met. Microscopic morphology ceased to be the central feature of academic training in the early 1900s, and epidemics of "microtome-mania" ceased to rage through student populations. General biology displayed the kind of ecumenical spirit that the new naturalists had envisioned. In 1917 Charles C. Adams commented on the "greater variety and . . . broader outlook" that had developed since the mid-1900s:

62. Kenneth M. Ludmerer, *Learning to Heal: The Development of American Medical Education* (New York: Basic Books, 1985). Robert E. Kohler, *From Medical Chemistry to Biochemistry: The Making of a Biomedical Discipline* (New York: Cambridge University Press, 1982). Steven C. Wheatley, *The Politics of Philanthropy: Abraham Flexner and Medical Education* (Madison: University of Wisconsin Press, 1988). Pauly, "Development of high school biology" (cit. n. 25).

63. John H. Comstock to Charles B. Davenport, 3 Nov. 1898, CBD-Gen.

64. Brown, "History of zoology teaching" (cit. n. 25).

65. Dora Otis Mitchell, "A history of nature-study," *Nature Study Review* 19 (1923): 258–74, 295–321, on pp. 295–311. Richard Olmsted, "The Nature-Study Movement in American Education," Ed.D. diss., University of Indiana, 1967. Anna B. Comstock, "The growth and influence of the nature-study movement," *Nature Study Review* 11 (1915): 5–11.

"A student is now permitted to study, in addition to anatomy and histology (which crowded aside taxonomy for a time), taxonomy, physiology, behavior, heredity, [and] ecology."[66] Perhaps that was really what most biologists wanted from the new natural history: not the thing itself but a more diverse discipline that no one specialty could dominate.

In sum, the combination of circumstances that had made the new natural history a plausible strategy of change in the 1890s no longer existed in 1910, and new circumstances gave the laboratory side an ideological and institutional weight that the field side had only briefly enjoyed.

However, the agitation for a scientific natural history had important and long-lasting consequences. It changed the cultural geography of the field sciences: physically, in the numerous field stations that dotted the continent, and intellectually, in the field practices of disciplines like ecology, as Charles Adams observed. There would have been practitioners of ecology without the public movement, but I doubt they would have been so numerous or have had such a distinct disciplinary identity. Without the impetus of a public movement, ecology would probably have remained a tendency within botany and zoology (as to some extent it was anyway), as experimental evolution was in zoology. But we are getting here into the realm of counterfactuals.

My point is that the movement to reinvent natural history opened up the potential for a border region of mixed laboratory and field practices. This was at first more an imagined reality than an actual one, as I suggested earlier; more a change in biologists' sense of what was possible than in what they actually did in their daily work. The ideological work of the new naturalists opened a zone of potential interaction, in which laboratory and field biologists could devise mixed forms of practice and reinvent their scientific identities. Two cultures do not mix easily when their practitioners operate in separate worlds, confident of the superiority of their own customs and practices. Mixing becomes more likely when it is first sanctioned ideologically and made part of an imagined common future. Individuals are inspired to act when their work is in novel ways charged with a greater ideological meaning or attached to some larger social movement.

Biologists in the lab-field border had already begun in the mid-1890s to mix lab and field practices: the first ecologists, for example; the experimental morphologists who became interested in variation and evolution; and the field naturalists who began to introduce experiment into life-history work. Some of these promising lines proved disappointing, like the classic Darwinian problems of mimicry and protective coloration, which seemed

66. Adams, "New natural history" (cit. n. 3), quote on p. 491.

too teleological to American biologists. But it was the new naturalists' ideological work that opened up possibilities of mixed practice, more than their early achievements. Once the border zone had an imaginable cultural geography, individuals could attempt to embody the idea of a scientific natural history in actual practices, though in ways that were not always quite what was anticipated or even desired.

CHAPTER 3

Border Crossings

Reform projects like the "new natural history" may expand what biologists imagine can and should be done, but do not by themselves create new mixed practices, or alter the relative standing of field and laboratory standards of what constitutes good science. The new natural history was an inspiration but not a practical guide to action. It was individual practitioners and their experiences of success and failure that shaped the border culture that in time emerged. In the last chapter we saw how a border science came to be imagined. We turn now to the initial occupation of the border region and examine how the balance tilted from field to laboratory values.

The relative authority of lab and field culture is the crucial issue. "Class feelings" may have moderated in the 1910s, as Charles C. Adams observed, but standing was still an issue for biologists who wished to work in the field and also live up to laboratory expectations. Field and lab disciplines in the 1910s and 1920s, though not antithetical, were still distinct scientific cultures with distinctive customs and standards of judgment. Field naturalists valued natural settings and locational ways of knowing—mapping and classification—as much as they did precision and causal analysis. They were more tolerant of complex problems and circumstantial evidence than experimentalists were, and more willing to sacrifice rigor to make a start on a complex problem. It was easy to believe that shared practices were a good thing, but it took time and experience actually to create them.

How lab and field values balanced out in this cultural mixing depended in part on who was doing the mixing: immigrants from labs, or field naturalists. Participation—who and how—defined the shape of the emerging border zone. Customs of participation differed in lab and field culture: laboratories were accessible only to experts, the field to people of various sorts. This difference became important when the movement for a new natural history expanded the possibilities for participation by people with a range of scientific standing—lab experts, field biologists, students, amateurs, sportsmen. In the absence of unambiguous rules of practice, the standing of new practices will depend more on the social standing of the practitioners, and participation by those of lesser scientific standing can discredit new fields of study, as we will see.[1] In labs, who participated was hardly an issue because restricted access meant that participants were much alike. But in border life, with its social fluidity, participation was a more determining cultural element. It was not just who got there first, but what they brought with them and what they did with it.

The most energetic advocates of a "scientific natural history" were laboratory-trained biologists, as we have seen, but naturalists had the field experience and knew where things were and how they worked. Who would predominate in the border zone: those on the "scientific" laboratory side, or the field naturalists? Where would shared practices develop, in labs or in the field? Given the prestige of lab methods and the restrictive rules of access to labs, there was little likelihood of experimentalists importing field methods, so it is not surprising that mixed practices were invented mostly in the field, by importing laboratory methods. The cultural geography of borrowing was decidedly asymmetrical.

Although its public advocates were mostly on the laboratory side, the idea of a scientific natural history was especially inviting to field biologists who thought their field experience would stand them well and hoped to raise their scientific credibility. Others drawn to the idea were inexperienced students who, encouraged by their new-naturalist teachers, were eager to join newly fashionable fields like ecology. As noted earlier, their enthusiastic but often ineffective pursuit of new-naturalist ideals ended up

1. On the connection between credibility and standing, see Steven Shapin, *A Social History of Truth: Civility and Science in Seventeenth-Century England* (Chicago: University of Chicago Press, 1994).

discrediting traditional field practices and causing a general retreat to stricter ideals of laboratory practice. As ideal became practice, the new natural history began to look to many ecologists like the old natural history, which was not what they had in mind.

There was a cycle of hope and disappointment in the early 1900s, and as a result the balance between lab and field cultures shifted decisively to the lab side in the following decade. Mixed practices came to be judged by laboratory, not field standards, even by field biologists themselves, who criticized themselves even where by field standards their work did not deserve it. Deference to laboratory values became a kind of reflex for field-workers, a cultural cringe that took a generation or more to get over. It was likely but not inevitable that laboratory values would predominate in the field; it was the unexpected consequence of the historical process by which biologists of varied sorts occupied a newly envisioned region of border practice. That is my thesis.

We can see this dynamic at work in two episodes: the brief fashion for biometry, and the rush into the new science of ecology, which waxed and waned between the mid-1890s and about 1910. These episodes had a boom-and-bust quality that alarmed the very biologists who initially had encouraged them and caused a general retreat to more restricted modes of participation and more traditional modes of practice. As a consequence of these cycles, promising mixed modes of fieldwork were discredited. A natural-history style of ecology was arguably more successful at first than more rigorously experimental modes, yet ecologists put more trust in laboratory methods. Biometry might have proved quite useful for studying natural populations and the origin of species, but systematists avoided it for a whole generation before rediscovering it in the 1930s. It took almost as long for many ecologists to become comfortable once more with natural-history methods.

This cultural dynamic made evolutionists and ecologists less tolerant of mixed practices, because they judged themselves by laboratory standards. It is like the inner Asian frontier of China as Lattimore describes it: the stringent social and political requirements of irrigation society made mixed agriculture seem a dangerous compromise with barbarian ways, though environmentally it worked fine. One could accept irrigation society totally or give it up for a nomadic way of life. Likewise in the border zone of biology: cultural imperatives meant adopting laboratory standards or doing natural history. It was either-or.

The Rise and Decline of Biometry

Biometrics (the statistical study of variation) originated in Britain in the 1890s in reaction to increasingly sterile and frustrating fights over theories of evolution. Its empirical treatment of variation appealed in the late 1890s and early 1900s, and collecting data on variations became quite the rage among evolutionary biologists. The most zealous and mathematically sophisticated of the English biometricians was the polymath Karl Pearson, but Pearson's evident ignorance of biology and insistence on mathematical rigor made him less influential among naturalists than were biometricians who were biologists first and mathematicians second, like William Bateson and W. F. R. Weldon. Bateson and Weldon were disenchanted morphologists who saw biometry as an escape from the maze of speculation. They worked in similar but diverging ways (then famously fell out over theoretical issues of Mendelian versus Galtonian heredity). Bateson essentially remained a laboratory morphologist (an early project to study evolution in the salt lakes of inner Asia was a costly failure), and his influential book of 1894 dealt mainly with laboratory materials. Weldon combined experimental and field approaches in his pioneering study of changes in the population of crabs inhabiting an estuary that was in the process of silting in.[2]

Although some biometric techniques required statistical training, others were quite simple and accessible. Basically it was a matter of measuring a large sample of some natural population and constructing graphs, or "polygons of variation," in which the quantitative values of characters like length or weight were plotted against their frequencies. Comparison of the polygons of different populations or the same population sampled at different times might then reveal the patterns and perhaps the causes of evolutionary change. Biometricians did little that was very different from what taxonomists did—collect, count, measure, compare; they just did it more precisely and on a larger scale.

Nor did biometric practice necessarily entail extensive fieldwork, if large collections were at hand in museums or herbaria, or even labs. It was essential to observe animals in their native habitats only if one wished to correlate variations with environmental factors, or to test biogeographers' "laws"

2. William B. Provine, *The Origins of Theoretical Population Genetics* (Chicago: University of Chicago Press, 1983), chap. 2. Lindsay Farrell, "W. F. R. Weldon, biometry, and population biology," unpublished paper, circa late 1970s. Donald A. MacKenzie, *Statistics in Britain 1863–1930: The Social Construction of Scientific Knowledge* (Edinburgh: Edinburgh University Press, 1981), chaps. 4–6. A good contemporary overview of experimental evolution is Charles B. Davenport, "A summary of progress in experimental evolution," 5 March 1903, CBD-Gen f. CSH beginnings.

of variation—for example, the rule that animals in colder regions were larger than those in warmer regions (Bergman's law) and had shorter extremities (Allen's law), or that geographical proximity correlated with phylogenetic relatedness (Jordan's law), or that animals in humid regions were darker than those in dry ones (Gloger's rule). Field biologists saw these rules as evidence that environment drove evolution. On the other hand, laboratory biologists were more inclined to believe that the motive force of evolution was internal to animals and plants—the idea of "determinate variation"—and was best studied by experiments on individual creatures in labs. Biometrics thus straddled the lab-field border. It did not automatically incline its practitioners to either the lab or the field, and it could be performed either with lablike mathematical rigor or in the "good-enough" way more typical of field practice. How it was used in actuality depended on local contexts of use.

Charles B. Davenport was the first and most influential American naturalist to take up biometrics, as a young instructor at Harvard. In 1892 he collected potato beetles during a stay at Alexander Agassiz's Newport lab and

3.1 Charles B. Davenport, circa 1903–4. American Philosophical Society Library (CBD, print collection no. 2). Courtesy of the American Philosophical Society.

gave them to two students to measure and graph. (For one student, William Tower, it was the beginning of a lifelong and disastrous obsession.) In 1899 Davenport published a textbook of biometric methods for biologists and inaugurated a large project on variation in scallops. He also inspired young biologists to study variation quantitatively, among them Frank Lutz, Roswell Johnson, Reuben Strong, and Robert Yerkes. Wherever he worked—Harvard, Chicago, Cold Spring Harbor—little groups of apprentice biometricians sprang up like flowers after rain. The zoologist Carl Eigenmann thought that the "remarkable interest in variation" among American naturalists was largely Davenport's doing.[3]

In fact, Eigenmann also deserves credit. He had long been intrigued by the unusually large variation in fishes, but he did not pursue the problem until he fell under Davenport's biometric spell during a stay at Harvard in 1894–95. Though he found Karl Pearson's works impenetrable, Eigenmann was hooked, and upon returning to Indiana he set in motion a scheme for a large-scale study of variation in nature.

Eigenmann's plan was to measure the variation of animals living in and around Turkey Lake, the site of the University of Indiana's field station, and other lakes nearby. Comparing polygons of variation from different places or in a changing population would, he hoped, turn up meaningful patterns. It was vital that he and his students work with specimens that they had themselves collected in the field, because without knowledge of the exact conditions in which creatures lived it would be impossible to draw inferences about environmental causes of variation.[4]

Collecting and measuring on such a scale required a steady supply of workers, but that Eigenmann had in abundance, in the students who cycled through his courses each year. They were eager, renewable labor (and free), and student participation in research was valued for its educational as well as its scientific fruits. American academic culture, especially in land-grant universities, placed a high value on grass-roots participation in elite forms

3. Charles B. Davenport to Robert S. Woodward, 25 Nov. 1905, CIW. John H. Gerould to Davenport, 29 May, 27 July 1894, CBD-Gen. Charles B. Davenport to Charles O. Whitman, 17 Jan. 1899, CBD-Gen. Davenport, *Statistical Methods with Special Reference to Biological Variation* (New York: John Wiley, 1899). Davenport, "On the variation of the shell of *Pecten irradians* Lamarck from Long Island," *American Naturalist* 34 (1900): 863–77. Reuben M. Strong, "A quantitative study of variation in the smaller North-American shrikes," *American Naturalist* 34 (1900): 271–98. Carl H. Eigenmann to Davenport, 17 Feb. 1901, CBD-Gen (quote).

4. Carl H. Eigenmann to Charles B. Davenport, 3 Jan., 29 Aug. 1895, 31 Dec. 1896; Davenport to Charles O. Whitman, 17 Jan. 1899; all in CBD-Gen. Eigenmann, "A new biological station and its aim," *Proceedings of the Indiana Academy of Sciences* 10 (1894): 34–35. Eigenmann, "The study of variation," *Proceedings of the Indiana Academy of Sciences* 11 (1895): 265–78.

3.2 Carl H. Eigenmann at the mouth of Donaldson's Cave, Indiana, where he performed experiments on the evolution of blind cave fishes. Courtesy of the University of Indiana Archives.

of culture like research, and rigorous observing, counting, and measuring were good discipline as well as healthy outdoor work.

Biometrics was ideal for Eigenmann's project. It did not require large investments in laboratories and expensive instruments, and the work, though laborious and "time-killing" was something even beginners could do. Students collected whatever creatures were abundant in the last week of June, repeated the process in late August, then arranged specimens by size to ready them for measurement by experienced hands. A new crop of student recruits would repeat the process each summer, gradually building up a picture of changing populations in a particular and intimately known place.[5] It was the new natural history in action.

The entomologist Vernon L. Kellogg embarked on a similar project at Stanford University in 1901. Stanford was not yet well endowed with labo-

5. Eigenmann, "The study of variation," (cit. n. 4), pp. 268–69.

3.3 Students doing ecological fieldwork on Syracuse Lake, Indiana, 1895. Carl H. Eigenmann, "Turkey Lake as a unit of environment, and the variation of its inhabitants," *Proceedings of the Indiana Academy of Science* 11 (1895): 204–17, plate 8.

ratories, but Kellogg had students eager to participate in vanguard research, and he had the Stanford campus, some fifteen square miles of undeveloped California landscape. Kellogg and his students collected and measured large series of some abundant insects—sixteen species in all—and constructed polygons of variation. The idea was to repeat the process annually and look for trends in local populations. Kellogg believed that variation was driven by some internal mechanism in determinate directions, and hoped that biometrics would make these trends visible.[6]

Enthusiasm for biometric projects was widespread around 1900. At the University of Minnesota, Henry Nachtrieb launched Francis B. Sumner on his career in experimental evolution in 1894 by giving him a collection of Philippine snails to measure. He also envisioned a long-term study of variation in the fishes of the upper Mississippi River system, making use of the Minnesota Biological Survey's floating laboratory boat, nicknamed "Megalops." Unfortunately, he lacked a museum, and without a place to keep large collections he could not begin his project. Raymond Pearl got his first experience in biometrics as a student assistant with Jacob Reighard's Great Lakes Survey, determining biometric "place modes" for local races of whitefish. ("Place modes" are ranges of variation typical of local populations.) In

6. Vernon L. Kellogg and Ruby G. Bell, "Studies of variation in insects," *Proceedings of the Washington Academy of Science* 6 (1904): 203–332. Mark A. Largent, "Bionomics: Vernon Kellogg and the defense of Darwinism," *Journal of the History of Biology* 32 (1999): 465–88.

3.4 Henry Nachtrieb and the floating field laboratory "Megalops," 1900. University of Minnesota Archives (HFN, photograph album, Department of Animal Biology).

1900 the zoologist Henry V. Wilson proposed to use the new U.S. Fish Commission's station at Beaufort, Maryland, for variation work, and Davenport planned to mobilize the summer's crop of students at Cold Spring Harbor to measure things. Reuben Strong reported in 1899 that almost everyone at the Marine Biological Station at Woods Hole that summer seemed to be measuring and graphing. In 1902 Roswell Johnson set his students to work calculating place modes for crickets, plants, frogs, and whatever came to hand.[7]

This vigorous grass-roots activity produced a bulge of biometric publications, culminating in 1900–1902 when over a dozen publications or abstracts appeared in two leading journals. A striking feature of this literature is the substantial proportion (about half) that were written by individuals who did not pursue careers as professional biologists; presumably they were undergraduates lured by their professors into a brief summer romance with bio-

7. Francis B. Sumner, "The varietal tree of a Philippine pulmonate," *Transactions of the New York Academy of Sciences* 15 (1895–96): 137–41; and Henry F. Nachtrieb, "The 'Megalops,'" *Minnesota Magazine* 6 (1899): 18–22. Raymond Pearl to Herbert S. Jennings, 11, 29 July, 2 Aug. 1901, HSJ. Henry V. Wilson, "Marine biology at Beaufort," *American Naturalist* 34 (1900): 339–60. Davenport, "The importance of establishing specific place modes," *Science* 9 (1899): 415–17. Reuben Strong to Davenport, 25 July 1899; Roswell Johnson to Davenport, 16 March 1902; both in CBD-Gen.

metry.[8] Whereas in England biometry was a top-down movement of a laboratory elite, in America it was grass-roots, based in the field, and fueled by the participation of inexperienced students. These features would ultimately discredit it, but while it lasted, the fashion for biometry was one of the main ways that biologists tried to practice the ideals of a scientific natural history.

However, the fashion was short-lived. Davenport abandoned his scallop project to get into the hot new field of Mendelian genetics. When he left Chicago in 1904, he left Frank Lutz stuck with a biometric project in a department that was "not in love with Biometry."[9] Biometric publication declined rapidly, and by 1904 journals were filled instead with cytology and genetics, the new fashion. It was boom-and-bust in less than a decade.

Why this sudden disenchantment? Scattered evidence points to the biometric work itself: it was irksome to do and seldom yielded conclusive results. Biologists shepherding flocks of inexperienced acolytes tended to choose animals that were easily collected and had characters that were easily measured, like size or color, without giving much thought to the biological significance of what was being measured. Convinced that variation was not merely random, enthusiasts had faith that measuring any character would reveal underlying "laws." In fact what they produced was masses of data that revealed no consistent patterns. Apparent trends would begin, only to reverse the next year, and a regular trend in one character would be contradicted in others. Doing biometrics felt Sisyphean.

Robert Yerkes's experience was perhaps typical. Davenport had persuaded him to extend his work on the behavior of fiddler crabs to include their variation. His initial enthusiasm waned, however, as the work proved tedious and unrewarding. He did not know what he was looking for, and the masses of data gave him no clues. After ten thousand measurements of crabs from Falmouth Harbor, he admitted that the work did "not *seem* to justify the time they have cost," though he hastened to add that someone else might see meaning where he could not. But when Yerkes then realized that all his data were worthless because he had not collected the crabs at one time, he went back to experimental behavior for good.[10] Alfred Mayer undertook a biometric study of butterflies and moths in the Brooklyn Museum's collec-

8. *Proceedings of the American Association* 49 (1900) to 53 (1903). *American Naturalist* 34 (1900) to 35 (1901). Charles B. Davenport, "The advance of biology in 1896," *American Naturalist* 32 (1898): 867–73.

9. Frank Lutz to Davenport, 30 Nov. 1907, CBD-CSH-2.

10. Robert M. Yerkes to Charles B. Davenport, 22 July, 26 Dec. 1899, 25 Apr., 18 July, 14 Oct. 1900, CBD-Gen.

tion (in part to prove that natural-history collections had modern scientific uses), but he dropped it within a year because he kept having to reinterpret his data: it was no better than metaphysics, he thought. Roswell Johnson, Reuben Strong, and Frank Lutz came to the same conclusion and turned to the more reliably productive methods of laboratory work.[11]

At Stanford, Vernon Kellogg too became disenchanted when nature failed to behave in a regular and meaningful way. His hopes rose in 1905, when the variation polygon of the Stanford colony of flower beetles continued a trend established in 1901, though others did not, which worried him. His fears were confirmed in 1910, when the Stanford population reverted to the baseline polygon of 1895. No determinate variation, just random fluctuation, and Kellogg canceled plans for genetic experiments, as there was nothing to test experimentally.[12]

Another reason for the decline of biometrics was the harsh criticism meted out to its leading practitioners. No one tried harder than Frank Weldon to live up to rigorous scientific standards, yet he was damned by biologists like William Bateson and even by Karl Pearson, his chief ally, for failing to live up to their benchmarks of purity and rigor. To laboratory biologists, experiments on variable natural populations seemed improper science. And mathematicians could not condone biologists' rough-and-ready use of statistics, however appropriate it might be to their material. Watching what happened to those who ventured into a border practice, what naturalist would wish to join them? The biometrician J. Arthur Harris reflected bitterly on Weldon's fate: "the daily care of hundreds of animals, the thousands of measurements and the drudgery of calculation," rewarded only with "sterile and hostile criticisms."[13] No wonder the fashion was short-lived.

But perhaps more than anything else, it was the boom-and-bust cycle itself that discredited biometrics, by making it seem an area that was merely fashionable, easy to get into, and attractive to opportunists and second-raters. Raymond Pearl, one of the few who did not bail out, perceived the social logic: "It is pretty easy to measure or count any old thing and write a paper about it," he wrote, "but that kind of work will never produce any sig-

11. Alfred G. Mayer to Davenport, 8 Jan. 1901, 2, 26 Feb. 1902; Reuben Strong to Davenport, 20 May, 21 Dec. 1900, 7 June 1901; Roswell Johnson to Davenport, 13 June 1902; all in CBD-Gen. Frank Lutz to Davenport, 30 Nov. 1907, CBD-CSH-2.

12. Vernon L. Kellogg, "Is there determinate variation?" *Science* 24 (1906): 621–28. Kellogg, "Is there determinate variation?" *Science* 32 (1910): 843–46. Kellogg to Charles B. Davenport, 20 Sept. 1909, 21 Oct. 1912, CBD-Gen.

13. J. Arthur Harris, "The measurement of natural selection," *Popular Science Monthly* 78 (1911): 521–38, p. 529.

nificant results. . . . In too many cases it appears as if a worker had collected a lot of data of a quantitative character, and then hunted around (often without success) to find a problem to which the data were in any way or fashion related." It did not help that few biologists took the trouble to master biometric technique and so made statistical errors.[14]

In fact, many biologists probably did bank on the high credit rating of mathematics and took up biometrics with little thought as to what purpose might be served. For example, when Roswell Johnson was stuck teaching the "cat course" (vertebrate anatomy) to undergraduates, he used the opportunity to do a biometric study of cat viscera (it was short-lived). A few years later the physiologist Reid Hunt wrote Davenport to inquire whether something could be done biometrically with the 500 guinea pigs that his lab ran through every week; it seemed such a waste not to use them.[15]

The fact that beginning students carried out much of the biometric work may also have contributed to the bust. I know of no open disapproval, but work done by green students corralled into group projects was unlikely to win respect. Even if the work was not superficial, and much of it was, it was bound to be devalued because of who produced it and how. In hindsight, harnessing biometrics to mass education probably did it more harm than good. A research area that seemed so inviting to participation proved, for that very reason, to be one in which reputations were more likely to sink than sail.

The boom-and-bust in biometry deprived field biologists of what was potentially a most useful way of analyzing natural populations. Taxonomists' brief flirtation with biometrics illustrates the point. They had good reason to take biometry seriously. The system of subspecies and trinomial nomenclature, which was then transforming taxonomic practice in zoology (especially ornithology), was quite amenable to biometrics. Unlike species, subspecies often looked so much alike that some could be distinguished only by differences in their range of variation. Despite their reputation for conservatism, taxonomists were in fact a sophisticated vanguard discipline, and the odd bit of evidence suggests that some at least regarded biometrics with wary interest.

For example, in 1899 Reuben Strong, one of Davenport's young disciples, presented his biometric work on shrikes to an audience of ornithologists.

14. Raymond Pearl to Herbert S. Jennings, 4 Feb. 1906 (quote), 27 July 1904, 8 Oct. 1905, 4 Feb. 1906, HSJ.

15. Roswell Johnson to Charles B. Davenport, 16 March, 13 June 1902; Reid Hunt to Davenport, 25 April 1905; all in CBD-Gen.

Expecting a hostile reaction, Strong was pleasantly surprised when his talk was cordially, or at least politely received. "[Frank] Chapman, C. Hart Merriam, Dr. [Jonathan] Dwight and others spoke very pleasantly of the work to me," Strong wrote Davenport, and Chapman made a point of telling him that he thought ornithologists "ought to thank me for bringing this subject before them but that it was 'very new' to them." Some older taxonomists were less enthusiastic: Joel Aseph Allen, for example, collared Strong after his talk and "showed some spleen toward statistical investigations and argued with more venom than wisdom against the possibilities of such work." But generally the ornithologists "seemed to admire the precision of quantitative study," though Strong doubted they would actually do it themselves. "They have neither the training nor the ambition to do such work when they can get a committee of their own members to recognize such new varieties as they can concoct as handles for their own names." But Strong concluded that it would be worth his time to push biometrics among ornithologists whenever he had the chance.[16]

Yet by 1910 taxonomists had rejected biometrics as an unsuitable, even alien intrusion. Why did a promising alliance collapse so completely? Pending a thorough historical study of the question, I suggest that the answer lies in the imperial attitude that a few of the more ardent biometricians adopted regarding taxonomy. They came not as potential partners but as carpetbagging reformers of what they saw as an inferior field. (Reuben Stone's crack about taxonomists wanting nothing but their names on species was a standard laboratory canard.) They tended to offer biometrics not as a complement to accepted taxonomic practices but as a replacement, and that was threatening.

Charles Davenport was probably as responsible as anyone for this diplomatic breach. Davenport made no secret of his low opinion of systematics, regarding it "as perhaps the least valuable sort of scientific work" because it was the furthest removed from experiment. Cataloging and revising seemed to him no more than necessary but profitless housekeeping chores, and he did what he could behind the scenes to divert research funds from taxonomy to experimental fields.[17] Publicly, he went so far as to propose—in all seriousness it appears—that the whole taxonomic system of genera and species should be abandoned for a system of biometric place modes. Types of animals would be defined biometrically and assigned a number, rather

16. Reuben M. Strong to Charles B. Davenport, 25 Nov. 1899, CBD-Gen.

17. Charles B. Davenport to Charles D. Walcott, 11 Oct., 18 Nov. 1904, CIW f. Genetics director.

like the Dewey decimal system of cataloging books. The skunk, for example, instead of being *Mephitis mephitis,* would be "74," New England skunks 741, New York skunks 747, and so on. It was that, he thought, or a further mindless expansion of taxonomists' trinomials and subspecies to quadrinomials or worse.[18]

Obviously this went beyond the ritual exchange of ethnoscientific insults; it was demeaning and struck at the central practice of systematics, and taxonomists took the threat seriously. Joel Asaph Allen singled out Davenport's work on place modes for his bitterest criticism, Reuben Strong reported.[19] And taxonomists responded in kind when they could. For example, C. Hart Merriam advised the Carnegie Institution that Davenport's work "has become so mathematical that it is entirely beyond the comprehension of ordinary non-astronomical zoologists" and quipped that his proposal of a laboratory for experimental evolution should be reviewed by the mathematics and astronomy committee.[20]

This was not a conflict of ancients and moderns, but a case of a potentially fruitful alliance between two vanguard groups derailed by cultural imperialism. Some taxonomists at least were prepared to see the new spirit of quantification as progress. But what biometricians offered was more like a takeover. Davenport did not want to improve taxonomy but to transform it radically and to put a different sort of scientist in charge. No wonder taxonomists felt that the smart thing was to steer clear of statistics altogether, and that is what they did, for almost thirty years.[21]

Biometry went down in field biologists' collective memory as a momentary infatuation: wearisome to perform, its promise overinflated by mathematical zealots, its results seldom worth the effort. That estimation was not entirely true, but memories of disappointed hopes can temporarily discredit potentially worthwhile practices. In fact, biometry proved to be a powerful tool for studying natural variation when evolutionary taxonomists revived it in the late 1920s. The boom-and-bust in biometry thus deprived field biologists of an effective tool for analyzing natural populations. It made the lab-field border less hospitable and drove potential settlers back to the lab-

18. Charles B. Davenport, "Importance of establishing specific place modes" (cit. n. 7). Davenport, "Zoology of the twentieth century," *Science* 14 (1901): 315–24, on pp. 316–17.

19. Reuben M. Strong to Charles B. Davenport, 25 Nov. 1899, CBD-Gen.

20. C. Hart Merriam to Henry F. Osborn, 6, 7 June 1902, CIW f. Advisory Committee Zoology. But see Merriam to Davenport, 18 Feb. 1896; Carl Eigenmann to Davenport, 3 Jan. 1895, 31 Dec. 1896; Alpheus Hyatt to Davenport, 23 Feb., 2 March 1901; all in CBD-Gen.

21. Ernst Mayr, *Systematics and the Origin of Species from the Viewpoint of a Zoologist* (New York: Columbia University Press, 1942), p. 135. Carl L. Hubbs, "Racial and individual variation in animals, especially fishes," *American Naturalist* 68 (1934): 115–28, on pp. 115–16.

oratory or to pure fieldwork. Thus Strong made his career in microscopic anatomy, and Yerkes, in experimental animal behavior; Frank Lutz, after a spell of genetics, returned to systematic entomology, and Raymond Pearl went into human biostatistics, a discipline more hospitable to biometrics.

One might suppose that the bust in biomathematics would have strengthened the natural-history side of border biology. In fact, it may have had just the opposite effect. By depriving field biologists of an appropriate and effective tool for treating natural populations, the biometry bust left the methodological field open for lablike conceptions of populations as quasi individuals or superorganisms. Such conceptions invited border biologists to use experimental methods designed for individual animals in labs and not for populations in nature. The collapse of biometry did not cause biologists to believe in superorganisms; such ideas already came naturally to those trained in laboratory morphology. But it did discredit the only viable alternative for conceiving of natural aggregations, and thereby strengthened lab-inspired approaches that in the end proved less suited to field conditions than biometry.

Ecology: Physiology or Natural History?

A similar pattern of historical development can be seen in ecology. Ecology was arguably the clearest manifestation of the new natural history in the early 1900s. (Frederic Clements observed that ecology was "as natural a rebound from the intensive laboratory research of the last twenty-five years as this was a logical reaction from the more diffusive studies of natural history.")[22] Ecologists too were plagued by chronic uncertainty over the identity and location of their discipline between lab and field. Most ecologists regarded their discipline as a kind of physiology of the field. In practice, however, ecology was a good deal more like traditional systematic botany and biogeography. Self-image and reality were thus unusually split: hence the insatiable soul-searching and self-criticism that is such a striking feature of ecologists' early history.

Ecology, like biometry, originated in Europe. It arrived in the United States in the form of two treatises on biogeography, one by Eugenius Warming (1895), the other by A. F. W. Schimper (1898). Both books were expressions of a movement in German botany to extend the concepts of experi-

22. Frederic E. Clements, "Scope and significance of paleo-ecology," *Bulletin of the Geological Society of America* 29 (1918): 369–74, on p. 369.

mental physiology to plant geography, that is, to take laboratory culture into the field.[23] The models that inspired American botanists to become ecologists were physiological in spirit but biogeographical in practice; not a compound of lab and field but a mixture of disparate elements.

The definition of ecology that proved most influential was formulated at the 1893 meeting of the International Congress of Botany, by a committee of plant physiologists chaired by an American, Joseph C. Arthur. Ecology, the committee ruled, was the branch of physiology that dealt with "the interrelations of organisms and their mutual adaptations." (Probably they had in mind topics like symbiosis, parasitism, and mimicry, which could be approached experimentally in labs.) A year later Arthur expanded this definition to include the relations between organisms and their environments, which was in fact the problem that became the core of the new discipline of ecology. Arthur called the new subfield "sociological physiology," which kept it firmly within the expanded domain of physiology but distinguished it from experimental physiology, which dealt with individual organisms not groups. In other words, ecology would be a branch of plant physiology, then an exclusively laboratory science, but a branch that would be practiced in the field on complex objects that most experimentalists shunned. Arthur anticipated that the methods of "sociological" physiology (that is, ecology) would "appeal especially to the lover of nature, [but] without losing their value as problems of the deepest scientific import," that is, without ceasing to be proper laboratory physiology.[24]

Thus conceived, plant physiology was a discipline that straddled the lab-field border, a community of practitioners sharing a devotion to laboratory values but working in different places on different objects. Roscoe Pound and Frederic Clements offered a similarly schizoid view of ecology as "simply that particular phase of physiology which is manifested in the structure and habits of plants in their various homes," and, in the same breath, as "preeminently the division of phytogeography which seeks the connection be-

23. Eugenius Warming, *Lehrbuch der ökologischen Pflanzengeographie: Eine Einführung in die Kenntnis der Pflanzenvereine* (Berlin: Borntraeger, 1896), (the original Danish edition appeared in 1895); and A. F. W. Schimper, *Pflanzengeographie auf physiologischer Grundlage* (Jena: G. Fischer, 1898). For their reception see Joel Hagen, *An Entangled Bank: The Origins of Ecosystem Ecology* (New Brunswick, N.J.: Rutgers University Press, 1992), chaps. 2–3. See also Robert P. McIntosh, *The Background of Ecology, Concept and Theory* (New York: Cambridge University Press, 1985), chap. 2; and Eugene Cittadino, *Nature as the Laboratory: Darwinian Plant Ecology in the German Empire, 1880–1900* (New York: Cambridge University Press, 1990), pp. 112–15, 146–57.

24. *Proceedings of the International Botanical Congress* (Madison: International Congress of Botany, 1894), pp. 37–39. Joseph C. Arthur, "The development of vegetable physiology," *Botanical Gazette* 20 (1895): 381–402, on p. 389.

tween causes and effects" (that is, environmental causes of distribution).[25] But could ecology be simultaneously a subdivision of two such different scientific disciplines? Generations of ecologists would struggle with this ambiguous, if not contradictory, cultural geography.

The ideal, of course, was a harmonious blend of laboratory and field methods. Volney Spalding observed that a fact created in "the Dismal Swamp or in the Sahara" was just as much a fact as one made in a laboratory; but he also acknowledged that many field problems would in the end have to be solved in physiology labs. (Ecology without physiology, he thought, was as anomalous as physiology without physics and chemistry.) Thomas Kearney made the same point: "Ecology . . . is field work, *par excellence,* yet, if one would go far in the line of investigation, he must also be at home in the laboratory and the herbarium."[26] But could ecologists reconcile such different kinds of practice as physiology and biogeography? That remained to be seen, as they worked to translate ideals into practices.

While ecologists elsewhere shared Americans' interest in a physiology of the field, none embraced the idea so wholeheartedly or suffered so acutely from the problems of treating communities experimentally. In Germany it was mainly physiological anatomists who occupied the territory that would become ecology. They worked happily with individual organisms, usually in labs, and regarded the study of vegetation in nature as someone else's concern, not theirs. French "plant sociologists," in contrast, pursued a floristic field practice close to traditional botany and showed little interest in making it experimental.[27] Thus in Europe, changes in disciplinary boundaries preserved the traditional division of labor between laboratory and field. American ecologists, in contrast, imagined a border culture that straddled lab and field, in which experimental practices would be applied equally to individual organisms in laboratories and populations in nature, and that was to cause problems.

Creating a new border discipline was especially difficult in the United States because the ground it would occupy had no clear owners. In France

25. Roscoe Pound and Frederic E. Clements, *The Phytogeography of Nebraska* (Lincoln, Neb.: Botanical Seminar, 1900), p. 161.

26. Volney M. Spalding, "The rise and progress of ecology," *Science* 17 (1903): 201–10, on pp. 207–8. Thomas H. Kearney, Jr., "The science of plant ecology," *Plant World* 2 (1898–99): 158–60, quote on pp. 158–59.

27. Cittadino, *Nature as the Laboratory* (cit. n. 23), pp. 146–48. Robert H. Whittaker, "Classification of natural communities," *Botanical Review* 28 (1962): 1–239, pp. 9–23. Rudy Becking, "The Zürich-Montpellier school of phytosociology," *Botanical Review* 23 (1957): 411–88.

and Germany large numbers of biologists with well-defined skills and identities stood ready to occupy any new subdivision of botany, and they did. But who in the United States would become "ecologists"? Physiologists? Arthur and others seemed to assume so, though plant physiology was little developed in the United States, having had to grow in the German shadow. There were not many in that quarter to colonize new terrain. Or would it be plant geographers and botanists who occupied the ground where ecology would grow? Or field naturalists, a large community with an honored history and developed traditions of field craft (though little touched by experimental culture)? What ecology would become depended on who first occupied that cultural terrain, and in North American that was up for grabs.

If Joseph Arthur and others expected physiologists to be the first ecologists, they were disappointed. Nor were ecologists trained plant geographers, though many were influenced by geography. In fact, the leading figures of this first generation, the opinion makers, were mainly botanists who were trained in the laboratory morphological disciplines (anatomy, embryology, cytology) and systematics but had a particular taste for biogeography and fieldwork. Henry Cowles (Ph.D. University of Chicago, 1898), Frederic Clements (Ph.D. Nebraska, 1898), William Ganong (Ph.D. Munich, 1894), Edgar Transeau (Ph.D. Michigan, 1904), and Henry Gleason (Ph.D. Columbia, 1906) exemplify this type. (As does Conway MacMillan, who did graduate work at Johns Hopkins and Harvard in the late 1880s without getting a degree.) A second type comprised ecological botanists who worked mainly on systematics and distribution but did not call themselves ecologists: for example, John Harshberger (Ph.D. Pennsylvania, 1893) and George Nichols (Ph.D. Yale, 1909). Besides these opinion leaders there existed a numerous rank and file of varied backgrounds—students, teachers, and amateur naturalists among them—who liked fieldwork and called themselves ecologists but lacked the elite's concern with laboratory methods. It was a very mixed lot, with a distinct difference between leaders and followers, and that gap had some unanticipated consequences.

What happened, as I suggested earlier, was that the new "ecologists" did what they knew how to do, which was descriptive natural history, and called it "ecology." As a result, this incongruity between claims and realities discredited natural-history approaches and drove better-trained ecologists toward unrealizable experimental ideals that were in hindsight less appropriate than older field practices. In this way laboratory values came to dominate an area of field biology that was initially equally open to laboratory and field practices. It was the same story as biometry, in reverse, but with the same end result: a cultural gold rush from the field side that ended up discrediting the very practices that were best suited to the study of communities and successions. Let us see in more detail how it happened.

Up "Brush Creek" and Back Again

Early ecologists' reflections on their standing and future exude a pervasive sense of crisis. With practically one voice, leading ecologists chastised themselves for the poor quality of their work, which threatened to discredit ecology, and deservedly so, they thought. Too many ecologists simply compiled species lists and described plant formations, failed to analyze their causes, and on the whole acted as if anyone, whether trained or not and competent, could be an ecologist. William Ganong, one of the best of his generation, complained that ecological papers were too often marked by a "vast prolixity" but little substance, and by pretentious but meaningless terminology. American ecologists excelled only in description, while in interpretation, "the very soul of ecology, we continue to kaleidoscope the old and familiar matter." Ecological method, he lamented, was "little better than a series of huge guesses . . . [with] no distinct methods of ecological experiment nor principles of ecological evidence." Howard Reed, a young veteran of the Michigan ecological survey and an aspiring plant physiologist, echoed Ganong's lament. "Ecology" was proclaimed as something new, Reed wrote, but what was actually published were "descriptions and photographs of the flora of 'Brush Creek' and 'Driftwood Lake.'" Reed found American ecologists to be ill trained in physiology and morphology, unoriginal, and uncreative; moreover, their work was not up to German standards—all in all, a poor show.[28] The derogatory association of ecology with casual outings in rural countrysides is clear enough: "Brush Creek," "Driftwood Lake."

Frederic Clements likewise condemned most ecological work as little more than descriptive floristics. Henry Cowles went even further, stating that most current American ecology was simply false. Burton Livingston later recalled how "the first fine frenzy of ecological interest" had produced a large descriptive literature. "[I]t is almost startling," he wrote, "to realize how ecology has come into being and grown to large estate with little more

28. William Ganong, "The cardinal principles of ecology," *Science* 19 (1904): 493–98, quotes on pp. 493–94; and Ganong, "Advances in methods of teaching: Botany," *Science* 9 (1899): 96–100, quote on p. 99. Howard S. Reed, "A brief history of ecological work in botany," *Plant World* 8 (1905): 163–70, 198–205, on pp. 163 (quote) and 204.

3.5 "Driftwood Lake"—actually, First Sister Lake in the Huron River Valley just west of Ann Arbor, Michigan—a favorite site for student ecological fieldwork. Lewis H. Weld, "A survey of the Huron River valley, II: A peat bog and morainal lake," *Botanical Gazette* 37 (1904): 36–52, on p. 37.

purpose than that of classification."[29] Senior botanists were also critical. Charles E. Bessey, Clements's mentor at the University of Nebraska, welcomed a report by the National Education Association criticizing ecology for haziness and guessing. "That is good," Bessey wrote a colleague. "That is exactly what nine-tenths of the ecological stuff is now-a-days. It is 'haziness and guessing.' Of course . . . there is such a thing as good, solid work in Ecology, but there is mighty little of it being done." John Merle Coulter, Cowles's department chief at Chicago, regarded ecology as promising but still "vast and vague" and short on solid morphology and physiology. The sentiment was widespread. Ecology, one high-school teacher warned, was a "dangerous" subject, tempting students to speculate and thereby undoing the good effects of rigorous, honest anatomy. Hard words! And these published statements, it seems, were just the visible bit of a critical groundswell that was, according to Ganong, "unpublished but wide-spread."[30]

29. Frederic E. Clements, *Research Methods in Ecology* (Lincoln, Neb.: University Publishing Co., 1905), pp. 1–4. Henry C. Cowles, "Research methods in ecology," *Botanical Gazette* 40 (1905): 381–82. B. E. L[ivingston], review of *Animal Ecology,* by Charles Adams, *Plant World* 17 (1914): 161–63.

30. Charles E. Bessey to A. F. Woods, 24 June 1903, quoted in Ronald Tobey, *Saving the Prairies: The Life Cycle of the Founding School of American Plant Ecology, 1895–1955* (Berkeley:

But who exactly were the guilty parties who were allegedly dragging ecology into well-deserved disrepute? Nowhere in any published critique are individual ecologists named or their works identified. Moreover, the published works of early ecologists do not seem to warrant such lambasting: descriptive, yes, but sound and in no way contemptible. Who were the sinners, and where is the evidence of their sins?

Amateurs and botanizers took some of the blame for indulging in a bit of fieldwork on their summer vacations and calling it "ecology." Clements, for example, wrote that "Floristic [ecology] . . . lends itself with insidious ease to chance journeys or to vacation trips, the fruits of which are found in vague descriptive articles, and in the multiplication of fictitious formations." The ease of describing landscapes and taking snapshots, plus the appeal of a voguish and accessible new science, was a fatal attraction for local botanists. "The bane of . . . ecology," Clements concluded, was "the widespread feeling that anyone can do [it]." Henry Cowles likewise condemned the "many 'contributions' to ecology which consist of a hasty gathering together of notes made in leisure moments during summer holidays." Real ecology, he asserted, was hard work, not "dilettante ecology," and he hoped that books like Clements's handbook would prevent it from falling into "a swift and merited disfavor" by keeping faddists out.[31]

Yet it was not just amateurs who were to blame, but also professional ecologists themselves. Ganong commented on the "'get-rich-quick' spirit" that drew ill-trained people into the field and admitted that he himself had been "one of the chief of sinners." Henry Gleason recalled how the work of Clements and Cowles had drawn a crowd of imitators, including himself: "Attracted by the novelty of their results and doubtless influenced also by the apparent ease with which results could be obtained, a host of others undertook ecological investigations, mostly by purely observational methods. Throughout the country, young ecologists, of whom I was one, descended on the dunes, the shores, the marshes, and the bogs and presently returned to the laboratories to write voluminous accounts of their observations."[32] If it

University of California Press, 1981), pp. 38–39. John M. Coulter, "Development of morphological conceptions," *Science* 20 (1904): 617–24, on p. 624. Elma Chandler, "The relative emphasis to be given physiology, morphology, ecology, and other phases of botany and zoology," *School Science and Mathematics* 6 (1906): 393–97, on p. 396. Ganong, "Cardinal principles" (cit. n. 28).

31. Clements, *Research Methods* (cit. n. 29), pp. 6–7. Henry C. Cowles, "Research Methods" (cit. n. 29).

32. Ganong, "Cardinal principles" (cit. n. 28), pp. 493–94. Henry A. Gleason, "Twenty-five years of ecology, 1910–1935," *Memoirs of the Brooklyn Botanical Garden* 4 (1936): 41–49, on pp. 41–42.

3.6 Henry C. Cowles (top right) and Homer L. Shantz in the field, clowning with students. University of Chicago Archives (negative no. AEP-ILP 243). Courtesy of the Department of Special Collections, The University of Chicago.

sounds as though Gleason and Ganong were talking about their students, they probably were.

The place to see "Brush Creek" ecology is not in the work of people like Gleason and Ganong, but in a spate of papers read at meetings of the American Association for the Advancement of Science and the Indiana Academy of Science. Beginning with a few per year about 1896, they peaked between 1898 and 1902 (no fewer than fifteen) before tapering off. (By 1910 it was back to experimental papers, with the occasional one on biogeography or systematics.) These papers display the qualities deplored by critics: descriptive and repetitious, some mere lists of species tarted up with the language of Warming and Cowles.[33]

Some are written as stories of field trips, with landscape description and local animal lore, geography, settlement history, and meteorological data.

33. Items in *Proceedings of the AAAS,* 1895–1910, and *Proceedings of the Indiana Academy of Science,* 1896–1909. A sampling of other state academy proceedings turned up few ecology papers.

Some are written in the first person and open with an itinerary and a description of the countryside surrounding the object of study—often a lake—as "a sort of frame for the picture of the lake itself," as one author put it. Verbal landscape pictures are supplemented with photographs.[34] They are reports of summer excursions, in the narrative mode of sporting and vacation magazines—definitely "Brush Creek" ecology.

Who the authors of these papers were we do not know (they do not appear in *American Men and Women of Science*), but most likely they were participants in summer projects like Carl Eigenmann's at Turkey Lake. Undergraduates enjoying a scientific vogue, high-school biology teachers brushing up on nature study—were these the sinners that Clements and Cowles and Ganong feared would discredit ecology? If they were—and I think they were—it would explain why the critics did not like naming names, since they would hardly disparage the students whom they themselves had encouraged to play ecologist. Hence, too, the sense of personal guilt that ecologists like Ganong obviously felt.

Ecologists had in fact actively encouraged local naturalists to join them, pointing openly to the need for descriptive work. "The first great need of Phytobiology, . . ." William Ganong wrote in 1894, "is accurate observation of fact in the field. . . . Field study of how plants behave in relationship with the external world is the great aim which the local botanist should keep clearly before him." Because it did not require long-distance travel, as modern taxonomy did, ecology was an ideal occupation for local botanists. Thomas Kearney likewise valued "the making of word pictures of vegetation" as the foundation of ecology and saw it a job for local botanists. Kearney, a botanist and self-taught ecologist with the U.S. Department of Agriculture, exhorted teachers, students, and amateur botanists to help build the new discipline. "It should be the task of each student of a local flora to present such a picture," he enjoined. Botanists like C. Stuart Gager encouraged high-school students to study and map local plant formations. Local naturalists asked Stephen Forbes to suggest good "ecological" projects: C. Walton Clark, for example, was "anxious to take up some small lake, stay with it the year round and follow out a few of the many lines of investigation it would suggest."[35] For "ecologists" whose range was limited to their home locales, Driftwood Lakes were the easy or the only choice.

34. H. Walton Clark, "Flora of Eagle Lake and vicinity," *Proceedings of the Indiana Academy of Science* 17 (1901): 128–92. E. B. Williamson, "Biological conditions of Round and Shriner Lakes, Whitley Cty., Ind.," *Proceedings of the Indiana Academy of Science* 15 (1899): 151–55.

35. William Ganong, "An outline of phytobiology," *Bulletin of the Natural History Society of New Brunswick* 12 (1894): 3–15, on pp. 3–5, 9 (quote), 10; see also Ganong, "Notes on the natu-

Ecologists' encouragement of students and amateurs was a habit deeply rooted in botanists' professional culture and history. Except perhaps for ornithology, botany had the longest undefended border with the world of amateur naturalists of any field science. Professional botanists have always valued and depended on amateur participation. *Botanical Gazette* and *Plant World* are full of specific advice to local botanists on how (and how not) to contribute to the science. The terms of this partnership were changing in the 1890s and early 1900s, as the increased technical demands of the new systematics made it almost impossible for local botanists to participate; indeed, amateur species-hunters were becoming a serious nuisance to professional taxonomists. And the collapse of this traditional social relation made ecology an appealing substitute. Describing and cataloging local vegetations was something useful for local botanists to do that would keep them from muddying the taxonomic waters.

The partnership between professional and amateur botanists was not a one-way street. Professional botanists sometimes adopted the literary voice of botanizers and sportsmen. Reports of botanical collecting excursions took the form of "tramping" stories, with an itinerary, details on the company and guide, campsites, Indians encountered in the outback, and description of landscape and terrain, as well as tips to good collecting places and botanical finds. This genre of botanical travelogue was not appropriate for scientific reports by then, but professional journals like *Botanical Gazette* continued to publish them in order to keep paying subscribers. As one editor (probably John Coulter) noted, "a very large number of our readers are chiefly interested in the unprofessional notes that come, free from the smell of the laboratory, with the freshness of the open air about them." "Fresh-air" pieces were featured in *Plant World,* which was inaugurated in 1897 specifically for an amateur audience, and leading botanists presented the latest science there in popular form. John Harshberger's narrative of a "tramp" up Mount Ktaadn is typical. Other pieces segued from tramping into the latest ecology, as Charlotte King did, for example, in a report of a "summer outing" along prairie rivers in Iowa, where "the student of ecology may see examples of the progress of vegetation formation, from those of the mud-flat and water's edge to the more stable formations of the higher islands."[36] Ecologists uncertain of their scientific standing might well come

ral history and physiography of New Brunswick," ibid. 17 (1899): 122–35, on pp. 131–33. Kearney, "Science of plant ecology" (cit. n. 27), quote on p. 160. C. Stuart Gager, "Botanical geography for schools," *Plant World* 8 (1905): 81–82. C. Walton Clark to Stephen Forbes, 1 Aug. 1904, INHS box 9.

36. Editorial [by John Coulter?], *Botanical Gazette* 13 (1888): 62–63; a typical "fresh-air" piece is L. N. Johnson, "A tramp in the North Carolina Mountains, II," *Botanical Gazette* 13 (1888): 318–21.

to regret this connection with a popular narrative genre and condemn their own efforts in the genre as amateurish.

Ecologists' scientific practices also resembled those of amateur naturalists in some ways. The first step in ecological research was usually to survey and list the species of a place and carefully observe and record the environment, which was essentially what naturalists and botanizers had always done. Ecologists, of course, did not stop with listing and describing, but went on to work out the dynamics of plant associations and successions. And there were important differences of aim and practice: local botanizers had an eye for rare species, while ecologists were interested in common species and their distribution. Collectors confined their work to counties and states, while ecologists studied natural biotic units across political boundaries. Traditional county or state floras listed species in Linnaean order, whereas ecologists grouped species by types of community or environment. The one group wanted to know what species were in their locality and where to find them; the other wanted to why plants grew where they did. But the actual work was similar, and it is no accident that the first major works of American ecology were regional or state floras or faunas, recast and adapted to ecological ends.[37]

Charles C. Adams noted how natural a transition it was for taxonomists, who knew species and their habits, to become ecologists. "To those who like the descriptive aspect of taxonomy," he wrote, "ecological studies also offer a new field for further description and classification." However, he also discovered, as leader of a biological survey of Northern Michigan, how difficult it was to get inexperienced field assistants to think and act ecologically, to resist the habits of collectors, pass by rarities, and keep detailed notes on distribution and habitat. "Such a change," he observed, "requires a modifica-

John W. Harshberger, "A botanical ascent of Mount Ktaadn, Me.," *Plant World* 5 (1902): 21–28. Charlotte M. King, "A summer outing in Iowa," *Plant World* 5 (1902): 222–25. For another such piece, see Louis H. Pammel, "Some ecological notes on the Muscatine flora," *Plant World* 2 (1898–99): 182–86.

37. Regional and state ecological surveys include Conway MacMillan, *Metaspermae of the Minnesota Valley,* Report of the Geological and Natural History Survey of Minnesota, Botanical Series 1 (Minneapolis, 1892). Pound and Clements, *Phytogeography of Nebraska* (cit. n. 25). Thomas H. Kearney, Jr., "The plant covering of Ocracoke Island: A study in the ecology of the North Carolina strand vegetation," *Contributions of the U.S. National Herbarium* 5, no. 5 (1900): 261–319. Henry C. Cowles, "The physiographic ecology of Chicago and vicinity: A study of the origin, development, and classification of plant societies," *Botanical Gazette* 31 (1901): 73–108, 145–82. Victor E. Shelford, *Animal Communities in Temperate America as Illustrated in the Chicago Region* (Chicago: University of Chicago Press, 1913). Charles C. Adams et al., *An Ecological Survey of Isle Royale,* Report of the Board of the Michigan Geological Survey, 1908 (Lansing, Mich., 1909).

tion of the habits of the mind [and] . . . reversion to the older attitude of mind is very easy." John Coulter likewise observed how different it was to make a list of species and to analyze a place ecologically, "a difficult bit of field work, calling for training and good judgment."[38]

But the similarity of the actual work of botanizing and ecology made it easy for students and amateur naturalists to think that listing and describing was ecology and to overlook real differences in practice. They thought they could do "ecology" even if they did not yet grasp the science, and that they could publish their work as the descriptive prologue of a future ecological project. The result was a literature that consisted too often of species lists and landscape descriptions without analysis of biological mechanisms and principles—in short, the "Brush Creek" ecology that so alarmed people like Frederic Clements and Henry Cowles.

Ecologists responded to this unexpected turn of events by retreating from the ideal of a scientific natural history and seeking salvation in quantitative and experimental methods. Within a few short years ecologists who had been promoting student and amateur participation were calling instead for strict laboratory method and rigor. William Ganong declared in 1904 that observational work had reached a point of diminishing returns—just eight years after he declared it to be ecologists' greatest need—and he argued that further progress would come from precise experiments on the physics of the environment and physiological life histories of plants. What was needed was new instruments, new measuring practices, and field laboratories, he announced, and it was a general chorus. John Coulter asserted that the progress of ecology would be measured by ecologists' "experimental works conducted upon a definite physiological basis." Field description and classification came to be seen as mere preliminaries to experiment. Howard Reed looked forward to the day when "experimental work is the regular thing, not only in the laboratory but in the field." Field laboratories and full-time experimenters would replace spare-time ecologists who did their work on "hasty midsummer vacation trip[s]." Frederic Clements's handbook of ecological methods (1905) was all counting, measuring, and experimenting, and it was hailed by ecologists as a blueprint for their future discipline.[39]

Where once fieldwork was seen as a good place to begin a career in sci-

38. Charles C. Adams, *Guide to the Study of Animal Ecology* (New York: Macmillan, 1913), pp. 8–9. Editorial [by John C. Coulter?], *Botanical Gazette* 21 (1896): 303–4.

39. Ganong, "Cardinal principles" (cit. n. 28), pp. 493–94. John M. Coulter, "Development of morphological conceptions," *Science* 20 (1904): 617–24, on p. 624. Reed, "Brief history of ecological work in botany" (cit. n. 28), on p. 204. Clements, *Research Methods in Ecology* (cit. n. 29).

ence, now it was the laboratory that seemed less risky. Henry Cowles stated that direct experience of nature, without prior laboratory training, was likely to mislead. Witnessing the striking adaptations of plants and "the panorama of succession," ecologists undisciplined by laboratory work found it all too easy to see vital forces and teleology. And when they did, what wonder that "some of our best biologists have seen naught in ecology but superficial vaporings or scientific nonsense?" Cowles looked to experimental physiology for prophylaxis: a good dose of laboratory culture would immunize ecologists against the intellectual diseases that could be caught from direct experience of nature.[40] Things had indeed changed since the mid-1890s, when too much laboratory work seemed the problem and new naturalists sought authentic knowledge in immediate, personal experience in the field.

Ecologists' embrace of laboratory ideals and practices also signaled a change in the social customs of recruitment. The laboratory—not the open door of fieldwork—became the well-guarded point of entry into ecological careers. Excursions to "Driftwood Lake" no longer sufficed. Laboratory culture, ecologists hoped, would attract a better quality of recruits and discourage those with a taste for outdoor life. William Ganong, who had once encouraged amateurs to emulate Charles Darwin's backyard experiments, now had second thoughts. By the early 1900s he was insisting that proper experiments had to be done in proper labs. Ecologists turned labwards in part because labs controlled access to the discipline and ensured a level of work (and workers) that laboratory biologists would respect. Quantitative and experimental methods, because they were arduous, would keep out the tyros whose presence was tarnishing ecologists' credibility. The turn to the laboratory was in part a response to a social problem: the discipline had grown too fast and attracted the wrong sort of people. For ecology to be a physiology of the field, it would need lab-trained biologists.

Frederic Clements, with his characteristic lack of guile, put it bluntly: "The employment of instruments of precision . . . demand[s] much patience and seriousness of purpose upon the part of the student. As a consequence, there will be a general exodus from ecology of those that have been attracted to it as the latest botanical fad, and have done so much to bring it into disrepute." Clements and John E. Weaver put it even more bluntly in 1924: "It is certain that much work of purely superficial or local character will continue to be done under the name of ecology, but the touchstones of

40. Henry C. Cowles, "The trend of ecological philosophy," *American Naturalist* 43 (1909): 356–68, on p. 358.

instrument, quadrat, and experiment afford a ready means of eliminating such papers from consideration. No study deserves to be called ecological that does not deal with the cause-and-effect relation of habitat and organism in a quantitative and objective manner."[41] Strong words: investigators who did not use instruments and do experiments were not ecologists, and their work need not be taken seriously. People could not be prevented from doing what they liked, but they could be declared irrelevant. Amateurs and recreational ecologists might be at home on Driftwood Lake, but they were trespassers on ecologists' disciplinary terrain. Here is a nice piece of boundary work, an exorcism of deviant practitioners.[42] Those who failed to use laboratory methods or who camouflaged old-fashioned floristics as "ecology" were not ecologists—and thus someone else's embarrassment.

Not all ecologists retreated so precipitately from the idea of a scientific natural history. Charles C. Adams kept an open mind about recruiting amateur naturalists. In his view, people who lived intimately with nature were instinctive ecologists because they experienced nature as interacting communities. "The natural history which a farmer, a fisherman, a summer vacationist, or a sportsman acquires . . . is the natural method of approach," Adams wrote, "and is generally of most permanent value, except possibly to some professional teachers or zoologists."[43] But the social logic of discipline building pointed ecologists the other way, and by 1910 most of them believed that laboratory experience was the more natural training.

What we are observing here is a reorganization of the cultural geography of the lab-field border zone, and a decisive shift of authority from the field to the laboratory side. The boom-and-bust cycle in ecology, like the gold rush in biometrics, was a turning point. It discredited natural-history practices and drove a generation of ecologists to embrace laboratory ideals and methods with what proved to be an excessive ardor. As the bust in biometry deprived evolutionists of an effective tool for studying variation in natural populations, so the reaction against a natural-history style of ecology discredited methods of intensive observation that were well suited to complex and changing biotic communities. Border life was restructured so that individuals choosing a line of work were more likely to choose a laboratory than

41. Clements, *Research Methods in Ecology* (cit. n. 29), p. 20. Frederic E. Clements and John E. Weaver, *Experimental Vegetation: The Relation of Climaxes to Climate,* Carnegie Institution Publication no. 355 (1924), p. 3.

42. For another example see Thomas F. Gieryn and Ann E. Figert, "Scientists protect their cognitive authority: The status degradation ceremony of Sir Cyril Burt," *Sociology of the Sciences Yearbook* 10 (1986): 67–86.

43. Adams, *Guide to the Study of Animal Ecology* (cit. n. 38), pp. 13–14, quote on p. 14.

a field approach, not because it was more effective in the field but because the alternative was associated with embarrassment and wrecked careers.

Guardians of the Faith: Genetics and Physiology

Border life in evolution and ecology were also shaped by what was going on in genetics and physiology. Looking to the lab for legitimacy, field biologists in effect gave laboratory practitioners a voice in deciding what was proper field practice. Who could say what ecologists should count as physiology in the field with more authority than laboratory physiologists? And who were better qualified than laboratory geneticists to decide what should count as a proper genetics for evolutionary study? At first there was reason to be optimistic that laboratory standards of rigor would not be incongruent with field objects and conditions. In the late 1890s plant physiology and genetics were open to field problems, but in the first decade of the new century they moved decisively toward a more exclusive laboratory culture and shed concepts and practices that field biologists would have found the most useful for work on populations in nature.

To understand this dynamic fully we would need histories of genetics and plant physiology written from the point of view of the field, a vantage that no historian has yet adopted. However, it is possible to get a general lay of the cultural land from what historians have done, and that will have to do for the present.

The separation of genetics from evolution in the Mendelian revival of 1900–1915 is a familiar story. It was also a separation of laboratory and field as a place for genetic practice. Early students of experimental heredity worked with populations using biometric methods, or crossed varieties of domesticated plants and animals and mined their recorded histories for insights into the origin of species. This mix of practices changed dramatically when cytologists discovered the role of chromosomes in inheritance. That and Hugo de Vries's work on "mutations" in *Oenothera* gave laboratory experiment a clear edge over historical and field methods. As Daniel MacDougal saw it, while evolution had been "monopolized by naturalists engaged in studying 'undisturbed nature' to very little profit[,] . . . the interrupted task of bringing evolution within the scope of experimental science so fairly begun by Darwin has been forcefully completed by de Vries." At just this time biometricians were decisively discredited by Mendelians, and in the early 1910s the invention of mapping genetics by Thomas Hunt Morgan and his students established a new mainstream genetics, one in which natural populations and fieldwork had no role at all. Cytogenetics dealt with transmission of individual traits, not genetic variability, and de Vries's "mutations" held out the promise that the origin of species could be studied in the laboratory without having to deal with messy populations and environments.[44] The celebrated achievements of cytogenetics meant a sharper separation of lab and field in the border zone, and a diminished authority for mixed, border modes of practice.

How cytogenetics ceased to deal with populations and became a science of individuals and standard organisms is a long story.[45] But the crucial event was the invention by the Danish botanist Wilhelm Johannsen of the distinction between "genotype" and "phenotype," that is, between variations that are inherited and those that are merely accidents of development or environmental modification. This concept grew out of Johannsen's "pure line" experiments with garden beans in 1901–3, which showed that selection for extreme variations had no effect on the offspring of plants descended from a single, self-fertilizing parent. As Johannsen first conceived of it, a pure line was a population, but that changed as Mendelian "factors" became identified with chromosomes. Soon "genotype" was applied to an individual's germ plasm and "phenotype" to its soma, the one inherited unchanged (except for rare mutations), the other a product of both inheritance and development.[46] This distinction gave geneticists an unambiguous criterion for distinguishing traits that were safe subjects for genetic experiments and those that were likely to give false results. Mutations, which could be physically located on chromosomes, were safe; variations in populations were decidedly unsafe and therefore off limits to proper science. For geneticists, Johanssen's invention of genotype was a historic turning point, when the muddle of experimental heredity was cleared away like so much brush and experimental work began confidently on secure foundations.[47]

44. Provine, *Origins of Theoretical Population Genetics* (cit. n. 2), chaps. 3–4. MacKenzie, *Statistics in Britain* (cit. n. 2), chap. 6. Robert Olby, "The dimensions of scientific controversy: The biometric-Mendelian debate," *British Journal for the History of Science* 22 (1989): 299–320. Bert Theunissen, "Closing the door on Hugo de Vries' Mendelism," *Annals of Science* 51 (1994): 225–48. Daniel T. MacDougal, "Darwinism and experimentation in botany," *Plant World* 12 (1909): 97–101, 121–27, quote on pp. 122–23.

45. Ernst Mayr, "The role of systematics in the evolutionary synthesis," In *The Evolutionary Synthesis: Perspectives on the Unification of Biology,* eds. Mayr and William B. Provine, (Cambridge: Harvard University Press, 1980), 123–36, on pp. 128–29.

46. Frederic B. Churchill, "William Johannsen and the genotype concept," *Journal of the History of Biology* 7 (1974): 5–30. See also Kyung-Man Kim, "On the reception of Johannsen's pure line theory: Toward a sociology of scientific validity," *Social Studies of Science* 21 (1991): 649–79.

47. Raymond Pearl, "The selection problem," *American Naturalist* 51 (1917): 65–91, on pp. 72–74.

The view was less rosy from the field side—from the point of view of those whose work had been declared mere brush. For them the detachment of phenotype from genotype and of genetics from the study of variation in populations was a distinct loss. The biometrician J. Arthur Harris lamented that genetics had "temporarily brought the study of evolution to a very onesided stage of development." Herbert Spencer Jennings, whose work on *Paramecium* established the pure-line orthodoxy in the United States, also worried that geneticists had lost sight of "the great evolutionary problems" and attempted "to stifle and exclude other lines of work bearing on evolution."[48]

This episode rankled and lingered in field biologists' collective memory. The ichthyologist Carl Hubbs, who was a schoolboy when it all happened, recalled how the study of variation in nature "was growing with a promise of becoming a major subfield of zoology" until it was swept away by the fashion for genetics: "The promise of a new field . . . , the greater satisfaction in the newer experimental as contrasted with the older circumstantial evidence which had been gained in biometry, the greater definiteness of individual over population analysis—such factors attracted biologists from biometry into genetics. And with the natural pride and satisfaction in their epochal progress, the geneticists developed a feeling which often approached scorn toward the variation studies of the preceding generation."[49] Viewed from the field, the taboo against genetic experiments on natural variation was not a take-off into continuous progress but progress stalled for a generation.

The invention of genotype and phenotype destroyed investments in mixed field practices and border careers. Vernon Kellogg, for example, pointed in 1904 to the costs in lost time and damaged reputations for those who failed to distinguish between "blastogenic" and "ontogenic" variations. Most studies of variation had not done so, he noted, and so were worthless to evolutionary science.[50] Studies of the inheritance of variations in nature were discredited, not because geneticists were biased against fieldwork but because natural populations, their objects of study, fell on the wrong side of the genotype-phenotype line and so were declared unsound and unsafe.

Experiments on natural selection were on the wrong side, too, because

48. J. Arthur Harris, "A neglected paper on natural selection in the English sparrow," *American Naturalist* 45 (1911): 314–18, on p. 314. Herbert S. Jennings to Charles B. Davenport, 17 July 1911, CBD-Gen.

49. Hubbs, "Racial and individual variation" (cit. n. 21), quote on p. 115.

50. Kellogg and Bell, "Studies of variation in insects" (cit. n. 6), on pp. 207–8.

selection operated on phenotypic variations that might or might not be inherited and so were unsafe material for experiments. Biometric studies of variable populations also lost caste because they measured phenotypic characters, which were inadmissible genetic objects. Evolutionists were thus deprived of two practices that could have taken them afield to work on natural populations. No wonder they preferred to stay indoors and do genetics. It was a deeply felt and long-remembered object lesson for those whose careers were deconstructed or saw it happen to others.

In the laboratory, too, fear of being caught on the wrong side of the genotype-phenotype line made geneticists wary of practices that might have taken them into the field. Genetic studies of wild organisms and varietal crosses of domesticated plants or animals were devalued because they did not deal with pure lines. Traits crucial to evolution, like size or fertility or rates of growth, became unfit objects of experimental study because they were phenotypic and sensitive to environmental modification. It became an article of faith among geneticists to study only simple characters that could be identified with chromosomes, as mutations. As Charles Davenport put it, so long as evolutionists took species as their units they "floundered helplessly in the quicksand of unclearness and complexity." But take characters as the units of study, and "we move on surer ground." Experiments on chromosomes were safely on the genotype side of the line; those that dealt with complex phenotypic traits were not. No wonder that T. H. Morgan gave up experimental evolution to pursue mapping genetics, and that genetics put more complex varieties of experimental heredity in the shade.[51]

The distinction between phenotype and genotype sharpened the line between laboratory and field (safe and scientific on the one side, risky and unscientific on the other) and narrowed the intermediate terrain where mixing could occur. By excluding variation and populations from laboratory work, geneticists made it harder for field biologists to mix experimental genetics with biometric studies in the field. Biologists interested in evolutionary genetics found themselves chasing a moving target: if they worked by rules of laboratory genetics, their objects of interest—populations—were

51. Charles B. Davenport, "Light thrown by the experimental study of characters upon the factors and methods of evolution," *American Naturalist* 46 (1912): 129–38, on pp. 129–30. See also MacDougal, "Darwinism and experimentation in botany" (cit. n. 44), on p. 125; and Robert E. Kohler, *Lords of the Fly:* Drosophila *Genetics and the Experimental Life* (Chicago: University of Chicago Press, 1994), chap. 2. For an acute reflection on the slight of hand involved in separating genotype and phenotype, see Edgar Anderson, "Hybridization in American *Tradescantias*," *Annals of the Missouri Botanical Garden* 23 (1936): 511–25, on pp. 512–14.

not admissible; if they devised practices appropriate for natural populations, they risked attack from geneticists. A sector of the border zone that for a time had seemed a place where lab and field could mix, became a place of either-or.

It was like the ecological frontier between Asian irrigation and nomadic societies, where irrigators' intricate social system made it impossible to operate in a mixed mode and be properly Chinese. Mixed modes of subsistence were viable ecologically, but not ideologically and politically. So, too, in evolution and genetics: mixed practices might have worked, but the events of the early 1900s made survivors overly wary of pursuing them. Memories of ruined careers discouraged border practices, reinforcing the differences between lab and field and devaluing the similarities.

Something of the sort occurred in plant physiology. As geneticists retreated to laboratories and put up warning border markers, so plant physiologists also retreated from a vision of their discipline that was open to fieldwork to a more closed conception of laboratory practice. The result for ecologists, as for evolutionists, was a narrowed middle ground. Those who wished to practice ecology as a physiology of the field found themselves chasing a receding laboratory standard. There seemed to be no way to do physiology in the field, not because it was impossible to devise physiological practices appropriate to field conditions, but because for laboratory physiologists—and thus the world—only lab-style work was admissible science.

The evidence for this assertion is indirect but suggestive. Certainly, ecologists who inclined to physiology saw the relation between the two disciplines as a reciprocal, mutually beneficial relation. Henry Cowles remarked that physiology had as much to gain from ecology as ecology did from physiology, in softening the artificiality of pure lab work. "In the old days," he recalled, "when physiology was a mere laboratory science, and therefore artificial, no experimental test could be too bizarre to be applied to plants." ("*Mere* laboratory science"—a bravura touch!) Physiologists needed ecologists to remind them that "the contortions of a puny plant in our illy-lighted and gas-ridden laboratories" solved none of the important problems of biology. Victor Shelford agreed: what value was there in experiments done by people ignorant of the natural life of the animals they studied? Physiologists needed ecologists to tell them that. Frederic Clements likewise saw a mutual benefit: if field ecology was too descriptive, physiology would give it analytical edge; if physiologists confined their work to pathological behavior in unnatural environments, ecology would inject a dose of fresh air. Clements was confident that physiologists, as well as ecologists, were capable of changing their ways. So was Cowles, who saw a new sanity in plant physiology, which he ascribed to the influence of ecology. As late as 1918 Shelford could write that "the day of the naturalist physiologist is at hand."[52]

Was this optimism simply wishful thinking, or were American plant physiologists really open to field methods and ecological influence? We do not know, but for a time, I think, they were. When agitation for scientific natural history was at its peak, many plant physiologists were employed in agricultural schools and stations, where a field orientation might appeal, and animal physiologists were not yet firmly entrenched in medical schools. They had reason to be open to any alliance that could give support, even field biology. After all, the plant physiologists who defined ecology in 1893 must have seen some benefit to themselves in defining it as a part of their discipline. Ecologists who looked to a truly mutual relation were too optimistic, but I think they had some cause for optimism. We will not know until the history is written.

As it turned out, physiologists did not need ecologists to build a discipline. The laboratory movement in agricultural and premedical education established physiologists on their own terms. They had no need to ally with ecologists or to compromise experimental ideals, and every reason to shed associations with fieldwork. The evidence of this cultural shift is evident in the careers of people like Daniel MacDougal or Burton Livingston, who in the 1900s worked actively in the field at the Desert Lab and in biological surveys, but in the 1910s became exclusively laboratory biologists. Charles Adams saw the handwriting on the wall in 1913, when he noted that ecological physiologists were leaning more to work on individual organisms than on groups. He hoped it was a passing phase.[53] It was not.

Reviewing the recent history of plant physiology in 1917, Livingston observed how the "older reverence for natural or 'normal' phenomena [had] largely disappeared." And a good thing too, he thought, for what was usual in nature was not necessarily normal. "We have learned," he wrote, "that the range of conditions offered by nature does not generally happen to be great enough to allow adequate experimental interpretation of plant pro-

52. Cowles, "Trend of ecological philosophy" (cit. n. 40), quote on p. 357. Victor E. Shelford, "Physiological animal geography," *Journal of Morphology* 22 (1911): 551–617, on pp. 591, 610. Clements, *Research Methods* (cit. n. 29), pp. 1–4, 11. Shelford, "Physiological problems in the life histories of animals with particular reference to their seasonal appearance," *American Naturalist* 52 (1918): 129–54, quote on p. 150.

53. Adams, *Guide to the Study* (cit. n. 38), p. 9. See also Sharon E. Kingsland, "The battling botanist: Daniel Trembly MacDougal, mutation theory, and the rise of experimental evolutionary biology in America, 1900–1912," *Isis* 82 (1991): 479–509; and Burton E. Livingston, "Some conversational autobiographical notes on intellectual experiences and development: An autobituary," *Ecology* 29 (1948): 227–41.

cesses." Better to make one's own conditions than take those that nature provides. What was most needed for progress in physiology was not more fieldwork but better methods for creating controlled environments: "[I]f a student has not a liking and talent for creating physical and chemical conditions such as never have occurred in nature," Livingston concluded, "he should not cast his lot with plant physiology, for the next generation."[54]

Here was a chilling message for any physiologist inclined to fieldwork, or for ecologists who aspired to do physiology in the field. What Livingston envisioned was not a reciprocal relation between ecology and physiology but a hierarchical one, in which the borrowing and accommodation were all one way. Laboratory physiologists even appropriated the virtues of "naturalness" for lab environments. "Normal" behavior, Livingston implied, cannot be observed in the unpredictable environments of nature but only in the controllable environment of labs. A student of Robert Yerkes defined "naturalistic" data as observations made "in the natural cage conditions."[55] So powerful was laboratory culture that it could make the lab seem more natural than nature itself.

This hardening of disciplinary rules left ecologists who pursued mixed practices in a painfully narrowed space. On the one side, Frederic Clements ordains that no work may be called ecology that is not causal, quantitative, and "objective"—that is, experimental. Anyone who does not like instruments had better stay out of ecology. Yet on the other side, Burton Livingston declares that anyone who likes instruments and experiments had better expect to stay out of the field and in the lab. Between the upper and the nether millstones, what prospect was there for a physiology of the field or a scientific natural history? To study nature as physiologists thought proper, ecologists would have to abandon fieldwork; to physiologists mixed practices were impure. But to pursue natural-history methods meant losing credibility among ecologists who looked to laboratory methods to secure their future. As with genetics and evolutionary biology, ecologists' uncritical embrace of laboratory ideals, plus physiologists' retreat from the field, made the ideal of a mixed border practice harder to attain.

54. Burton E. Livingston, "A quarter-century of growth in plant physiology," *Plant World* 20 (1917): 1–15, on pp. 9–10.

55. Richard W. Burkhardt, Jr., "Ethology, natural history, the life sciences, and the problem of place," *Journal of the History of Biology* 32 (1999): 489–508, quoting N. Utsurikawa, on p. 500. For an earlier appropriation of the idea of "natural" by microscopists, see Graeme Gooday, "'Nature' in the laboratory: Domestication and discipline with the microscope in Victorian life science," *British Journal for the History of Science* 24 (1991): 307–41.

Conclusion

Cultural border zones are created when two cultures meet at a common boundary in circumstances that permit transactions across the border. In biology around 1900 the movement for a new natural history created the potential for such a zone between laboratory and field biology. But only the potential: ideology can justify cultural mixing as a good and useful thing, but it is not necessarily a precise blueprint for what can be mixed and how. As when ground is newly cleared in nature, so, too, with new cultural space. What fills the vacant space depends on history: how it is populated, how much from one side and how much from the other, and what the new occupants do there.

The lab-field ecotone of the early 1900s was mainly occupied from the field rather than the laboratory side. Field biologists possessed skills and practices that were more adaptable than lab practices to field conditions, and they had a larger pool of potential occupiers. They were also less inhibited about cultural mixing than their laboratory cousins. For these same reasons, however, their occupation resulted in a decided shift toward laboratory ideals and a narrowing of possibilities for cultural mixing. The boom-and-bust in biometry deprived field workers of an instrument that was well suited to the study of variation in natural populations, leaving the way clear for laboratory models. Similarly, the rush of students and local botanists into ecology and the reaction against Brush Creek modes of practice turned ecologists against natural-history methods and drew them to laboratory ideals that were difficult to realize in the field. At the same time, both geneticists and physiologists retreated to their laboratories and to a stricter ideal of experimental work that was harder for field biologists to apply in real situations in nature. In short, field biologists were pushed toward laboratory practices at the same time that these practices became less appropriate and accessible.

The dynamics of cultural frontiers are highly context dependent: that is the point. Mixed practices may be encouraged in cases where both sides want something particular and limited from each other, as in trading frontiers. But where the authority of an imperial culture is at stake, and the borrowing is one-sided, mixed cultural practices may be discouraged because they compromise dominant practices and bring borrowers little approval from those they emulate. The frontier between Chinese irrigators and mixed agriculturalists is an example of this dynamic. So, too, is the lab-field frontier in biology in its early years in North America. Field biologists wanted to adopt laboratory practices, but the mixed practices that would have suited their objectives earned no respect on the laboratory side and so were effectively off limits. The simplest choices were to leave the field for the laboratory, or to give up lab culture and be a field naturalist. Stay put and grow paddy rice, or follow the wandering herds.

Ultimately it was the difference in standing between the laboratory and field sciences that drove the cultural dynamic of the border zone, but it did so indirectly through individuals' choices of where to work and how. Ecologists sought to become physiologists of the field not because physiologists had higher status, but because botanists' custom of open participation encouraged a gold rush that threatened the standing of the fledgling discipline. It was the social and cultural dynamic that gave reality to the symbolic difference between laboratory and field. So, too, with evolution: it was not the higher status of experimental genetics that caused field biologists to give up on natural populations, but the cycle of boom-and-bust in biometry. Potentially viable alternatives to experimental genetics were discredited by the rush of untrained practitioners into the border zone.

The tensions thus created between field problems and laboratory ideals were not evenly distributed along the lab-field border; rather, there were hot spots, where memories of particular losses or failures made biologists more sensitive to apparent threats to disciplinary standards. Taxonomists long remembered Mendelians' assertions that mutations were the raw material of new species, when it was obvious to anyone who had studied species in nature that they could not be. Taxonomists remembered and resented biometricians' attempt to substitute place modes for species. Geneticists who experienced the winnowing of the early 1900s were likewise quick to reject any mode of practice that reminded them of the bad old days of phenotype genetics. (It was this particular hot spot that Dobzhansky inadvertently encountered when he followed *Drosophila obscura* into the field.) Ecologists were similarly oversensitive to modes of practice that reminded them of their excursions down Brush Creek to Driftwood Lake. Field biologists' deference to laboratory values was neither an imposition by imperialists of experiment, nor false consciousness, but the result of the social dynamics of a newly opened and occupied border zone.

CHAPTER 4

Taking Nature's Measure

We have seen how border biologists looked more to the laboratory than to the field for guides to an imagined future. That future began to be a present, a lived reality, as biologists took laboratory artifacts and practices into the field. That is the subject of this chapter and the next. Values and expectations shape what people do and what communities become; but so do actions and their outcomes, what works and what does not. Future projections prove accurate or misguided as people act them out. Following laboratory objects and practices into the field and seeing how they fare tells us a good deal about the cultural dynamics of the emerging border culture. Counting and measuring are considered in this chapter; experiment in the next.

Material culture is always a powerful vehicle of cultural mingling and transformation. Material things and practices carry as much cultural baggage as theories or theologies, but more unobtrusively. They can serve as the entering wedge for a larger cultural complex, when the whole complex would seem threatening to recipients' own culture. Moreover, things and ways of handling things can be detached from their cultural contexts and so acquire new meaning in different cultural contexts. Trademark items of a dominant culture become items of prestige in borrowing cultures; or vice versa, if marginal cultures are seen to have an authenticity untarnished by participation in a discredited history. New forms of culture may also arise out of creative misunderstandings. These effects are well documented in anthropology and history and need not be belabored; they also operate in science.

Tools and practices of production are especially powerful because they are the essential instruments of ways of life—agricultural, technological, sociopolitical, scientific. They have different meanings and uses in different cultural contexts—there is no simple determinism. The point is that tools of livelihood and production travel well because they are designed to solve practical problems shared by most societies: making a living from nature, using resources, fashioning a human environment. The cultural baggage of items of daily use may thus be subtly but strongly transforming.

Tools and practices are especially important in the border relations of sciences because they are the defining elements of scientific disciplines, the stuff of daily concern, the objects of methodological flirtations and commitments, the material and practical core around which scientific cultures accrete. Tools and practices are valued as prestige items—precision instruments or fancy statistics, for example; but their usefulness matters even more. Counting and measuring make publications and win respect, and that is why they have proved such powerful vehicles for mixing lab and field.

Field scientists have always borrowed from the lab side; it is not a modern phenomenon. The physical sciences of the field have been heavily invested in instruments and quantitative measurement for a long time: meteorology and topography especially, but more recently hydrology and geophysics. Counting and measuring also have a deep history in the biological field sciences. Alexander von Humboldt was famously emphatic about the importance of instruments and measuring in biogeography.[1] Indeed, before the era of laboratory building, instruments and measurement were not necessarily thought of as peculiar to laboratories or as items to be "borrowed"—they were of the field. Lab and field were a common context. This changed when laboratories became the preferred locus for precision measurement, and when instruments and measuring practices became too delicate and exacting to be of use to field biologists. In laboratory culture precision measurement became an end in itself, an emblem of scientific high culture; and the association of precision with status and reward drove labo-

1. Michael Dettelbach, "Global physics and aesthetic empire: Humboldt's physical portrait of the tropics," in *Visions of Empire: Voyages, Botany, and Representations of Nature,* ed. David P. Miller and Peter H. Reill (Cambridge: Cambridge University Press, 1996), 258–92.

ratory scientists to devise ever more precise and refined tools and practices.[2] It ceased to matter whether or not they might be useful in the field.

Thus it was not really until the late nineteenth century that the idea of borrowing laboratory tools for use in the field became problematic—or, indeed, that it was seen as borrowing. (If the "field" was a by-product of the laboratory revolution, so too was transcultural borrowing: no boundary, no borrowing.) True, instruments and quantification had remained elements of field biology since Humboldt's day. But the material culture of laboratory experiment had become far more specialized and precise since then, and it far outstripped what field biologists were accustomed to using. Two cultures evolved at different rates and achieved, to borrow a biological metaphor, something closer to reproductive isolation. The high status of experimental instruments increased the potential rewards of borrowing by field biologists, but their refined designs made appropriation more difficult. Precision instruments need simplified and controlled environments to show their technical and symbolic superiority. The unnatural uniformity of laboratories guarantees that instruments produce uniform results; and we trust instruments and experiments in part because we know that they are deployed in the right kind of place.[3] Could a material culture designed for laboratories work in places as variable and uncontrolled as dunes, floodplains, prairies, or bogs? Did laboratory ideals of precision have meaning beyond the laboratory threshold? The culture of precision instruments of counting and measuring coevolved with laboratories, and from the laboratory point of view, the very idea of precise measurement or experiment in nature could seem a contradiction in terms.

And if instruments could be modified for use outdoors, could field biologists take them on without losing their naturalists' souls? Might field biologists adopt laboratory artifacts merely for their cultural prestige even if they were less effective than naturalists' homegrown methods? Laboratory practices and paraphernalia had become powerful and potentially threatening vehicles of a dominant culture, like the material culture of imperial powers—French arms and measures, British steel and machines, American mass production and marketing. Could field sciences survive a successful assimilation of laboratory tools and cultural values without disappearing into the dominant culture? In the 1890s these became real questions for the first time.

2. M. Norton Wise, ed., *The Values of Precision* (Princeton: Princeton University Press, 1995).
3. Steven Shapin, "The house of experiment," *Isis* 79 (1988): 373–404.

Trust in Numbers: Quadrats

Of all the elements of laboratory culture, enumeration and quantification are the ones that travel most readily. Counting things and manipulating numerical data are generally easy to do, and the payoffs in credibility are potentially great, thanks to our apparently unshakable trust in anything quantitative. Counting is a familiar skill in everyday life, is flexible, and requires little expensive machinery (at least until computers came along). The most important quantitative method in ecology—the quadrat—was invented in the summer of 1897 by Roscoe Pound and Frederic Clements, about the time that ecology was officially identified as a new discipline.[4] Within a decade or two it was so widely adopted as to be taken for granted, a trademark of ecological practice. I doubt that any method has done more to give ecology standing as a proper science.

Essentially, the quadrat is a way of sampling the uncountable number and variety of things in nature. It is easy to do: stake out a square plot of appropriate size, count every plant in it (or animal, if sessile), and repeat the process until further sampling does not significantly change the averaged frequencies of species present. Assuming that the quadrats have been located in representative parts of the formation being studied, the resulting table of frequencies will describe it as precisely as one wishes.[5]

In practice, of course, it is not quite that simple. Knowing where to locate quadrats to get biologically meaningful results requires skill and experience. Deciding how big to make them and how many to do also requires judgment: smaller and fewer quadrats are quicker to do, but more or larger ones give more precise results, and precision, even if not strictly required for the problem in hand, affords valuable epistemological cover. Pound and Clements's original quadrats were five meters square, but one-meter squares became the standard for grasses and forbs, two-meter squares for shrubs, and ten-to-fifty meter squares for trees.

Where exactly in the lab-field border zone did the quadrat arise? The historian Ronald Tobey has pointed to the laboratory side, observing that quantification was much in fashion at the time in the up-and-coming botanical schools of the Middle West—especially in Charles Bessey's school at the University of Nebraska, where Clements was a student. (Bessey looked

4. Ronald C. Tobey, *Saving the Prairies: The Life Cycle of the Founding School of American Plant Ecology, 1895–1955* (Berkeley: University of California Press, 1981), chap. 3.

5. Roscoe Pound and Frederic E. Clements, "A method of determining the abundance of secondary species," *Minnesota Geological and Natural History Survey Botanical Series* 4 (1898): 19–24, pp. 20–22.

4.1 Counting a quadrat on Mount Garfield, Colorado. Frederic E. Clements, *Research Methods in Ecology* (Lincoln: University Publishing Co., 1905), on p. 165.

to physiology to make botany scientific; ecology was too like children's nature study for his comfort.)[6] However, a quantitative spirit was also abroad among field botanists at the time. Mass collecting was becoming standard practice in the late nineteenth century, mainly in zoology but also botany, and species descriptions commonly included abundances, which were a useful aid to collecting and identifying. The categories were rough and ready—"social," "gregarious," "copious," "sparse," "rare," and "solitary," with finer distinctions indicated by compound terms or numerical subscripts. This practice was based on naked-eye estimation, not sampling, and hardly qualifies as precision measurement. But it was implicitly quantitative and invited refinement, and its popularity shows that quantification was as much a feature of field as of laboratory culture in the late 1890s, though the two groups had quite different ideas of what it meant in practice. And because collecting was already a field practice, its quantitative techniques were more accessible to ecologists than the more exacting and puritanical practices of laboratory science.

In fact, the idea of actually counting a quadrat sample, rather than just eyeballing a patch of vegetation, was almost certainly conceived by Roscoe Pound the amateur botanist and biogeographer, not by Clements the trained laboratory man, and for floristic not ecological purposes. Pound's aim in counting samples of vegetation (remarkably no one had done it be-

6. Tobey, *Saving the Prairies* (cit. n. 4), pp. 53–57.

fore) was to create more precise and minutely subdivided categories of abundance for their book on Nebraska's flora. Counting samples was especially useful for secondary species, which were harder to estimate by inspection than species that were either very numerous or very rare.[7]

The quadrat method was thus initially an intensified form of an established field practice; but it had unexpectedly subversive effects. Pound and Clements did not expect that counting would greatly alter estimated abundances, just make them more accurate. So they were surprised when the counted abundances of many secondary species turned out to be very different from naked-eye estimates. The human eye, they realized, is an imperfect scientific instrument, because it is easily misled by the aesthetic biases of the mind's eye (or is it the eye's mind?). Plants with prominent and attractive flowers seem more abundant than they are in fact, and humble and inconspicuous plants are in reality more abundant than they appear even to experienced field men like Pound. The observer's gaze is selective and serves several purposes at once: visual pleasure and the desire to admire or possess striking forms, as well as precise science. Worse yet, long experience in the field only makes observers more prone to error, by stamping an erroneous picture more indelibly on their minds. Ronald Tobey has observed that the invention of the quadrat marked a fundamental change in the way field biologists perceived nature: "Ecology had 'taken leave of its senses,' and hitched its intellect to mathematics."[8] This is overstated, I think. The quadrat method did embody laboratory values of precision, true, but it came out of field practice.

It was almost certainly Clements, not Pound, who first recognized the potential of the quadrat as an ecological tool. He was the aspiring ecologist, lab-trained and eager to put respectable scientific foundations under the new discipline. Quantitative knowledge of the species composing plant communities was relevant to all the key items on ecologists' agenda: classifying types of vegetation, explaining zonal patterns and migrations, and succession.

Although early ecologists liked to think of themselves as physiologists of the field, what they mainly did was classifying plant associations, as taxono-

7. Tobey, *Saving the Prairies* (cit n. 4), pp. 55–53. Roscoe Pound, "The plant-geography of Germany," *American Naturalist* 30 (1896): 465–68. Roscoe Pound and Frederic E. Clements, *The Phytogeography of Nebraska, I: General Survey,* 2nd ed. (Lincoln, Neb.: Botanical Seminar, 1900), pp. 56–63. Pound and Clements, "Method of determining" (cit. n. 5), pp. 19–20. Clements, *Research Methods in Ecology* (Lincoln, Neb.: University Publishing Co., 1905), pp. 166–67.

8. Pound and Clements, "Method of determining" (cit. n. 5), pp. 19–20. Tobey, *Saving the Prairies* (cit. n. 4), pp. 65–69, quote on p. 68.

mists classified species. (Clements compared quadrats to taxonomists' type specimens—an apt and revealing comparison.)[9] Though essential, classification was a relatively low status activity. However, the quadrat method promised to make it more precise and scientific by defining types of community in terms of exact percentages. The ideal was quantitative classification, as some hoped biometrics would make taxonomy one of the exact sciences. Quadrats seemed an ideal way to infuse quantitative ideals into ecology without compromising its fundamental concern with classifying.

Quadrats could also be applied to the dynamics, as well as the statics, of vegetation. Quadrats permanently staked across the boundary between two associations could reveal movements that occurred too slowly to be perceived except by quantitative counts over a long period of time. The same method could also be used to characterize precisely the vanguard patches of an invading association or the rearguard patches of a retreating one and to watch, in slow motion as it were, the succession of vegetation in a single place.[10] No wonder that such a versatile tool caught on so fast.

Technically the quadrat method was simple and inexpensive, almost a household technology. Its physical paraphernalia consisted of measuring tapes with eyelets sewn in at decimeter intervals (any cobbler could make them), and metal stakes to secure the tapes—or lacking those, hatpins, meat skewers, or nails would serve well enough, common household items. Simple procedures were developed for counting and recording: small and inconspicuous plants were counted first, decimeter by decimeter, then the easier ones. In difficult cases plants were broken off or pulled up as they were counted to avoid confusion.

The quadrat rapidly evolved into a small family of specialized techniques. The basic "list" quadrat—a list of species and frequencies—sufficed to characterize plant communities. The "chart" quadrat mapped the precise location and identity of every individual plant and was useful for studies of competitive or nurturing relations between species. The areas occupied by foliage could also be drawn in. These schematic maps were often supported by photographs. The "transect" quadrat (Clements loved giving things names) was a line of quadrats laid out across a valley or mountain slope; it gave precise definition to the boundaries between vegetation zones. And if production of biomass was the object sought, plants could be clipped, dried, and weighed to make a "clipped" quadrat, or carefully dug up from a series of

9. Clements, *Research Methods* (cit. n. 7), p. 163.

10. Frederic E. Clements, "The development and structure of vegetation," *Botanical Survey of Nebraska* 7 (1904): 5–175, on pp. 5–7, 84–85, 105–7.

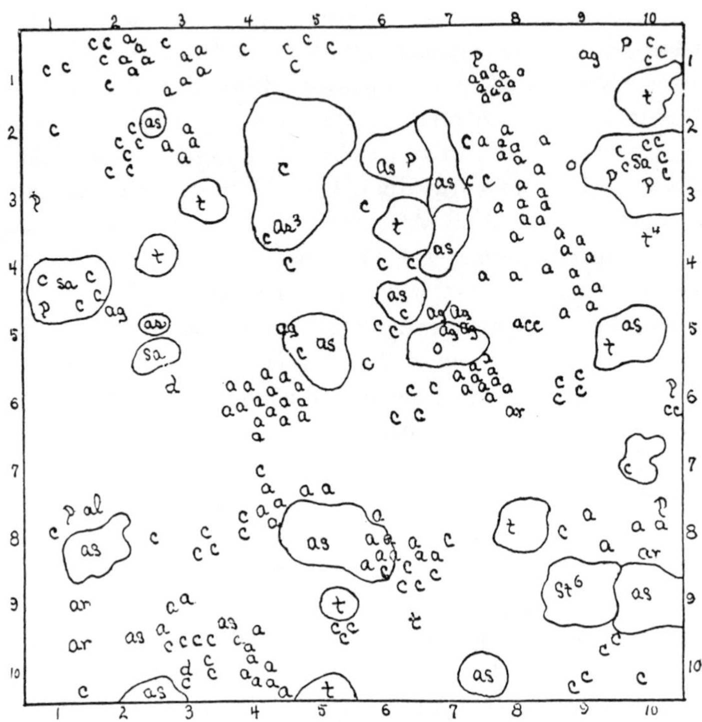

4.2 A typical chart quadrat. Clements, *Research Methods in Ecology,* on p. 169.

adjacent quadrats as the growing season progressed. "Permanent" quadrats, precisely located with respect to topographical survey markers, were indispensable for studying invasions and retreats. William Cooper installed a set of permanent quadrats at Glacier Bay, Alaska, to obtain "accurate, unimpeachable data" on the movement of plants into bare ground vacated by the retreating ice. Cooper also produced what may be the oddest type of quadrat, bits of ancient forest floor that had been covered by silt, then ice, and brought again to the light of day. Washed of the protective silt, the mosses stood up as if in life—a "paleoquadrat."[11]

Quadrats can also be made into instruments of active experimentation. Cleared of all vegetation by herbicides or excavation, "denuded" quadrats can be used to study the process by which plants invade and resettle vacant ground—a simulated primary succession. Or they can be burned, mowed,

11. Clements, *Research Methods* (cit. n. 7), pp. 161–76. William S. Cooper, "The recent ecological history of Glacier Bay, Alaska," *Ecology* 4 (1923): 93–128, 223–46, 355–65, on pp. 355, 112–14.

or pruned of selected species and then observed to see how these interventions affected species competition and replacements. These experimental quadrats are simple models of the changes caused by fire, disease, or human interventions. Or quadrats can be removed en bloc and transplanted to different environments to see how the composition of species changes—a simulation of invasion and shifting ecotones. Experimental quadrats enable ecologists to mimic and examine natural processes quantitatively and in the environments where these phenomena naturally occur rather than in a greenhouse or laboratory.[12]

Although the quadrat is best suited for terrestrial plants, which stand still to be found and counted, it can be adapted to mobile animals and subaqueous environments. Streams and rivers were a special problem, but even there it is possible to count. Victor Shelford devised a kind of denuded quadrat for fast-flowing streams by removing all plants and animals from a small area, installing meter-square pieces of concrete or rock roughened or covered with stones to simulate the stream bottom, and watching how the quadrat was recolonized. The environment of this artificial quadrat could also be manipulated experimentally, by building dams or diversions to alter rates of flow.[13]

For those without research assistants and grants there were ways around the practical demands of quadrat counting. One such compromise was the "walking" quadrat devised by Henry Gleason in his study of the Illinois sand region. The quadrat method gave excellent results, Gleason thought, but it was too labor intensive for a one-man operation, and he found that he could obtain satisfactory results simply by walking along a line two paces at a time and listing species in the order of the space they occupied (not their number) in the imaginary two-by-two meter quadrat directly in front of him (or larger ones for trees). Roland Harper devised a similar method for his botanical survey of northern Florida. Quadrats were out of the question for large surveys, but estimates of the volume of vegetation served Harper's purpose and could be done from a moving train or auto. For many purposes, Gleason thought, simple observation was quite good enough.[14]

12. Clements, *Research Methods* (cit. n. 7), pp. 173–75, 306–13.

13. Victor E. Shelford and Samuel Eddy, "Methods for the study of stream communities," *Ecology* 10 (1929): 382–91, on pp. 384–88.

14. Henry A. Gleason, "The vegetation of the inland sand deposits of Illinois," *Bulletin of the Illinois State Laboratory of Natural History* 9 (1910): 23–174, on p. 40. Roland M. Harper to Arthur G. Vestal, 31 May 1914, AGV box 2. The walking quadrat closely resembled the time-honored method of timber cruisers for getting reliable, if not exactly quantitative, estimates of standing board feet.

But counting quadrats simply is more labor and time intensive than observation, and for field biologists it represented a major escalation in the investment required to participate in vanguard ecology. It upped the ante of commitment to the discipline and gave a competitive advantage to ecologists who could get funds for equipment and assistants. Raising the threshold of participation did not bother Clements. He liked the fact that counting quadrats was time consuming, for him it was a sign of progress: modern science was modern because it was demanding. The extra labor would discourage tyros and second-raters, he hoped, and prevent them from bringing the new discipline into disrepute. "Every ecologist . . . that has the interests of his field at heart and deprecates the present slipshod work," he wrote, "will appreciate the necessity of methods which seem like drudgery to the mere dabbler."[15] Counting and quantifying would secure respect for ecology by making its practices more like those of laboratory science. They also made the recruitment and labor practices of ecology, with its long and open border with the world of amateur naturalists, more like those of the laboratory sciences, whose borders with public culture were regulated and vigilantly patrolled.

Not all ecologists shared Clements's enthusiasm for quadrats. Henry Gleason and Roland Harper, for example, thought that the number of individuals per unit space was a less meaningful indicator of a community than the physical space each occupied, or their biomass. In their view, counting overvalued rare species and represented associations as more heterogeneous biologically than they were in fact. Harper thought his estimates of the volume of trees in board feet were more strictly quantitative than mere numbers of stems. For precise measurement of grassland formations a better method than counting would be to mow a few acres, separate the species, and weigh each separately.[16]

Henry Cowles also regarded quadrats as impracticable and not as good a representation of natural vegetation as simple description. "Personally, I do very little quadrat counting," he confessed to Arthur Vestal, "even when I am working intensively on a small area. I believe I have rather more confidence in a subjective method than in an arbitrary method like the quadrat. After all, one must select one's quadrats, and that brings in the subjective element."[17] He had a point: one did have to decide that a community existed

15. Clements, *Research Methods* (cit. n. 7), pp. 160–63, quote on p. 163.

16. Gleason, "Vegetation of the inland sand deposits" (cit. n. 14), p. 40. Roland M. Harper to Arthur G. Vestal, 8 Aug. 1914, AGV box 2.

17. Henry C. Cowles to Arthur G. Vestal, 26 June 1914, AGV box 2.

and what a "typical" example was before one could stake a quadrat. Characterizing these preselected stands in quantitative terms made them no more objective, but only gave a specious air of objectivity to a practice that was in fact not much different from old-fashioned taxonomic intuition.

Vestal came to a similar conclusion in 1912. After some hours of counting quadrats in the dry-mesa plains near Tolland, Colorado, it dawned on him that the tables of numbers he was compiling did not express the composition of the association as he was actually seeing it. To capture the character of a type of vegetation one had to know what each plant looked like and how it grew: "One cannot tell from looking at a table of figures what the important plants are without knowing which plant is large, which has a certain growth-form, which can resist competition for space, etc." What mattered, in short, were the qualitative features of vegetation that were easily observed but lost in counting. As an experiment Vestal lumped species into life forms (grasses, bushes, interstitials), estimated their abundance and did a simple cross-tabulation. In ten minutes he got a picture of the mesa association that seemed truer to nature than any table of figures. For Vestal's purpose the eye was the superior instrument. Numbers were all very well, but they had to be supplemented by descriptions, and descriptions without the numbers would do well enough for a preliminary survey, Vestal thought. He left the counting for the next ecologists to come that way.[18]

Trained in departments dominated by labs, but field naturalists by predilection, Gleason, Cowles, and Vestal valued quantification but also recognized that the mathematical eye was as limited and prone to bias as the observer's naked eye. What, after all, do numbers mean to plants? They experience competitors by the space they occupy and the resources they consume, not their numbers. It is the visible dominants what matter most biologically, not the more numerous but less obtrusive secondaries. Perhaps the human eye, with its preference for visual prominence, is more like a plant than the counting eye of the quadrat grid, and so a more appropriate way of perceiving nature.

Ecologists did not "take leave of their senses" when they embraced the quadrat; they created a mixed practice that combined traditional field observation with quantification. Counting did not replace the senses but amplified them. Ecology remained a classifying science, with quadrats serving as an analogue of taxonomists' "type" specimens. Choosing typical places to count remained a matter of individual judgment—"subjective" if you will, but not unscientific. Moreover, first-generation American ecologists

18. Arthur G. Vestal to Henry A. Gleason, 4 Sept. 1912, AGV box 2.

remained indifferent to the sophisticated statistical techniques of the European plant sociologists. Americans were interested in plant communities, and for defining and comparing communities the quadrat was the ideal tool. (The one American who did work on statistical sampling—Henry Gleason—was also the most outspoken opponent of Clementsian concepts of community and succession.)[19] Experienced observation by eye (and all the senses) remained an essential element of ecological practice even as ecology became a counting discipline.

In sum, the quadrat was in many ways an ideal field method for the new naturalists. It was simple and flexible and brought contemporary laboratory ideals of precision to field methods that were already semiquantitative, but nonviolently, so to speak. And the quadrat method remained in part reassuringly dependent on field experience and know-how. Quadrats provided a measure of credibility at relatively minimal cost. And, of course, they worked: Clements remarked that he had never "listed or mapped a quadrat without discovering some new fact or relation, or clearing up an old question."[20] The quadrat enabled first-generation ecologists to participate in the culture of quantification in a way that was congruent with their naturalist yearnings and the realities of fieldwork.

Taking Nature's Measure: Instruments

Measuring was less simple and straightforward for field biologists than counting was. Ecologists could lay out a quadrat anywhere and start counting; measuring in the field required specialized physical instruments, most of which were initially borrowed from other disciplines and designed for other purposes. Some were borrowed from laboratory disciplines. The practical difficulties of transporting and maintaining instruments in field conditions were immediate and daily reminders that measuring, unlike counting, is most easily done in a steady and predictable place (hence labs). Yet instruments have the cachet of laboratory culture, and measuring was generally acknowledged to be a sign of a progressive science. Thus for field biologists, taking nature's measure was well worth the effort, or so many believed.

Among ecologists no one preached the gospel of instruments and exact

19. Rudy Becking, "The Zürich-Montpellier school of phytosociology," *Botanical Review* 23 (1957): 411–88. Malcolm Nicolson, "Henry Allan Gleason and the individualistic hypothesis: The structure of a botanist's career," *Botanical Review* 56 (1990): 91–161, on pp. 119–28.

20. Clements, *Research Methods* (cit. n. 7), p. 163.

measurement with greater zeal than Frederic Clements. Substantial parts of his much-used handbook of ecological field practice (1905) were devoted to instruments and measuring. Instruments were as essential for field ecology, Clements believed, as they were for laboratory physiology. They were to ecology "what the library and herbarium are to taxonomy," and he went so far as to say that any work that did not use instruments and quantitative methods should not be considered ecology.[21]

Clements practiced what he preached. As head of botany at the University of Minnesota, he spent so much of the department's budget on instruments that his colleagues were on the point of open rebellion. His inventory of instruments for the 1918 field season at Pike's Peak included 3 hygrothermographs, 3 soil thermographs, 3 "selagraphs" (recording photometers), 1 rain gauge, 4 anemometers, 5 photometers, 1 barometer, 24 atmometers, 12 Green's thermometers, 1 soil thermometer, 5 cog psychrometers, 2 soil tubes, 144 soil cans, 3 scales, and 1 water bath, plus the usual cameras, microscopes, and minor paraphernalia. Each of the Alpine Lab's three measuring stations had its own outfit of machines, turning them into little laboratories under the open sky. Clements's devotion to instruments was perhaps a little out of the ordinary, but it is fair to say that most ecologists shared it to some degree. And those who did not, like Henry Gleason, felt obliged to apologize or explain why.[22] Instruments for most "physiologists of the field" were an emblem of a modern, progressive science.

Limnologists were another group who embraced field instruments, especially Edward Birge and Chancey Juday, who created the first American school of limnology at the University of Wisconsin. In the 1910s and 1920s they transformed descriptive hydrobiology into a science of quantitative physical and biological measurement. Their studies of Wisconsin's lakes included quantitative measurements of temperature, plankton populations, dissolved respiratory gases (oxygen and carbon dioxide) and nutrients, solar radiation and absorption, and so on. (Their aim was to understand the distribution of aquatic plants and animals in terms of the physical structure of lakes.) An inventory of field instruments at the main field station at Trout Lake, Wisconsin, in 1930 included several kinds of underwater thermome-

21. Clements, *Research Methods* (cit. n. 7), chaps. 2, 4, quote on p. 20. Clements to Otto Rosendahl, 6 Jan. 1914, UM-Bot box 1 f. 2. Frederic E. Clements and John Weaver, *Experimental Vegetation: The Relation of Climaxes to Climate,* Carnegie Institution Publication no. 355, (1924), p. 3.

22. Frederic E. Clements to John E. Weaver, 30 April 1918, FEC box 62. Clements, "Experimental methods in adaptation and morphogeny," *Journal of Ecology* 17 (1929): 356–79, on pp. 367–68. Gleason, "Vegetation of the inland sand deposits" (cit. n. 14), p. 35; see also Bertram W. Wells to Frederic E. Clements, 15 April 1925, EFC box 65.0

ters, pyrlimnometers (light and heat absorption), solarimeters and pyrheliometers (solar radiation), tow and plankton nets, precision plankton traps and dredges, water samplers and testers, chemical equipment, one crank and two pump centrifuges, plus boats, autos, and other gear.[23]

Field biologists borrowed or adapted instruments from various disciplines. Although we think of instruments as preeminently laboratory artifacts, they were also central to the practices of certain field sciences, especially the physical sciences of the field like geodetics, geophysics or meteorology. Hydrobiologists, too, used machines for dredging and netting, and biogeographers might include thermometers and barometers in their field kits. A few field instruments were common household items, like the photometers of various sorts that had been developed for the amateur photography market. But when ecologists began to take up instruments in the early 1900s, the physical sciences of the field were the richest and most accessible resource.

The instruments most commonly deployed by early ecologists came from these quarters, especially meteorology. Meteorological instruments had been originally designed for field use or were already adapted from laboratory instruments, so the threshold for cultural borrowing was low. These instruments were designed to survive field conditions and could be purchased from catalogs in standard, field-proven designs, the simpler ones relatively cheaply. (Instruments that recorded automatically were the most convenient and desirable but tended to be too expensive for ecologists in the days before research grants.) Moreover, of the environmental variables recorded by weather stations—temperature, rainfall, wind speed, relative humidity—several seemed likely causes of the distribution of vegetation.

But why, one might wonder, did ecologists need to take instruments afield at all; why did they not simply use official weather data? The reason is that weather data are collected at widely dispersed "standard" places—usually near towns, where the people live who use the data—but these are quite unlike the environments in which plants and animals live. Furthermore, meteorologists report data as averages, which again serves the purposes of human users but not of biologists, who need a more particular and local knowledge. What matters to plants and animals is very local weather, the extremes and rare events that spell life or death. "We must get away from means, averages month, year etcetera *to the extreme,*" Charles

23. Chancey Juday and Edward A. Birge, "The highland lake district of northeastern Wisconsin and the Trout Lake limnological station," *Transactions of the Wisconsin Academy of Sciences* 25 (1930): 337–52, on pp. 347–52.

C. Adams wrote, "and get to storms, storm tracks, fronts, local and general winds etcetera—more to understand the relations of life to climate. It is no surprise that the usual weather data in ecol[ogy] paper[s] cannot be related to the biota."[24] Ecologists were thus obliged to do their own measurements in the field, with whatever instruments were available or could be modified to suit their purposes.

Many standard meteorological instruments served well enough, though some were simply not designed for the places where ecologists worked. For example, the standard weather bureau instrument for measuring relative humidity was the sling hygrometer, a dry-bulb–wet-bulb thermometer swung around at the end of a longish rod to ensure good air flow. It was portable, easy to use, commercially available, and affordable (five dollars in 1905). However, it was designed for use in open places, where vegetation did not alter "normal" ambient conditions, and it was precisely those local environments that ecologists wanted to measure, because that is the climate that plants and animals actually experience. In dense vegetation and small places the sling hygrometer could not be used at all, and in brush or forests it was liable to get smashed against a tree. Much better for ecological use, Clements found, was a gadget improvised by mounting wet and dry thermometers on a seventy-five-cent kitchen egg beater—household technology, like the quadrat.[25]

Limnologists likewise made use of hydrobiologists' collecting technology. However, hydrobiology was a qualitative science, unlike meteorology, so limnologists were obliged to modify and adapt existing equipment for quantitative work. Ordinary nets and dredges were standardized and fitted with simple shutters that could be quickly opened and closed to get precise, standard samples. Thermometers were redesigned to permit accurate measurements quickly from precisely known depths.[26] These modifications were quite simple and mechanical. In both ecology and limnology, borrowing from adjacent field disciplines was relatively straightforward, certainly simpler than borrowing from laboratory science.

For much field-biological work, however, instruments designed for other field sciences simply did not suffice. It was necessary in such cases to devise instruments specifically for the needs of field biologists, often by adapting

24. Charles C. Adams to Alexander Ruthven, 11 March 1907, AGR box 51 f. 1 (emphasis in original).

25. Clements, *Research Methods* (cit. n. 7), pp. 37–44.

26. Chancey Juday, "Limnological apparatus," *Transactions of the Wisconsin Academy of Sciences* 18 (1916): 566–92. Edward A. Birge, "A second report on limnological apparatus," ibid., 20 (1922): 535–52.

4.3 The home-made eggbeater psychrometer. Clements, *Research Methods in Ecology,* on p. 39

instruments from physics and physiology labs. That was not as easy as modifying instruments already adapted to outdoor use. Laboratory instruments are designed for precision work in constant, controlled environments and must be redesigned for use in the inconstant, even violent conditions of nature.

Some of the difficulties of translation were purely physical. Laboratory instruments do not have to be weatherproof, because laboratories have no weather. They do not have to be portable, or quick and simple to use, because conditions are always the same, and measurements can be repeated until they are done right. In the field, however, it is imperative that instruments be lightweight and portable, robust, and easy to use. Fieldwork is expensive and difficult to carry out, and every minute counts. In laboratories, instruments are tended by in-house expert mechanics. Fieldwork is usually performed by lone workers or small groups far from any shop. It is not a question of convenience: in the field heavy, delicate, and complicated instruments do not work at all.

Other problems of translation arose when field biologists began to measure variables that meteorologists and hydrobiologists had not. For measuring incident heat and light, for example, Birge and Juday had to turn to the precision electromechanical instruments of the University of Wisconsin's physicists. If hydrobiologists' nets and dredges had to be smartened up for quantitative work, the physicists' photoelectric cells and galvanometers had to be dumbed down—made less exact and tougher. They had to be precise but not too precise; portable and robust enough for use in boats on rough water; and fast, so that large numbers of readings could be fit into tight survey schedules.

Improvements were incremental. For example, Birge and Juday's first instrument for measuring radiation absorption in lake water was a simple black-bulb thermometer fitted with a shutter, which was serviceable but slow because it had to be lifted to the surface for each reading. It was succeeded by a version that could be read underwater through an extendable brass telescope, but that was awkward and was replaced by an array of thermocouples wired to an onshore galvanometer. Eventually a robust gal-

4.4 Edward Birge and Chancey Juday deploying a plankton trap, Lake Mendota, Wisconsin, 1917. State Historical Society of Wisconsin (negative no. WHi(X3)37040).

vanometer was devised for onboard use. For the design and construction of these instruments, Birge relied on the physics department's skilled mechanic, J. P. Foerst.[27] For the heavier technology of filtering large volumes of lake water, Birge and Juday turned to another local resource, the dairy industry, when it turned out that centrifugal milk clarifiers were ideal for separating nanoplankton fast and in bulk.[28] Lucky for Birge that the tourists' "Land o' Lakes" was also the Dairy State.

The advent in the 1910s of the automobile as an instrument of fieldwork, though one might think it would ease the problems of taking laboratory instruments afield, in fact made them worse. The violent jarring of autos on unpaved rural roads (which most roads were at the time) or on cross-country travel was far harder on delicate instruments than the gait of horses or pedestrians. For example, Clements's fine platinum-wire spectrophotometer did not survive an auto trip in the field and had to be rebuilt to be hardier.[29]

But the problems of taking laboratory instruments into the field went beyond ease of use and survivability. Those were relatively easy to solve. More complex and difficult were design issues like precision. How precise should an instrument be to suit ecological or limnological purposes? Here the differences between laboratory and field became harder to negotiate, but crucial because the credibility of field measurement was at stake. In laboratories instruments tend to evolve toward greater and greater precision, because in lab culture displays of precision are taken as a visible sign of competence and credibility and deflect doubt and criticism. In the disciplined environments of labs it is possible to do such performances with relatively little fuss, even if they are not strictly necessary for the job in hand, and rewards for incremental investment in precision are virtually guaranteed.

In the field the calculus of investment and reward is not so simple. The costs of precision are greater where there is no temperature and humidity control and no artificial light, and the rewards are less certain, because good results in such conditions cannot be guaranteed. Precision instruments fail, ruining runs of data that might not be repeatable. And many features of nat-

27. Birge, "Second report" (cit. n. 26), pp. 539–47. Edward A. Birge, "Transmission of solar radiation by the waters of inland lakes," *Transactions of the Wisconsin Academy of Sciences* 24 (1929): 509–80, on pp. 510–14, 564–69. Birge to T. E. Brittingham, Jr., 18 June 1930; and Charles E. Mendenhall to Birge, 9 Aug. 1932; both in Edward A. Birge Papers, University of Wisconsin Archives, Madison, Wis., box 6.

28. Juday, "Limnological apparatus" (cit. n. 26), pp. 582–88. Chancey Juday, "A third report on limnological apparatus," *Transactions of the Wisconsin Academy of Sciences* 22 (1926): 299–314, on pp. 299–303. Edward A. Birge, correspondence with H. C. Beckman and the De Laval Separator Co., 1914–15, UW-LIM box 5. Birge to Henry F. Moore, 21 April 1915, UW-LIM box 6.

29. Frederic E. Clements to Arthur G. Tansley, 29 March, 12 May 1923, FEC box 63.

ural phenomena simply cannot be precisely measured. Burton Livingston advised ecologists not to waste their time measuring some environmental factors with great accuracy if they could not measure all to the same degree.[30] But precision is what assured credibility in the scientific world, and even in the field measurement is likely to be judged by laboratory standards. There is no precise answer to the question, "how precise."

Ecologists were often torn between conflicting goals of precision and practicality. Clements, for example, advised ecologists to use standard laboratory instruments, which were capable of greater precision than was required in the field but could always be used less precisely if one wanted. But he also did not hesitate to use cruder instruments when it was clear that they were good enough. For example, Glen Goldsmith tested photochemical reactions and photographic emulsions as photometric instruments because they were simple and practicable in the field, though less exact than mechanical instruments. Burton Livingston likewise valued precision highly but did not pursue it as an end in itself; his practical ideal was to improve instruments for all environmental variables up to a reasonable level of precision, not to perfect those that were capable of high precision, as a lab man might do.[31] The symbolic and practical uses of precision measurement pulled field biologists in two directions, both toward the lab and toward the field.

Ecologists generally preferred instruments that gave a middling level of precision—a "good-enough" strategy that was more typical of the culture of engineering than of experimental science. The ideal, as Frederic Clements put it, was to measure with a precision appropriate to the factor being measured and to the observed effect. Glenn Goldsmith, his chief assistant at the Alpine Lab, favored "the simplest thing which will give results of the required accuracy." More precise instruments could be reserved for supplementary work in the lab.[32] What is "good enough" is a matter of personal judgment, of course; there are by definition no universal standards for it.

Yet while a good-enough philosophy might serve ecologists' needs, it was unlikely to win respect from plant physiologists, and often it was their opinions that counted. Clements was made sharply aware of this difference in outlook in 1930, when he and Goldsmith proposed to do a comprehensive reevaluation of available instruments, testing them in the field and if neces-

30. Burton E. Livingston, "Present problems of physiological plant ecology," *American Naturalist* 43 (1909): 369–78, on pp. 41–42.

31. Clements, *Research Methods* (cit. n. 7), pp. 21–22, 24. Glenn Goldsmith to Frederic E. Clements, 10 May 1930, FEC box 68. Livingston, "Present problems" (cit. n. 30), pp. 41–42.

32. Clements, *Research Methods* (cit. n. 7), pp. 21–22, 24. Glenn Goldsmith to Frederic E. Clements, 22 Feb. 1931, FEC box 86 f. heredity and environment.

sary redesigning them for ecological work. The idea was to devise instruments that suited the specific needs of ecologists, who were interested in biology and only secondarily in physical factors per se. For that reason Goldsmith planned to steer clear of instruments developed by physical scientists to measure single physical factors, because they were usually not meaningful biologically and were too finicky and cumbersome for field use. Goldsmith also insisted that ecologists perform the field tests, and not chemists or physicists, to ensure that the project would not be hijacked in a pursuit of precision for its own sake. "The advice of a man who is a specialist in a narrow field is seldom practicable," Goldsmith believed, "due to the fact that he is thinking of the measurement of a single factor in the most accurate possible way and not the total complex of factors, much less the simultaneous biological work which must be carried out." Goldsmith would do the design and construction in his home shop, and Clements would field-test the prototypes at Pike's Peak.[33] The results would be compiled in an updated edition of Clements's handbook.

Initially the project appeared to have won approval from Herman Spoehr, the head of the Carnegie Institution's newly consolidated Division of Plant Biology and Clements's (unwelcome) boss. Spoehr was a plant physiologist and a hard-nosed experimentalist with little sympathy for field ecology, and Goldsmith was surprised that he had approved an instrumentation project run by and for field biologists. He was right to be suspicious. Spoehr's interest in the project quickly evaporated. The reason, Clements learned indirectly, was Spoehr's belief "that ecologists lacked the necessary physical background and training and that instruments of precision probably could not be applied to biological problems anyhow." Goldsmith was disappointed but not surprised that Spoehr would insist that chemists run the tests to laboratory standards of precision.[34] Perfecting precision instruments had too much symbolic importance for experimental physiologists to be left to field ecologists. For people like Spoehr, "good-enough" was decidedly not good enough.

No less troubling than precision for those who took laboratory instruments into the field was the issue of simple and compound variables. Most physical instruments are designed to measure single physical variables, like temperature or relative humidity. However, plants and animals do not respond to single variables but rather to complex combinations. As a result,

33. Glenn Goldsmith to Frederic E. Clements, 28 March 1930, FEC box 68.

34. Frederic Clements to Glenn Goldsmith, 4 Jan., 5 Dec. 1930; Goldsmith to Clements, 20 Dec. 1930; all in FEC box 68.

ecologists often found it impossible to relate instrumental data to actual biological phenomena. But physical scientists and physiologists had little reason to develop instruments for measuring complex variables, because experimental protocol admitted only simple ones.

Take temperature, for example: a simple matter one might suppose, since instruments for measuring it were highly perfected—ordinary and recording thermometers, thermocouples, minimum-maximum indicators, and others. But plant germination and growth depend not on temperature alone but on a combination of temperature, humidity, sunshine, soil moisture, and so on. There was no proven method for correlating temperature records and plant behavior. It was the same with relative humidity. As a physical property, relative humidity was easy to measure, but the power of the atmosphere to evaporate water from plants depended not on relatively humidity alone but, again, on temperature, sunlight, and the like. So, too, with rainfall: easy to measure and abundantly recorded, but how much of the rain that fell was available to plants, that was the question. Runoff and groundwater levels were hard to measure, and meteorologists were not interested in these variables since they did not affect people.[35]

It was the same story for solar radiation. A variety of fine instruments were available, but plants use only particular parts of the radiant spectrum, and existing instruments had been designed to serve physical sciences or the various purposes of everyday human life, not plants. For example, photometers were useless to ecologists because they measured reflected, not incident, radiation—just the thing for photography, but not for plants. Physicists' bolometers measured radiant energy with precision but were too delicate for field use. No off-the-shelf photometer measured light as plants actually experienced it, a real problem for plant ecologists.[36]

Like precision, the issue of compound variables is deeply rooted in the different cultures and workplaces of lab and field science. Experiments operate on single variables; if they don't, they are by definition not proper experiments. Compound variables are a dangerous pitfall for experimentalists. Letting one slip undetected into an experimental design is a serious blunder, and taking one apart is seen as a signal achievement, a step from darkness to enlightenment. No wonder that laboratory scientists had little interest in devising instruments that measure them, however useful they

35. Livingston, "Present problems" (cit. n. 30), pp. 43–46. Clements, *Research Methods* (cit. n. 7), pp. 25–44, 64–75.

36. Livingston, "Present problems" (cit. n. 30), pp. 43–46. Clements, *Research Methods* (cit. n. 7), pp. 48–64. Frederic E. Clements and Glenn W. Goldsmith, *The Phytometer Method in Ecology,* Carnegie Institution Publication no. 356 (1924), pp. 4–5.

might be to field biologists. What paid off for them were more precise ways of measuring single variables.

For border biologists the logic of investment was more complicated. They valued causal analysis, but if they claimed too much for single variables as causes they opened themselves to attack as "reductionists"—an easily leveled and fearful allegation. The history of field biology is littered with the wreckage of single-factor theories. C. Hart Merriam's theory that vegetation zones are caused by differences in average temperature, for example, was especially galling to ecologists, who knew that other factors were involved. And there were the various schemes for classifying types of vegetation according to single environmental factors: Eugen Warming's put the causal onus on water, while others featured temperature, soils, or evaporation. The proliferation of these single-factor theories finally served only to discredit them all. These embarrassing failures in ecologists' recent past were a present reminder of the risks of putting too much faith in the kind of data produced by instruments designed by laboratory scientists for their own purposes.

Plant-Machine: Atmometers and Phytometers

There were ways of making instrumental measurement biologically meaningful, and field biologists tried most of them. A few put their hopes in mathematical algorithms that would combine single-variable weather data into complex variables that could then be correlated with plant distribution. Charles C. Adams hoped in this way to rework the vast accumulation of weather data to make it "dynamic," that is, a measure of the conditions that actually affect the life of animals and plants. Burton Livingston also saw a future in data reclamation, for biologists who had a mathematical bent. (Weather bureaus do the same when they combine temperature, wind, and relative humidity into compound variables like "wind chill" and "comfort index," which measure the physiological ability of human surfaces to stay warm or to cool off by sweating.) But alas, mathematical salvage of weather data proved clumsy and artificial. Adams soon lost heart, and Clements urged ecologists to do their own measuring and leave official weather data alone, as it could only bring them disrepute.[37]

37. Charles C. Adams to Alexander Ruthven, 11, 19 March 1907, AGR box 51 f. 1. Livingston, "Present problems" (cit. n. 30), pp. 43–46. Burton E. Livingston, "Atmometry and the porous cup atmometer," *Plant World* 18 (1915): 21–30, 51–74, 95–111, 143–49, on pp. 269–70. Clements, *Research Methods* (cit. n. 7), pp. 20–21.

A more fruitful strategy was to design instruments that could measure the complex variables that plants and animals actually experience. Make machines more like plants, in other words. The bias of meteorological instruments toward single physical factors would be designed out, and an ecological purpose built in instead. Alternatively, "standard" plants could be redesigned for quantitative measurement of the complex variables that affected them, turning them in effect into machines—"phytometers." We will look at both.

The porous cup atmometer is the instrument that most neatly illustrates the strategy of making machines more like plants. This little gadget measures the rate at which water evaporates from a surface. In its simplest form it is just an unglazed ceramic cup filled with water and connected by a glass tube to a reservoir of water. Water diffuses to the outside surface of the porous cup, where it evaporates; the loss of water (measured in volume per unit time) represents the rate of evaporation from the cup surface. Several features of this simple but elegant device suit it for biological work. First, it is relational: it does not measure an absolute property of the atmosphere, like relative humidity, but the relation between a particular surface and the atmosphere. Also, the rate of evaporation is not a simple variable but one compounded of relative humidity, temperature, exposure to the warmth of the sun, and air movement.

The atmometer, unlike meteorological instruments, thus measures a variable that plants experience: the power of the atmosphere to evaporate water from their leaves. In fact, the instrument is like an artificial plant, as Burton Livingston observed, though without the living plant's active power of regulating transpiration. The porous cup is a mechanical leaf, drawing up water through its stem and evaporating it from its surface. This is why atmometry data are directly relevant to plant activity; the evaporating power of the air is, along with light, the crucial factor that determines how good a place is for plants to grow.[38]

First invented in France in 1848 but then forgotten, the porous cup atmometer was rediscovered in 1904 in Germany and simultaneously in the United States, by Burton Livingston. The standard method at the time for measuring evaporation was crude: set out an open pan of water and weigh it from time to time to get the rate of evaporation. It was a household or

38. Burton E. Livingston, "Evaporation and plant development," *Plant World* 10 (1907): 269–76, on pp. 269–70. Livingston, "Evaporation and plant habitats," *Plant World* 11 (1908): 1–9. Livingston, "Paper atmometers for studies in evaporation and plant transpiration," *Plant World* 14 (1911): 281–89, on p. 99. Livingston, "Atmometery" (cit. n. 37).

backyard instrument: cheap, handy, foolproof, a model of simplicity. But for use in field conditions the pan was not a practicable instrument. It could not be used in rain, and it attracted birds and small animals, which drank the water, splashed it about, and left behind excreta and debris. Furthermore, the rate of evaporation turned out to vary with the shape of the pan and water level. A pan of water was much less like a plant than like a pool or pond, something limnologists might find useful, but not ecologists.

In its first incarnation the porous-cup atmometer was strictly a laboratory device and not field worthy, but its promise was evident, and it evolved rapidly, thanks mainly to the efforts of Livingston and the ecologist Edgar Transeau. By 1910 or so it had become a simple and robust field instrument, transportable, easy to use, and quite exact, so long as it was calibrated from time to time in a laboratory against a designated "standard" cup. It was small enough to be placed in the microenvironments in which plants actually lived and sufficiently inconspicuous to escape the notice of vandals. The first models could not be used in rain, but addition of an ingenious and simple arrangement of mercury valves made it all-weather.[39]

Dubious of the first models, ecologists took up the improved, rainproof type with alacrity. The "artificial plant" probably did much to get ecologists into the habit of taking instruments into the field, because for the first time an instrument produced data that could be correlated to the life of plants. The atmometer helped make field ecology into a measuring discipline, exactly as its developers intended. It carried an element of laboratory culture into fieldwork, but in a way that was appropriate to biologists' concerns.

Its resemblance to a plant is part of what made the porous cup atmometer so attractive to border biologists. Like plants, it responds to particular local places, with their particular and ever-changing patterns of light and shade, humidity and atmospheric circulation. (Laboratory instruments, in contrast, measure only weatherless, placeless places.) But, as Burton Livingston observed, the cup atmometer does not mimic particular plants, which differ in their physiological powers of transpiration, but a *generic* plant: "[T]he exposure, surface, etc., of the porous cup atmometer are . . . nearly similar to the corresponding features of plants in general." It was precisely this generic likeness of cup to "plants in general" that made the instrument scientific by laboratory standards. The atmometer was like a plant but lacked

39. Livingston, "Atmometery" (cit. n. 37), pp. 22–29, 59–62, 99–105, 110–11. Edgar Transeau, "The relation of plant societies to evaporation," *Botanical Gazette* 45 (1908): 217–31. Transeau, "A simple vaporimeter," *Botanical Gazette* 49 (1910): 459–60. Transeau, "The relation of plant societies to evaporation," *Botanical Gazette* 45 (1908): 217–31.

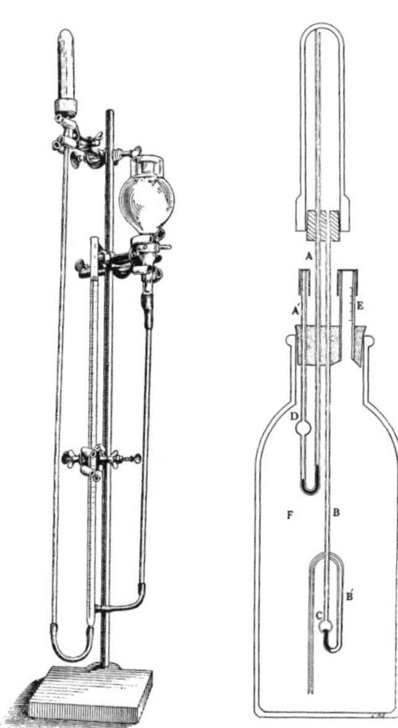

4.5 Two atmometers. *Left:* the original laboratory apparatus. Burton E. Livingston, *The Relation of Desert Plants to Soil Moisture and to Evaporation,* Carnegie Institution Publication no. 50 (1906), on p. 26. *Right:* the improved model, rainproof and fieldworthy. Livingston, "Atmometry and the porous cup atmometer," *Plant World* 18 (1915), on p. 61.

the particularity of plant species or individual plants; it was a kind of mechanical "standard" plant,[40] and that is why ecologists loved it. They always hoped that the U.S. Weather Bureau would one day carry out atmometry on a continental scale, but it never did. (Leaves do not appropriate funds, skins do.)

Livingston and others went to considerable lengths to make the porous cup atmometer more nearly resemble a leaf. One modification used a moistened paper disc: being thin, like a leaf, it responded more rapidly to changes in temperature than the porcelain cup. Unfortunately, it tended to dry out at the edges so was given up. Livingston tested cups blackened with lamp black or made of dark clay, which he thought would absorb sunlight more as green leaves do. William Tower apparently experimented with spherical cups, which caught the sun at the same angle throughout the day, as plants did.[41] Few other field instruments had this peculiar property of being simultaneously mechanical and plantlike, standard yet responsive to particular,

40. Livingston, "Atmometry" (cit. n. 37), pp. 110–11.

41. Livingston, "Paper atmometers" (cit. n. 38), p. 99. Livingston, "A radio-atmometer for comparing light intensities," *Plant World* 14 (1911): 281–89, on p. 281.

variable environments. It was quintessentially an artifact of the border, a blend of lab and field, machine and nature.

The reverse strategy, which in principle achieved the same end, was to make plants into quasi instruments for directly measuring the capacity for environments to sustain their growth. "Phytometers," as Frederic Clements called them, were the most natural imaginable measuring device, not plant-like cups but the very thing itself. They required no redesign or adaptation, as laboratory instruments did to be field worthy, and retained no apron strings of dependence on laboratory machine shops and standardizing labs. Although the idea of "standard" plants smacked of laboratory culture, the thing itself was not a portmanteau of hidden cultural baggage from the lab, as machines were bound to be. Phytometers promised to make ecology quantitative and lablike using (literally) homegrown objects.

Clements was the most enthusiastic advocate of phytometery (about the only one), and he spent considerable time and effort between 1906 and the mid-1920s trying to make the idea work in practice. "Plants," as he put it, "are the best measure of plants." His first attempts used cut plant shoots in potometers (sealed bottles of water), but the results were too variable to be of any use. (Dead plants made poor machines.) He then turned to whole living plants set out in pots in the environments he wished to measure. For these "standard" plants Clements preferred weeds or domesticated species, because they stood up better than native species did to constant handling. He then measured rates of growth, photosynthesis, and transpiration in the same way that physiologists did in greenhouses (by simply weighing sealed pots). These physiological rates he took to be quantitative measures of the ability of particular places to support plant life—from, so to speak, the plant's point of view. Clements also kept his phytometer test plots au naturel: small, untidy, and scattered about the mountain landscape of Pike's Peak. In photographs they could be mistaken for unkempt gardens or abandoned camp sites or other bits of nature that humans use briefly and carelessly.[42] Clements managed in the end to get some data that he was not afraid to publish, but with difficulty; it is not surprising that phytometry never caught on.[43]

In fact, the idea of phytometry was a scientific variant of venerable folk practices that also used plants to measure the suitability of places for growing plants. Farmers scouting the North American forests for homestead

42. Frederic E. Clements, "Ch. 4. Alpine Laboratory and Transplant Gardens," n.d., p. 90, FEC box 84 f. Colorado.

43. Frederic E. Clements to John E. Weaver, 10 Nov. 1922, FEC box 63. Clements and Goldsmith, *Phytometer Method* (cit. n. 36), pp. 51–53. Clements to Glenn Goldsmith, 14 Feb. 1919, FEC box 62.

4.6 "Phytometers" deployed in improvised and typically untidy experimental field gardens. Frederic E. Clements and Glenn W. Goldsmith, *The Phytometer Method in Ecology,* Carnegie Institution Publication no. 356 (1924), plate 5.

sites used to judge the agricultural potential of a place by observing the species of trees that grew there. (Black walnut and tulip meant good soil for wheat, sycamores and elms for corn, oaks and dogwood for cotton, and so on.)[44] An even older example of folk phytometery was the Native American practice of planting corn when the leaves on the oak trees were the size of a

44. Thomas H. Kearney, Jr., "Report on a botanical survey of the Dismal Swamp region," *Contributions of the U.S. National Herbarium* 5:6 (1901): 321–550, on p. 473.

squirrel's ear. The phytometers—oak leaves in this case—were a visible indicator of the combined factors influencing plant growth in a particular place and a particular season. Normal variations in locale and weather made planting by the calendar—the modern, universalistic practice—riskier. (Like politics, all agriculture is local agriculture, and all ecology is local ecology.) Clements's "phytometer" may be seen as another of these folk practices, an ethnoscientific one, and that was perhaps its problem. Such practices work because they are particular and not too precise and thus suited to making choices in local and variable environments—an ideal kind of knowledge for field biology, one might think. But where the universalistic standards of the placeless place are honored, as they were by many border biologists, local practices will be less honored.[45]

Instrumental Eye: The Camera

Not all the instruments of field ecology came from the world of professional science. One—the camera—came out of popular culture, and it was an important one. The development of cheap, easy-to-use cameras and roll film in the 1890s transformed a complex technology into a ubiquitous household item. Professional photographers had used wet-plate cameras in the field since at least the 1860s, but the equipment was clumsy and finicky, and as an adjunct to ecological fieldwork it was more trouble than it was worth. But the black-box Kodak turned ecologists, along with millions of others, into avid landscape photographers. Ornithologists were the first biologists to take cameras into the field, no doubt because of their photogenic subjects and their unusually active connection with the world of recreational naturalists. (Home and workplace cultures mingled intimately in that sector.) Frank Chapman was the first American ornithologist to use photography routinely in his work and lectures, starting in 1888, and was again the first to use motion pictures, about twenty years later.[46]

45. The history of weights and measures offers a paradigmatic example of the conflict between local and universal practices: Kenneth Alder, "A revolution to measure: The political economy of the metric system in France," in *Values of Precision,* ed. Wise (cit. n. 2), 39–71. See also James C. Scott, *Seeing Like a State: How Certain Schemes to Improve the Human Condition Have Failed* (New Haven: Yale University Press, 1998), chap. 9 and p. 312.

46. Reese V. Jenkins, *Images and Enterprise: Technology and the American Photographic Industry 1839 to 1925* (Baltimore: Johns Hopkins University Press, 1975). Beaumont Newhall, *The History of Photography* (New York: Museum of Modern Art, 1982; 5th ed. 1988). Alfred O. Gross, "History and progress of bird photography in America," in *Fifty Years' Progress of American Ornithology, 1883–1933* (Lancaster, Pa.: American Ornithological Union, 1933), 159–80.

4.7 Edith and Frederic Clements in the field with camera, digging, 1917. American Heritage Center, University of Wyoming (FEC box 120).

Ecologists, who came on the scene about the same time as the black-box Kodak, were also quick to take it up, and in their hands a recreational toy became a scientific instrument. Indispensable, Frederic Clements called it in 1905, no less so than meteorological instruments: "[I]t is as important for recording the structure of vegetation as the automatic instrument is for the study of the habitat. No ecologist is equipped for systematic field investigation until he is provided with a good camera."[47] And ecologists did so equip themselves. Ecological publications from the early 1900s are full of grainy photographs of fields, bogs, woods, lakes. Henry Cowles assembled a collection of some 5,000 glass-plate photos, rather humdrum rural scenes but now a unique record of landscapes devoured by urban sprawl. As part of his ecological survey of Minnesota, Conway MacMillan created a gallery show of thirty-by-forty inch photographs of typical formations—"plant portraits in their habitats" or ecological group shots. It was good science, and the public loved it.[48]

There are several reasons why ecologists took so avidly to what they might have seen as a mere toy. Most important, the camera helped ecologists to do what was already an obligatory but rather onerous field chore: namely, minutely observing and recording in notebooks the places being

47. Clements, *Research Methods* (cit. n. 7), pp. 188–96, quote on p. 188.
48. Conway MacMillan, "A botanical art gallery," *Botanical Gazette* 28 (1899): 430–31.

worked. Early ecological reports were elaborately, obsessively descriptive, and the most common complaint of leaders of early ecological surveys (apart from bad weather and biting insects) was the difficulty of getting field assistants to take regular field notes. Snapshots went some way to solving that problem; they were much faster and easier than note-taking—and more fun—and they were more compact in publications than lengthy word pictures. Never popular with field-workers, note-taking seemed antiquated when cameras became standard items. As one field-worker found when his camera broke in the field: "There are a good many things that could have been recorded quickly with the camera which could hardly be put in words as they came along. Now I must depend on my memory . . . aided by such notes as I could make at the time."[49] Ecologists' customs of description and note-taking predisposed them to adopt the camera as a mechanical substitute for these literary chores: the camera was preadapted. Its ubiquity as an item of popular entertainment also eased its translation into a scientific instrument. Cameras were cheap and accessible, and as items of mass consumption well supported by users' guides. As Charles Adams observed, students of wildlife photography in 1913 had four excellent how-to books from which to choose—more than there were guides to ecological field instruments and practices.[50] For that ecologists had to thank the large number of amateur bird and animal watchers.

The camera's association with recreation and amateur naturalists did give early ecologists some qualms. One anonymous zoologist welcomed the new rapprochement between professional zoologists and naturalist photographers, but noted potential dangers: "Of indiscriminate observation there has already been too much; what is needed is discrimination," he warned. "There is [a] danger that the camera may become as powerful a fetish as the microtome has been. . . . Pretty photographs are of no more value than pretty microscope slides."[51] Frederic Clements likewise made the point that the camera was a serious scientific instrument, not a plaything, and should be used accordingly. Photography, he warned, could easily reinforce the tendency of Brush Creek ecologists merely to describe and to claim any slight variant of vegetation as a new type. "The misleading definiteness which a photograph seems to give a bit of vegetation has been

49. Alfred C. Weed to Wilfred H. Osgood, 11 July 1928, WHO box 10.

50. Charles C. Adams, *Guide to the Study of Animal Ecology* (New York: Macmillan, 1913), pp. 37–39.

51. "Tendencies in zoology," *Popular Science Monthly* 60 (1901–02): 95–96.

responsible for a surplus of photographic formations, which have no counterparts in nature," he complained. "Indispensable as the photograph is to any systematic record of vegetation, its use up to the present time has but too often served to bring it into disrepute." Burton Livingston also warned that photographs were a superficial record; the real causes of vegetation were to be found by laboriously mapping the various types of soils.[52]

The camera's eye, unlike the human kind, did not discriminate between what was essential and what was incidental, as good ecologists did. Cameras recorded everything indiscriminately, rather as early biometricians counted everything, or as Brush Creek naturalists described all they saw. It is easy to see why ecologists would have mixed feelings about the camera, at a time when they were striving to make natural history "scientific" and distance themselves from the weekend "ecologists" who threatened to discredit the new discipline. In an ambiguous cultural borderland, the visual power of photographs and their slavishly faithful representation were ambiguous virtues. Worth a thousand words, no doubt, but they also could give spurious authenticity to ill-conceived ideas.

In practice Clements advocated a "good-enough" approach to ecological camera work, as he did with other field instruments. It didn't pay to spend hours getting the perfect shot, he advised. The camera proved most useful as a complement to analytical, literary description, not as a substitute for it. It was not a precision instrument but an instrument of observation, a mechanical help to note-taking—that most traditional and essential of field practices. It was an artifact well suited to border biology.

Making the Place Right: Forrest Shreve

Taking nature's measure requires more than instruments and laboratory techniques. Skills distinctively of the field are also required, skills of selecting or adapting natural places to make them suitable for machines and quantitative measuring. It takes experience to know exactly where and how to place instruments so that they produce good, steady data, and how to avoid the hundred little things that can go wrong in a place where the unexpected is expected. For long-term stakeouts it may be necessary to redesign

52. Clements, *Research Methods* (cit. n. 7), p. 7. Burton E. Livingston, "The distribution of the upland plant societies of Kent County, Michigan," *Botanical Gazette* 35 (1903): 36–55, on pp. 53–55.

natural places to make them more like experimental gardens, though not too much like gardens lest their natural authenticity be lost. In laboratories, obviously, there is no such difficulty, since there is no variation between one place and another. But in nature no two places are exactly alike, and procedures for finding the right ones to measure are a crucial part of instrumental practice.

Forrest Shreve's activities in the Santa Catalina Mountains offer good examples of the craft of exact measurement in the field. Although he is best remembered for his late studies of the geography and natural history of western deserts, Shreve spent much of his career making quantitative measurements on mountain slopes. He never put much stock in Clementsian theories of community and succession, but he did share Clements's conviction that plant distribution depended on environmental factors, which had to be measured with precision instruments. Mountains were a particularly good place to correlate vegetation and environmental variables, because their vegetation belts and environmental gradients are compact and thus easily recognized and measured: good places for intensive, long-term study.[53] (In flat country, climate zones are hundreds of miles wide, making the deployment of instruments logistically difficult, as Clements learned in the Midwestern prairies.)

Shreve chose his field sites for their convenience to his laboratory headquarters. In the 1900s he worked the Blue Mountains of Jamaica from the Cinchona Experimental Station; in the 1910s and early 1920s the Santa Catalina Mountains were his field, just twenty miles from the Desert Laboratory at Tucson, Arizona. Later Shreve worked the Santa Lucia mountains in California's coastal range, a day's auto ride from the Carnegie Institution's laboratory at Carmel. Shreve pursued a somewhat different mix of laboratory experiment and field observation in each place, according to his current interest and the opportunities of the terrain. In Jamaica he spent most of his time in the station's physiological laboratory doing experiments on transpiration, but the Blue Mountains' dramatically contrasting climates and vegetations (cold rain forests on the north slope and warm, dry subtropical forests on the south) enticed him into the field, laden with a kit of atmometers, hydrographs, psychrometers, thermographs, and soil thermometers. In the Santa Lucia Mountains, in contrast, Shreve did almost no instrumental work, in part because there were few roads and trails, and in part because the vegetation zones seemed to depend less on environmental

53. Janice E. Bowers, *A Sense of Place: The Life and Work of Forrest Shreve* (Tucson: University of Arizona Press, 1988).

factors than on contingencies of soils and fire history, neither of which could be measured with instruments.[54]

It was in the Santa Catalina Mountains that Shreve most fully realized his ideal of quantitative measurement in the field. The ecologists of the Desert Laboratory strongly supported such work (Burton Livingston did much of his work on the atmometer there), and the terrain of the Santa Catalinas was almost ideal for the deployment of instruments. This range is one of a number of small, isolated ranges rising abruptly out of the Arizona desert. Their small size and steep gradients of elevation (from 3,000 to over 9,000 feet) plus a climate regime of infrequent but violent rains, created a highly dissected topography of steep-sided ravines and sharp ridges. The vegetation in this compact vertical landscape varied from desert to alpine within a few miles and was arranged in well-marked zones ideal for instrumental study. (C. Hart Merriam did his classic work on life zones in just such a place, the San Francisco Mountains of northwest Arizona.) The Santa Catalinas had the additional advantage of being a forest preserve and thus provided with a good system of trails maintained by the U.S. Forest Service. Other island mountains, like Kits Peak, were wild, trackless, and far more difficult to work.[55]

Shreve's purpose in the Santa Catalinas was to disprove Merriam's simple temperature theory (which he thought few believed except Merriam's loyal followers in the Bureau of Biological Survey). Where Merriam had just mapped vegetation boundaries and recorded temperatures in a single season in the field, Shreve's project was long-term and heavily invested in a range of instruments. Suspecting that water was crucial to plant distribution, Shreve measured a range of environmental variables—including rainfall, soil moisture, relative humidity, evaporation, and air and soil temperature—over a wide range of terrain. What he hoped to find (but never did) were discontinuities in the gradients of single factors that matched those of vegetation zones.

Taking the measure of a mountain's life zones was in principle just a mat-

54. Forrest Shreve, *A Montane Rain-Forest,* Carnegie Institution Publication no. 199 (1914), pp. 5–9, 22–41, 41–54. Shreve, "Experimental work at Cinchona," *Science* 43 (1916): 918–19. Shreve, "The vegetation of a coastal mountain range," *Ecology* 8 (1927): 27–44.

55. Forrest Shreve, *The Vegetation of a Desert Mountain Range as Conditioned by Climatic Factors,* Carnegie Institution Publication no. 217 (1915), pp. 5–16. [Frank Lutz?], "Report on the 1916 expedition to Arizona and Texas," pp. 2, 11, Academy of Natural Sciences Archives, Philadelphia, series 113 box 3 f. 74. C. Hart Merriam, "Results of a Biological Survey of the San Francisco Mountains Region and Desert of the Little Colorado in Arizona," U.S. Bureau of Biological Survey, North American Fauna no. 3 (1890). Jane Maienschein, "Pattern and process in early studies of Arizona's San Francisco peaks," *BioScience* 44 (1994): 479–85.

ter of taking simple transects: place batteries of automatic recording instruments at intervals from base to peak. In practice, however, it took experience to decide what, where, and how. Shreve experienced a sharp learning curve and a rising investment in machines and the time spent caring for them. Burton Livingston had staked out three stations in the Santa Catalinas in 1906 for testing his new atmometer designs; Shreve expanded these to five in 1908–10. The data they produced, however, proved too discordant and full of gaps for publication, and in 1911 he had to start all over again with improved instruments at six stations, this time tending them assiduously on rigid schedules to ensure that every machine was working properly all the time. Shreve also expanded the range of instruments, installing rainfall gauges at four stations in 1908 and at seven or eight in 1911–14, with instruments for measuring soil moisture as well as precipitation. In 1921 and 1922 he installed soil thermographs at six stations between three thousand and nine thousand feet.[56] Getting lab-standard data in the field turned out to be an exacting and often disappointing experience.

It was not that Shreve's instruments were ill designed for outdoor work. The problem was making them work reliably when left alone in rough country. In the mountains, instruments can be visited only periodically and are subject to vagaries of weather and human and animal depredation. Only a strictly observed schedule of visits, readings, standardizing, and maintenance can make the material culture of labs work properly in variable and accident-prone environments. Field know-how and judgment were crucial. Laboratory training and skill in using instruments were not sufficient to make a good field-worker; it was too easy operating instruments in places without hills, weather, and bears. (Labs are meant to make it easy for beginners to get started fast.) John Weaver, Clements's chief coworker, did not trust even his most senior graduate assistant: "I find many men are all right in the laboratory," he wrote Clements, "but cannot quite put it across in the field. We want clear cut data and we want no question marks at the end of the season." The more complicated the instrument, the more training was required to use it skillfully in rough places.[57]

Placing transects and instruments also required knowledge and judgment that could only be acquired by experience in the field—almost a kind of scientific geomancy. For example, Shreve chose the south slope of the

56. Shreve, *Vegetation of a Desert Mountain Range* (cit. n. 55), pp. 46–64. Shreve, "Soil temperature as influenced by altitude and slope exposure," *Ecology* 5 (1924): 128–36, on p. 131.

57. John E. Weaver to Frederic E. Clements, 13 March 1923, FEC box 63. See also Shreve, *Vegetation of a Desert Mountain Range* (cit. n. 55), pp. 46–64; and Glenn Goldsmith to Clements, 28 March 1930, FEC box 68.

Santa Catalinas for his string of stations because it was steeper than the north slope, and thus stations at thousand-foot intervals of elevation were closer together and more convenient to work—a vital consideration for long-term work in difficult terrain. Moreover, the ridges and canyons of the south slope ran more or less east to west, creating distinct differences of vegetation on north and south aspects and extreme and easily measurable differences in light and temperature. (On the north slope, drainage patterns ran generally north to south, and canyon slopes had similar vegetation on both sides.) Finally and most important, the south slope was geologically homogeneous, with one kind of rocks and soils over the entire area, whereas the varied rocks and soils of the north slope were a complicating variable. Choosing the right place enabled Shreve to hold constant a variable that in many instrumental studies had made causal inferences uncertain.[58] But it required intimate personal knowledge of a large area, and an eye for the particular small places in which the signal of cause and effect was greater than the noise of natural variability.

The care and judgment that Shreve exercised in the Santa Catalinas paid off, confirming his expectation that water, not temperature, was the key to vegetation zones. Single environmental factors showed no discontinuities, just gradual continuous change across vegetation zones. However, Shreve found that a doubly compound variable, the ratio of daily evaporation to soil moisture, did show discontinuities that coincided with the boundaries of vegetation zones. It was good evidence of a causal relation. Temperature effects were more complicated, Shreve found, because plant survival depended on local vagaries like average and extreme values, seasonal changes, and duration of freezes, which cannot be reduced to simple mathematical expressions.[59] Some features of plant and animal life are just not amenable to the techniques of measurement: that is an inescapable fact for border biologists.

Take a drive with Shreve across the Sonoran desert from the Desert Lab to the foot of the Santa Catalinas. We park, then ascend from the Santa Cruz Valley up Soldier Canyon to Green Mountain at seven thousand feet and from there to the highest station on Mount Lemmon at nine thousand feet, stopping at intervals to read and check batteries of instruments; then back down to the desert playa and home to the Desert Lab. Taking this cross-country trip, we also perform a cultural transect of the lab-field border zone, from a lab in nature to field stations that are tiny patches of laboratory cul-

58. Shreve, *Vegetation of a Desert Mountain Range* (cit. n. 55), pp. 6–16.

59. Shreve, *Vegetation of a Desert Mountain Range* (cit. n. 55), pp. 92–94. Shreve, "Soil temperature" (cit. n. 56).

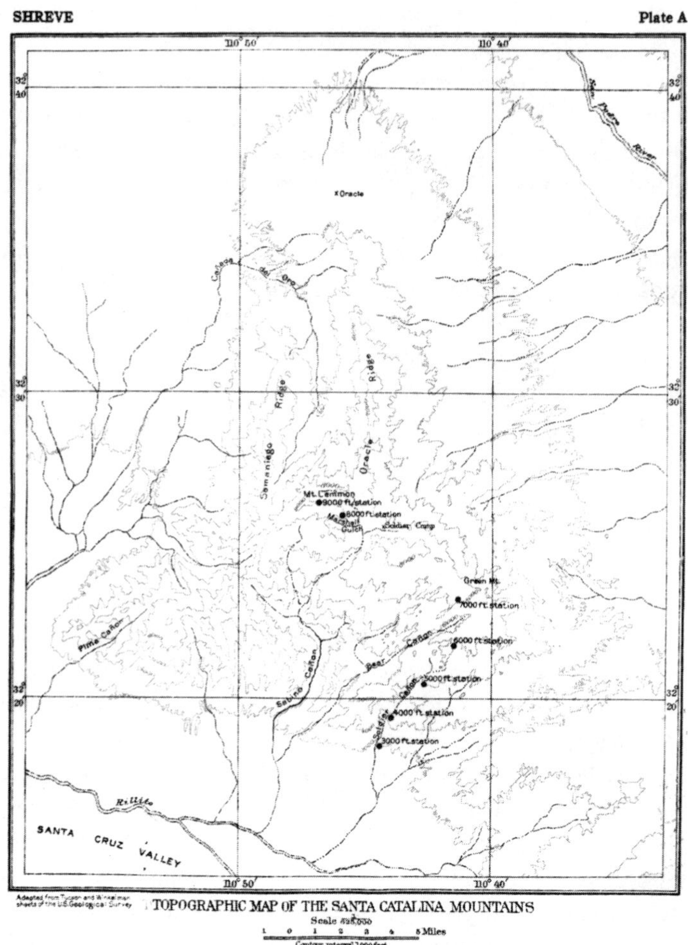

4.8 Topography of the south slope of the Santa Catalina Mountains, Arizona, showing stations where Forrest Shreve installed field instruments. Forrest Shreve, *The Vegetation of a Desert Mountain Range,* Carnegie Institution Publication no. 217 (1915), plate A, on p. 56.

ture in the midst of a vast natural world in which laboratory things and practices will always be out of place.

Conclusion

Following instruments and quantitative practices into the lab-field border zone, we see that they are powerful vehicles of cultural transformation, but also that they are not simple or one-way vehicles of influence. Laboratory instruments and practices change when they are taken afield. Ceramic cups become leaflike objects, and sharply precise tools are dulled to suit imprecise problems and environments. Places and skills of placing become a crucial part of instrumental practices in the field in a way that is quite alien to laboratory experience. And things and practices acquire new meanings when moved from places of greater to those of lesser cultural authority. These adaptations made precision instruments less threatening to field biologists' identity and authority.

The cultural geography of the lab-field border appears decidedly varied, and would be even more so if we were to extend our survey to biogeography and the physical sciences of the field. However, we may hazard some general points about the role of material culture and practice in shaping a cultural ecotone.

Cultural proximity is one point: field biologists adopt instruments and practices more easily from the physical sciences of the field than directly from laboratory physics or physiology. And no wonder; meteorology, hydrology, and topography were counting and measuring cultures long before the new naturalists came on the scene, and their tools and standards of precision were already adapted to field conditions. The threshold for borrowing is lower along the boundary between the biological and physics sciences where both are of the field than when one is of the lab. Place trumps subject.

A second point concerns the symbolic and practical calculus of investment. Field biologists use things and practices that pay off in insights and publications and that bring them credibility. That is true in every region of science: lab, field, and border zone. Where counting and measuring are seen to serve compelling intellectual agendas, even substantial investments in new things and skills will seem worthwhile, and problematic ideological baggage can be finessed or reworked. Praxis trumps ideology—often, though not always. But because the credibility of border practices are less secure than those of labs, field biologists who wish to be properly scientific may embrace precision instruments and quantification with an ardor that goes beyond any practical use, as ecologists did to keep out tyros. Or they may reject the practices and paraphernalia of lab culture for fear of being hijacked, as taxonomists rejected biometry. In the field symbolic meanings are powerful because the field is culturally diverse and close to the worlds of recreation and economy, and practices there are contested and ambiguous.

No wonder that the lab-field border in biology has a ragged cultural topography, some areas hardly influenced at all by counting and measuring, like systematics; others transformed, like limnology. Ecologists were in between; "autecology" (the side that deals with individual animals) became a branch of laboratory physiology, while "synecology," which deals with natural populations, remained a mixed border science. American ecologists embraced quadrats and instruments but were indifferent to the statistical plant sociology that captivated Europeans. Americans made the study of animal behavior a laboratory discipline, while Europeans made ethology an outdoors science.

But the fundamental point is this: the practices and material culture of counting and measuring were a powerful force in realizing the ideals of a scientific natural history and creating a distinctive border culture. Because counting and measuring were relatively simple to do and well suited to field problems—doable and productive—they were among the first practices to define a lived, as distinct from an imagined, border culture, that blended lab and field. (Translating the practices and material culture of experiment was more taxing and risky, as we will see next.) The adaptations of instruments to place, and of natural places to the requirements of counting and measuring, are more than indicators of a new cultural geography: they are what created it.

CHAPTER 5

Experiments in Nature

It took ingenuity to count and measure in the field, but there were precedents to build on; doing experiments there was more problematic. Experiment, unlike counting or mensuration, is an activity designed specifically for laboratory environments. Its power is realized only in simplified and controlled environments, and it is difficult (though not impossible) to replicate such places in nature. Experiment carries heavier cultural baggage, and its use in the field is more likely to be contested than counting or measuring. Following counters and measurers into the field, we experience a gentle transition from laboratory to field culture. Following experimenters, we feel a sharper discontinuity. The difficulties of taking experiments afield reveal an uneven cultural topography.

It has been customary to see the difference between laboratory and field as one of practice: on the one side observation, on the other experiment. Observation does well enough where experiments are not possible, but experiment is the real thing, and if one cannot take experiments afield, it is well to take one's work as quickly as possible into a lab. These assumptions were built into the institutions of modern science, and field biologists have had to reckon with them.

Promoters of the new natural history argued that

both experiment and observation were limited, and that the ideal was a combination of the two. Stephen Forbes, for example, foresaw that ecologists would turn more and more to experiment but hoped they would do so only after thorough field study. "It is thus always the field observation," he wrote, "or the laboratory observation made under conditions which involve the least practicable departure from natural conditions . . . , which must precede and suggest the experiment." Henry Cowles projected a similar balance between the analytical precision of experiment and the natural authenticity of fieldwork: "[T]he fundamental method of the future is to be field experimentation combined with observation . . . ," he wrote, "the exact methods of the laboratory carried into the field."[1]

Lab-trained evolutionists like Samuel J. Holmes likewise pointed to the benefits to both sides of combining taxonomy and genetics. "How many of the subspecies which have been so industrially [sic] manufactured by American mammalogists and ornithologists would melt away in the hands of the experimenter" Holmes feared to contemplate. But were geneticists' "mutations" or the variations induced in lab animals by experimental morphologists any more secure? Did they have any relation to variations in nature? It was risky to put too much stock in results with lab-bred animals in unnatural environments, Holmes thought, and essential to study mutations and variation in nature as well as in the lab.[2]

How field biologists attempted to combine laboratory work and fieldwork is the subject of this chapter. It was difficult. The history of such efforts in the first few decades of the twentieth century is largely a history of frustration and disappointment. However, these experiences tell us a good deal about the structure and dynamics of border life. Like good naturalists we must follow experimenters into the field and observe where they go and what they do, catching them on the move in their natural habitats as they did the creatures they studied. We must pay special heed to field practices that have lablike qualities, and to the actual places where these practices were performed. Much of our evidence is evidence of place.

A cultural-geographic approach is well suited to these tasks because in the field, deciding what to do is often the same as deciding where to do it. Practitioners in the zone of ambiguous identity and cultural mixing will be

1. Stephen A. Forbes, "Special report of the Biological Experiment Station," University of Illinois *Trustees Report* 18 (1895–96): 302–26, on p. 310. Henry C. Cowles, "Ecological aspect of the conception of species," *American Naturalist* 42 (1908): 265–71, on pp. 265–66.

2. Samuel J. Holmes to William E. Ritter, 20 Sept. 1912, WER box 11.

especially alert to the symbolic implications of where they work, because their standing depends in part on how they position themselves in relation to the lab-field border. A more experimental mode of practice will be read as situating a practitioner closer to the high-status laboratory side; a natural-history mode, closer to the lower-status field side. So for each variety of field-experimental practice we must ascertain the place where it was done and its position in the cultural ecotone.

Movement is especially revealing of cultural topography. Cultural borders are regions of both physical and cultural movement (like all frontiers), and trajectories in space have symbolic import. Movement from field to lab is a sign of a progressive career; a move from lab to field is more ambiguous. Experimenters have always gone afield to collect their "material," but collectors always return to the lab, and their excursions are generally brief and occasional. The movements of field experimenters are more varied. The ideal was a shuttling between lab and field: formulate problems in the field, experiment in labs, then back to the field to check results and get new problems—that was the pattern implicit in the ideal of mixed observation and experiment.

In practice, a more common pattern in these early years was an uneven lurching movement in careers, periods devoted to work in the field followed by abrupt shifts to experimental work, with the cycle sometimes repeating. This cycling movement was driven by hope and disappointment and is characteristic of a time when mixed cultural practices hardly existed; simply importing experimental methods from the lab was bound to disappoint. Or, acting on the common wisdom that field observation is a preliminary to laboratory experiment, naturalists moved to labs only to find that laboratory experiments were not very relevant to field problems.

Early border dwellers tended to apply experimental methods to natural populations as if they were dealing with individual experimental animals, and to natural places as if they were laboratory setups. Biologists were eager to realize the ideals of the new natural history, and in the absence of proven mixed practices, what choice did they have but to take experimental procedures off the shelf and try them out in the field? Predictably, the results of such attempts were often disappointing and occasionally disastrous. Stable synthetic practices did gradually evolve in the 1930s, as we will see in later chapters, but before that happened, instability and movement were the common experience of those who aspired to bring experiment to the field. For a decade or two, the lab-field border was a place of things and people in motion, a mosaic of varied practices, a shifty, patchy place—like nature itself.

Experimental Evolution

Experimental evolution was a popular pursuit *circa* 1900, though much of it was experimental morphology under an alias. The idea was simply to force the pace of variation by exposing creatures to chemicals or extreme physical conditions (cold, heat, wet, dry), to see if heritable changes in morphology were produced: mutations, perhaps, or Lamarckian adaptations, or determinate variations toward related species, or even the transformation of one species into another. T. H. Morgan exposed fruit flies to strenuous selection, hoping to cause heightened variability beyond the species' normal range.[3] Experimentalists thus sought to speed up nature's snail-paced experiments in speciation to the pace of human observers' brief lives. The techniques involved were the standard ones of experimental morphology but with an evolutionary gloss. No wonder this type of experimental evolution was popular.

Potentially, such work could impel practitioners out of labs and into the field, if they began wondering whether forced lab variations had anything to do with the actual variations of species in nature. Some morphologists did wonder, like Joseph Powers, and a few did go afield, as Victor Shelford did, to compare the modifications that he had produced experimentally in tiger beetles with those found in nature. But fieldwork was time consuming and unfamiliar to experimental morphologists, and most stayed put in their labs.[4]

Experiments on natural selection were another common mode, and because that mode dealt with populations and not just individuals, it took those who practiced it more regularly out of doors. Again, the basic idea was simple: collect the victims and survivors of some selection event and see if the two classes displayed measurable differences in size, weight, or protective features that might suggest degrees of fitness. In the field one could take advantage of natural disasters, as Hermon Bumpus did when he collected and measured 136 sparrows killed or brought to the brink of death by a freak ice storm; or Charles Davenport, who saw the bright side of the decimation by crows of a mixed flock of domesticated and wild fowl turned out to graze on the grounds of the laboratory at Cold Spring Harbor. (Wild ancestral vari-

3. Robert E. Kohler, *Lords of the Fly:* Drosophila *Genetics and the Experimental Life* (Chicago: University of Chicago Press, 1994), pp. 37–43.

4. Joseph H. Powers, "Morphological variation and its causes in *Amblystoma tigrinium,*" *Studies of the Zoological Laboratory of the University of Nebraska* no. 71 (1907). Victor Shelford, "Color and color pattern mechanism of the tiger beetles," *Illinois Biological Monographs* 3 (1917): 395–512, on pp. 395–97. A good overview is John A. Detlefsen, "The inheritance of acquired characters," *Physiological Reviews* 5 (1925): 244–78.

eties seemed more fit, but the numbers were too small to be sure.) Similarly, Henry Crampton studied the victims and survivors among the pupae of moths wintering over on trees that grew in Brooklyn. Others looked for effects of intraspecies competition, as J. Arthur Harris did in a deliberately overcrowded bean patch, or Frank Lutz with competing species of beetles in bottles. Protective or warning coloration was another appealing subject for experiment. Joseph Reighard, for example, studied the reactions to artificially colored bait fish of a colony of grey snappers living under the pier of the Tortugas Marine Station (a "laboratory colony" he called it).[5]

This type of experimental evolution was practiced mainly in domesticated places or on laboratory grounds. However, none were quite up to laboratory standards: the numbers were too small, there were no controls, the differences were barely perceptable, and so on. Selection experiments arranged by nature using crows or sleet were uncontrollable and unrepeatable events, hardly experiments. And measurable differences between quick and dead, if there were any at all, were usually in dimensional characters (e.g., length of body or limbs) that seemed unlikely to provide any selective advantage. Evidence was seldom more than circumstantial and usually annoyingly inconclusive.

It was in the context of such experiences that biologists began to imagine practices that combined laboratory work with studies of variation and distribution in nature. A few managed to get resources and institutional backing and spent the rest of their careers trying to achieve this aim: for example, Francis Sumner, William Tower, and Henry Crampton. Others had the same idea but for one reason or another were unable to act on it, among them Vernon Kellogg, Karl Eigenmann, and several of Davenport's young disciples: Arthur Banta, Roswell Johnson, and Frank Lutz. Usually it was personal experience of the shortcomings of lab or garden experiments that drove them to the field, and their achievements mostly fell short of expectations. Nonetheless, their activities began to give shape to a distinctive border culture.

5. Hermon C. Bumpus, "The elimination of the unfit as illustrated by the introduced sparrow, *Passer domesticus*," *Biological Lectures Delivered at the Marine Biological Laboratory* 6 (1898): 209–26. Charles B. Davenport, "Elimination of self-colored birds," *Nature* 78 (1908): 101. Henry E. Crampton, "Experimental and statistical studies upon *Lepidoptera*, I: Variation and elimination in *Philosamia cynthia*," *Biometrica* 3 (1904): 113–30. J. Arthur Harris, "The measurement of natural selection," *Popular Science Monthly* 78 (1911): 521–38. Frank E. Lutz, "Experiments with *Drosophila ampelophila* concerning natural selection," *Bulletin of the American Museum of Natural History* 24 (1915): 605–24. Jacob Reighard, "An experimental field-study of warning coloration in coral-reef fishes," Carnegie Institution Publication no. 103 (1908): 261–325. Useful overviews are Raymond Pearl, "The selection problem," *American Naturalist* 51 (1917): 65–91, on pp. 66–75; and George M. Cook, "Neo-Lamarckian experimentalism in America: origins and consequences," *Quarterly Review of Biology* 74 (1999): 417–37.

These early border inhabitants were zoologists and morphologists who believed in experiment but also had a taste for fieldwork. Six of the eight were youngish (late 1920s to early 1930s) and of the new-naturalist generation. They were well placed institutionally, all but one trained in major centers of experimental morphology. Sumner and Crampton got their Ph.D.s at Columbia, and all but one of the rest at either Harvard or Chicago. (Kellogg, trained in entomology at Kansas, was the one bootstrapper.) In their careers, too, they rose quickly to the top. Kellogg was a professor at Stanford, Eigenmann at Indiana, Crampton at Barnard College, and Tower at the University of Chicago. Johnson, Lutz, and Banta were employed at the Carnegie Station for Experimental Evolution. Experimental evolution was an outgrowth of lab morphology and a movement of the center, not the margin.

Exemplary, if not typical of a mixed practice, is Francis Sumner's study of the white-footed mouse, *Peromyscus,* which he began in 1913 and continued until forced to give it up in the early 1930s. Sumner's determination to combine genetics and biogeography was fired by almost ten years of frustrating laboratory experiments done with makeshift equipment at the Bureau of Fisheries Laboratory at Woods Hole. He measured fish that survived or succumbed in low-oxygen water, hoping to see evidence of natural selection, but with a null result. He subjected laboratory mice to extremes of temperature, hoping to induce variations that mimicked the biogeographical "law" of variation, that animals in colder climates had shorter extremities. The evidence was suggestive, but meaningless without study of variation in animals in nature. So, too, experimental study of protective coloration was worth little without knowledge of predation in nature. Sumner believed in experiment but was acutely aware of its limitations, unlike the cocksure physiologists he observed at the nearby Marine Biological Laboratory, who talked laboratory jargon and, he thought, had "all the earmarks of a cult."[6]

Sumner's fieldwork proved no more satisfactory. A comprehensive biological survey of Buzzard's Bay and Vinyard Sound was much admired by biologists, but it was basically an inventory, and Sumner had little interest in fieldwork that did not have definite theoretical purpose. So when he found himself in 1913 in San Francisco, again working for the Bureau of Fisheries

6. William B. Provine, "Francis B. Sumner and the evolutionary synthesis," *Studies in the History of Biology* 3 (1979): 211–40, on pp. 215–16; Cook, "Neo-Lamarckian experimentalism" (cit.n. 5), pp. 430–33. Francis B. Sumner to Charles B. Davenport, 18, 21 May, 5 July 1908; 11, 15 April, 18 Oct. 1909; 10 April, 27 July, 13 Sept. 1910; all in CBD-Gen. Sumner, "Some effects of external conditions on the white mouse," *Journal of Experimental Zoology* 7 (1909): 146–55. Sumner, *The Life History of an American Naturalist* (Lancaster, Pa.: Jaques Cattell Press, 1945), pp. 169–80, quote on pp. 170–71.

5.1 Francis B. Sumner, Woods Hole, 1901. Scripps Institution of Oceanography Archives (photograph file People).

as the naturalist on the research ship *Albatross* and preparing for an extended survey of California's coastal waters, Sumner determined to make one final effort to put the laboratory and field sides of his experience together. Learning that biologists at the University of California were thinking along the same line, Sumner made an appeal to the director of the Scripps Institute, William E. Ritter, to underwrite a project. To the Californians no one seemed better qualified than Sumner, and a few months later he was at Scripps at work on *Peromyscus.*[7]

Sumner's plan was to combine genetic analysis of species characters with biometric measurements of variation and distribution in nature. His first objective was to ascertain whether the characters that taxonomists used to differentiate subspecies really were hereditary, that is, whether there was any truth in the common belief that subspecies were species in the making.

7. Francis B. Sumner, "A biological survey of the waters of Woods Hole and vicinity," *Bulletin of the U.S. Bureau of Fisheries* 31 (1911). Provine, "Francis B. Sumner" (cit. n. 6), pp. 215–22. Charles A. Kofoid to William E. Ritter, 2 July 1913, WER box 13. Ritter to Sumner, 25 June 1913, FBS-Fam f. 13. Sumner to W. W. Campbell, 21 Aug. 1924, FBS-Fam f. 7.

If they were real, Sumner would ascertain by genetic experiments whether subspecific characters showed blending or Mendelian inheritance, that is, whether they arose by a process of mutation or through some physiological effect of local environments on the germ plasm. (He was inclined to a neo-Lamarckian view but kept an open mind.)

Transplantation provided a simple experimental test of subspecies. If subspecies retained their defining characters when reared for several generations in different environments, then their characters were inherited and they were real entities. Sumner chose three subspecies of *Peromyscus maniculatis* for his test: one occupying temperate mountain and coastal areas, a darker subspecies living in redwood forests, and a light-colored desert subspecies. He bred them in laboratories at Berkeley and La Jolla, using local stocks as controls. The results were quick and unambiguous: subspecies were indeed stable forms. Systematists were not surprised by the result, but pleased that they had been proven right in a way that their laboratory brethern might finally take seriously.[8]

Showing experimentally that the differences between subspecies were the result of exposure to different environments was less easy. Attempts to mimic and speed up environmental effects in the laboratory failed. Sumner built an ultra-dry, ultra-hot "desert room" in hope of turning the dark temperate-zone subspecies into the light-colored desert form, but the wooden apparatus disintegrated, the mice all died, and Sumner doubted that an experimental setup could ever mimic the environment in which these burrowing animals actually lived, so even a null result meant little.[9] He therefore turned to biogeographers' observational method of correlating natural variations with climate gradients. To make it a proper experiment, Sumner proposed to measure single environmental factors and correlate them one by one with variations in single characters—standard experimental protocol.

Sumner established four field stations near his main collecting areas, installing at each one a battery of instruments to measure sunlight, air and soil temperature, surface air and burrow humidity, rainfall, and barometric pressure. The operation was more expensive than he had foreseen but essential,

8. Francis B. Sumner, "Genetic studies of several geographic races of California deer-mice," *American Naturalist* 49 (1915): 688–701. Provine, "Francis B. Sumner" (cit. n. 6). William E. Ritter, Joseph Grinnell, Charles A. Kofoid, John C. Merriam, and Francis B. Sumner, "A proposed plan for the study of certain problems of evolution and heredity," April 1913, FBS-Fam f. 23. Sumner to Ritter, 1 Feb., 29 June 1913, FBS-SIO box 5 f. 530. Examples of systematists' reactions are: David S. Jordan to Sumner, 16 July 1923, FBS-Fam f. 7; Joseph Grinnell to Annie Alexander, 7 July 1923, AA box 2.

9. Francis B. Sumner, "The stability of subspecific characters under changed conditions of environment," *American Naturalist* 58 (1924): 481–505, on pp. 499–503.

he thought, to move beyond the "loose discussion in regard to the 'environment'" that had given rise to such dubious generalizations as Merriam's temperature zones. Correlations between gradients of humidity or temperature and variations in size, length of limbs, or color, would be definite (if still circumstantial) evidence that environmental factors had caused subspecies to differentiate. But the results were disappointing: each character seemed to vary in a different direction, some in accord with familiar biogeographical laws, others not.[10] Taking experimental, quantitative methods afield did not improve upon naturalists' qualitative practices.

Meanwhile, Sumner carried on experimental breeding at the Scripps mouse colony or "murarium," crossing subspecies to see how species characters were inherited. The results argued against mutations, he believed, despite the growing consensus among lab geneticists that "blending" inheritance was the result of normal Mendelian segregation in multiple-factor characters.[11] Sumner's genetic apostasy was the result of applying methods designed for standard animals to the complex traits of a wild species—the result, in other words, of bringing nature into the lab. (He would later grudgingly recant his unorthodox beliefs.)

As he pursued a genetics of the field, Sumner became a devoted field naturalist. He acquired an eye for behavior, habitat, and subtle differences of form, and he indulged a childhood taste for outdoor life. He established relations (symbiotic, he hoped, not wholly parasitic) with ornithologists and mammalogists, especially with Joseph Grinnell, director of the Museum of Vertebrate Zoology at Berkeley and the authority on California mammals. Grinnell gave Sumner tips about places to find unusual mice and taught him the art and craft of fieldwork. The two were often in the field together and became trusted friends and partners in peromyscology. It was a reciprocal partnership; as Sumner became a desert rat, Grinnell turned his hand to simple field experiments.[12]

Sumner probably did not expect to go quite that native. His original plan

10. Francis B. Sumner to William E. Ritter, 2, 11 Sept. 1913, 4 Jan., 15 May 1914, FBS-SIO box 5 f. 530. Sumner, "Genetic studies" (cit. n. 8). Sumner, "Geographic variation and Mendelian inheritance," *Journal of Experimental Zoology* 30 (1920): 369–402, on pp. 369–75.

11. Francis B. Sumner, "Results of experiments in hybridizing subspecies of *Peromyscus,*" *Journal of Experimental Zoology* 38 (1923): 245–92. Sumner, "Report to staff," 19 April 1917, FBS-SIO box 5 f. 532.

12. Sumner, *Life History* (cit. n. 6), pp. 216–17. Sumner, "Some biological problems of our southwestern deserts," *Ecology* 6 (1925): 352–71, on p. 353. Sumner to William E. Ritter, 7 April 1914, FBS-SIO box 5 f. 31 (on Grinnell's experiments). Joseph Grinnell to Annie Alexander, 17 April 1920, AA box 2. Sumner to Grinnell, 7 April 1916, 25 Feb. and n.d. [received 11 March] 1920; Grinnell to Sumner, 10 April 1916; all in JG-MVZ.

5.2 Francis B. Sumner with his field auto, "Perodipus," in the Panamint Mountains, California, 1920s. Scripps Institution of Oceanography Archives (FBS-Fam f. 51).

projected an initial period of collecting and observing in the field as a preliminary to an extended period of genetic experiment. But soon Sumner was shuttling to and fro between laboratory and field and spending as much time in the one as in the other. After discovering that museum collections, even Grinnell's, were not sufficient for biometric work, he collected extensively in the field himself (large collections at seventeen locations, smaller ones at a half dozen more). He spent two months in the field near Eureka, on Humboldt Bay in 1914, worked the area around Victorville in 1915, and checked out reports of unusually colored mice in the volcanic region of Big Basin in 1916. There were two major field trips in 1919, one to the lower Colorado River and another with Grinnell to the Mohave Desert and Death Valley.[13] In the 1920s he concentrated his efforts on fieldwork in special places—islands, peninsulas, light sand and dark lava areas—where "Nature's experiments" in speciation had left legible traces.

Sumner thus gradually developed an experimental natural history. Collecting excursions for laboratory material evolved into a continual shuttling between lab and field. Fieldwork was performed to laboratory standards,

13. Ritter et al., "Proposed plan" (cit. n. 8). Francis B. Sumner to William E. Ritter, 12 Feb. 1913, FBS-SIO box 5 f. 530. Sumner, "Some results of a twelve years' study of deer mice," n.d. [c. 1925], FBS-Fam f. 25. Sumner, "A naturalist's rambles in southwestern deserts," n.d., FBS-Fam f. 21. Sumner, "Work by F. B. Sumner 1914–1915"; "Work by F. B. Sumner, July 1915 to June 1916"; "Work of F. B. Sumner during academic year 1919–1920," all in FBS-SIO box 5 ff. 532–33.

and experiments were designed to be relevant to natural phenomena. He knew laboratory biologists would dismiss field observations if they were not quantitative or experimental.[14] But his experience in both field and lab gave him confidence that field observations were as likely as any experiment to reveal how evolution worked.

Shifting Ground: Field and Lab

Another attempt to combine laboratory and field was William Tower's study of variation in beetles of the genus *Leptinotarsa,* begun in 1903 and collapsing in scandal fifteen years later. Tower was the most flamboyant of Charles Davenport's young proteges: bright, energetic, and fiercely ambitious and independent. As a student at Harvard (where he antagonized his instructors and, typically, did not bother to get a degree) he plunged headlong into every new fashion: biometrics, experimental heredity, evolution. Like Sumner, he was drawn to fieldwork by frustration with laboratory experiment. (Did extreme heat and humidity produce heritable variations? No. Could a species be transformed into another by strong selection? No, again. Did predators select for adaptive traits? Apparently not. Could mutations be produced experimentally? Tower claimed he had produced them but could not have.) This apprenticeship showed Tower the limitations of experiment and the need to combine lab work with biogeography and experiments on animals in their natural haunts. "[T]he present furor for breeding 'mutants' . . . in laboratories and gardens" had little to do with how evolution actually worked, he concluded; one had to make experiments "direct in nature."[15] In 1904 a small grant from the Carnegie Institution enabled him to try it himself.

Tower's interest focused on the Mexican plateau, where *Leptinotarsa* was thought to have originated. A preliminary collecting trip in 1903 confirmed that Mexican species were in fact extraordinarily variable: a sign of an actively speciating form, it was believed. If nature could be caught in the act of making species, it would be in such a place. Tower mapped the distribution of Mexican species, measured their biometric "place modes," and scouted out sites for transplant experiments. The southern and southeast-

14. Francis B. Sumner, "Desert and lava-dwelling mice, and the problem of protective coloration in mammals," *Journal of Mammalogy* 2 (1921): 75–86.

15. William L. Tower, *An Investigation of Evolution in Chrysomelid Beetles,* Carnegie Institution Publication no. 48 (1906), pp. 59–112, 168–215, 271–74, 286–96, 314. Tower, report to Robert S. Woodward, 1 Nov. 1906, quotes on pp. 6, 12, CIW.

ern flanks of the great central plateau seemed ideal. A dense network of reliable railroads gave easy access to a wide range of climatic zones from the humid lowlands of coastal Vera Cruz to the alpine heights of the Sierra Orientale—dry and rainy forests, grassland plains, and deserts—everything required for experiments in nature.[16]

Like most American biologists at the time, Tower believed that new species evolved when a "plastic" species migrated out of its normal range into new environments. The founder species of the genus *Leptinotarsa,* for example, was thought to have lived along the southern edge of the Mexican plateau and to have budded off new species as it dispersed north, following its food plant as it spread north with the livestock industry into the North American plains (where one species became the notorious Colorado potato beetle). The main evidence for this picture was the orderly distribution of species: one in the moist coastal lowlands, a second on the escarpment, and a third and fourth on the plateau and the high plains—"like windrows along the slope of the Mexican highlands." Biometrically these species seemed to form an orderly series, each successive form slightly more extreme than its southern neighbor. The evidence suggested determinate variation triggered by exposure to extreme environments, but it was circumstantial.[17] Only experiment would prove it.

Tower hoped to show experimentally that a species transplanted to another's environment evolves to resemble the native form. Transplantation would thus mimic and accelerate the slow natural process of dispersal and speciation—that was the idea. From makeshift headquarters on the outskirts of Orizaba (chosen for its salubrious climate), Tower established five experimental sites at different elevations from the Gulf coast to the high peaks, from sea level to 10,000 feet in just fifty miles. He set up wire enclosures, planted *Solanum rostratum,* his beetles' favored food, and hired local Indians to tend the experimental plots when he was in Chicago.[18]

This network of makeshift field stations was a kind of dispersed outdoor laboratory, with nature providing a range of distinct environments and ex-

16. William L. Tower to Charles B. Davenport, 29 July, 14 Aug. 1903, CBD-Gen. Tower to Carnegie Institution, 15 Jan. 1903; Tower, "Report of work done," 1 Oct. 1904, both in CIW.

17. Tower, *Investigation of Evolution* (cit. n. 15), pp. 1–3, 9–48, 52–57, 113–20, 259–61, quote on pp. 113–14.

18. William L. Tower to Davenport, 20 Feb., 3 March 1904, 21 Sept. 1905, 27 May 1906, CBD-Gen. Tower, "Report of work done . . . ," 1 Oct. 1904; [Tower], untitled application, 1 Oct. 1905; Tower, "Report of progress . . . ," 1 Oct. 1905; Tower to Robert S. Woodward, 10 Aug. 1906; all in CIW. William L. Tower, *Mechanism of Evolution in Leptinotarsa,* Carnegie Institution Publication no. 263 (1918), pp. 24–25, 237–43.

5.3 Topographic map of Mexico showing sites of William Tower's fieldwork. William Tower, *Mechanism of Evolution in* Leptinotarsa, Carnegie Institution Publication no. 263 (1918), plate 19, on p. 340.

perimental populations. Fortunately, *Leptinotarsa* beetles live in dense, localized colonies, each one a more-or-less contained experimental unit, like an individual experimental animal in a laboratory cage. Tower could collect all the beetles in a colony, keep them in cages while he measured the patterns of their colored spots, then release them to multiply and be measured again the next year.

Tower was the first to combine field and lab practices, and his energy and remarkable claims brought him meteoric fame. He was a golden boy, given every advantage and sought after by universities, though his project made him too expensive for most. (Davenport was especially impressed by the way he bullied University of Chicago officials into fitting out an experimental greenhouse.)[19] Despite growing uneasiness about his dogmatic manner and suspiciously unrepeatable claims, in his heyday Tower was the exemplary new naturalist.

But his heyday was brief. Most of the intensive fieldwork in Mexico was done in 1905 and 1906, before the Carnegie Institution cut off its support. Tower carried on at a few stations, visiting twice a year, but they suffered neglect and depradation, and in 1910 the Mexican Revolution put an end to the project. Tower continued experimental work at Chicago and at the Carnegie Institution's Desert Lab, and may have collected in Panama, where

19. Charles B. Davenport to Robert S. Woodward, 25 Nov. 1905, CIW.

he hoped the Carnegie Institution would set up a field laboratory. But so far as one can tell he did no futher experiments in the field.[20]

Tower may not have found his retreat from the field entirely unwelcome. His reports hint at the unusual practical difficulties of his work in Mexico. Stations in the tropics and high mountains were too hard to reach and had to be abandoned. Others were destroyed by agricultural "improvement," a constant hazard everywhere. His most promising transplant colony, which seemed to be showing progressive variation toward the local species type, was wiped out by a freak heat wave. Locals hired to protect the colonies in Tower's absence refused to go into the countryside on strange missions and could not prevent vandalism of experimental cages.[21]

It also transpired that Tower's field "experiments" were far from meeting laboratory standards. Colonies were not as isolated as he had thought; beetles from one mingled with nearby colonies, compromising biometric and genetic data. (Any lab geneticist who discovered that his pure-line stocks had bred on the sly with wild intruders would throw the data out.) Tower had not actually measured environmental factors but used official meteorological data, which he knew were useless for local studies. He probably got his taxonomy wrong. Even when things went well the results were frustratingly inconclusive; transplants would begin to show determinate variation, then revert to type. Beetle colonies seemed to vary in all directions at once.[22] From about 1910 to the abrupt end of his career a decade later Tower devoted himself to genetic experiments and to the ideal that, as he put it, "[t]he rigid conditions demanded in physical science should be for the biologist the model of operation."[23] No jabs now at the limitations of experiment; as the fashion for a scientific natural history declined and cytogenetics became fashionable, Tower moved with the tide, back across the border to the laboratory.

There are many reasons for biologists to move from field to lab or the other way, and many in the early 1900s had careers that shifted fitfully across the lab-field border. Arthur Banta, another of Davenport's young disciples,

20. Daniel T. MacDougal to Robert S. Woodward, 13 June 1908, CIW f. Desert Lab MacDougal. Tower to Charles B. Davenport, 13 July 1908, 18, 20 Feb. 1909; Davenport to Tower, 20 Jan., 15 Feb. 1909; all in CBD-Gen. Tower to Woodward, 17 Sept. 1911, CIW. Tower, *Mechanism of Evolution* (cit. n. 18), p. 40.

21. Tower, "Report of progress," 1 Oct. 1905 (cit. n. 18), p. 7. Tower, untitled report to Robert S. Woodward, 1 Nov. 1906, p. 5, CIW. Tower, *Mechanism of Evolution* (cit. n. 18), pp. 47, 151, 293–98, 223–35.

22. Tower, *Mechanism of Evolution* (cit. n. 18), pp. 25–26, 237–43, 278, 302–7, 326–40. Harriet B. Merrill to William M. Wheeler, 9 March 1914, WMW box 23.

23. Tower, *Mechanism of Evolution* (cit. n. 18), pp. 25–26.

moved to the laboratory after a promising beginning in the field. He undertook a study of blind cave fauna in 1903 in Mayfield's Cave, Indiana, that ended twenty years later in a concrete laboratory "cave" at Cold Spring Harbor. Banta's first step was a natural-history survey of the species inhabiting Mayfield's cave and the streams of the surrounding countryside. He observed their favored haunts, life cycles, and habits of hiding and feeding; recorded the cave's temperature and air currents; and did a few experiments on the reactions of cave and surface animals in a makeshift laboratory in a nearby cement plant (the only building in that rural area with 24-hour electrical service).

Banta, however, seems to have regarded his natural-history survey as a preliminary to laboratory experiment. Carl Eigenmann, who had suggested the project, hoped that Charles Davenport would fit out Mayfield's Cave as a natural laboratory for Banta, but Davenport had no interest in a "laboratory" so far removed from centers of laboratory life. In 1909 Davenport enticed Banta to Cold Spring Harbor by building him an artificial "cave" for experiments on cave forms. This "cave" was a low underground gallery divided into three chambers, each fitted with concrete tanks with sloped bottoms and a continuous flow of water from a nearby spring. Stocked with a menagerie of cave and laboratory animals, this artificial cave was meant to simulate the real thing, but in a controlled environment.[24] The natural fauna and setting of Mayfield's Cave took a back seat to simplicity, convenience, and control of a laboratory space.

Banta's experiment was the obvious one: surface-living animals were reared generation after generation in the dark, and cave animals in the light, to see if they would develop heritable modifications such as unpigmented skin or degenerate eyes. Banta believed that evolution occurred when novel environments stimulated a built-in mechanism of determinate variation, and he hoped to mimic the process that had occurred naturally when creatures set up housekeeping underground. In fact, unpigmented forms did turn up, and an eyeless salamander, but neither modification was heritable. Year after year Banta kept generations of animals alive in their "cave" but always with null results, and gradually it became a sideline to Banta's experiments on sex determination, which yielded regular publications and were more congruent with the station's genetics program. Yet Banta never lost

24. Arthur M. Banta to Charles B. Davenport, 18 Sept. 1909, CBD-Gen. *Carnegie Institution Year Book* 9 (1910): 82–86; 10 (1911): 84–85. Carl H. Eigenmann to Daniel C. Gilman, 7 Oct. 1903, CIW. Eigenmann to Davenport, 18 Oct 1911, 1 Nov. 1911; Davenport to Eigenmann, 9 Dec. 1909, 24 Oct. 1911; all in CBD-Gen.

5.4 Arthur Banta with the staff of the Carnegie Station for Experimental Evolution, on the occasion of a visit by Hugo De Vries (seated, center) in 1911. Banta is at upper left, Charles Davenport at lower right. American Philosophical Society Library, Carnegie Institution Department of Genetics photograph collection (group no. 214). Courtesy of the American Philosophical Society.

hope: even when Davenport finally pulled the plug on the "cave" in 1924, he wanted to keep at it despite the "meager and disappointing" results.[25]

Ironically, it was Banta's natural-history survey that shed real light on the origin of cave forms, and by a simple act of sorting. Rather than compiling lists of species, he lumped cave-dwelling species into four functional types: occasional visitors, temporary residents (e.g., hibernating), permanent residents, and species that lived exclusively there, including blind forms. With a classification based on animals' reactions to their environments, Banta could compare the distribution of each type in cave and surface habitats and infer how cave dwellers got involved in nature's experiment in evolution.

What he found surprised him: there was a large overlap of cave and surface forms (three-quarters of cave species were strays and visitors), and related cave and surface species both habitually sheltered in dark, wet places. Banta

25. *Carnegie Institution Year Book,* 1911–1925. Arthur M. Banta, reports to Charles B. Davenport, 1911–1924, CBD-CSH-2.

concluded that the ancestors of cave forms had taken up life in caves voluntarily, because they preferred life in dark, wet places, and some subsequently became modified. With one simple stroke, Banta swept away speculative and often silly explanations that biologists had offered for the evolution of cave forms—for example, Ray Lankester's idea that animals with good eyes escaped, while those with deformed eyes could not and had to make do in a second-rate place.[26] It was a classic example of scientific natural history.

True, Banta was no closer to knowing the mechanism of modification, but at least he knew how nature's experiment had been set up and the material with which it had begun. He even had experimental support for his theory of preadaptation. Studies of the responses of cave and surface water fleas to light and touch suggested that the surface form would leave a cave if by chance it encountered outside light, while the cave form would stay. (Cave animals reacted negatively to intense light and were sensitive to mechanical stimuli, he found, while surface forms reacted negatively to low-intensity light but positively to light if kept in the dark.) It was a fine piece of work, linking laboratory experiment to field observation.[27]

Why, one cannot help but wonder, did Banta give up the ecological and natural-history side of his work, which yielded modest but real insights into evolution, while persisting in quite orthodox experiments on modification that gave no encouraging results at all? Banta never said, but he probably took for granted, as most biologists then did, that laboratory experiments were best and that proper careers moved from field to lab. We might conclude that experimental biologists, like cave animals, were preadapted to life in places with a roof overhead and an unvarying climate.

But the balance between lab and field could also tip toward the field, as it did in Henry Crampton's twenty-five-year study of variation and evolution in the genus *Partula.* Though less famous than Darwin's Galapagos finches, these Pacific land snails had been an object of interest since the mid-nineteenth century, when the Reverend John T. Gulick began to study Hawaiian species in the hope of making a case for isolation as a cause of speciation. Crampton's previous experiments on natural selection in New York City had been inconclusive, and he was looking for a variable species to study in more natural environments when his friend Alfred G. Mayer offered him Gulik's collection and unfinished project. *Partula* seemed ideal,

26. Arthur M. Banta, *The Fauna of Mayfield's Cave,* Carnegie Institution Publication no. 67 (1907), pp. 5–10, 89–106. Banta, "Outline of work in progress on cave faunas," to Charles B. Davenport, 15 Feb. 1905, CBD-CSH-2.

27. *Carnegie Institution Year Book* 9 (1910): 82–83. Arthur Banta to Charles B. Davenport, 15 Dec. 1910, 14, 20 Jan. 1911, CBD-CSH-2.

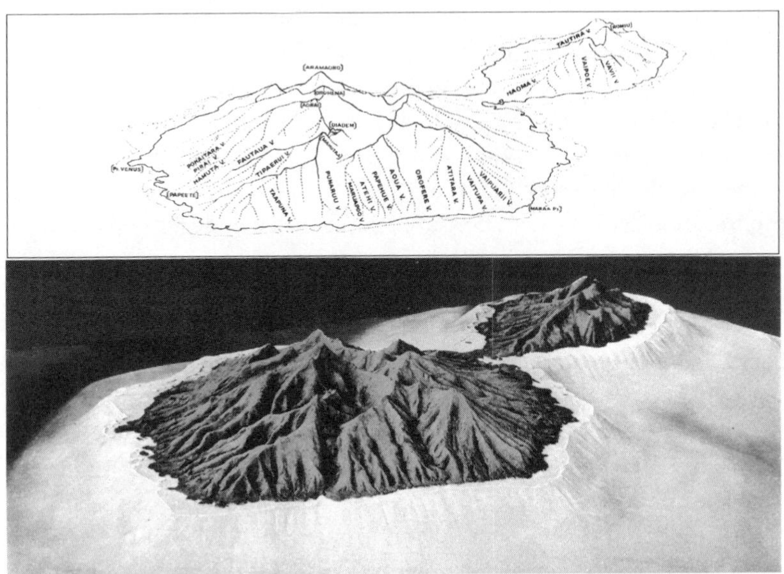

5.5 Model and topographical map of Tahiti, showing isolated valleys where Henry Crampton collected and observed the natural variation of land snails. Henry E. Crampton, *Studies on the Variation, Distribution, and Evolution of the Genus* Partula, Carnegie Institution Publication no. 228 (1917), Plate 12, on p. 34.

with its boundless range of variations and confined ranges in the isolated and steep-walled valleys lining the flanks of volcanic islands. If a case for environmental agency in variation was to be made, the South Sea archipelagos seemed the place to make it. Crampton's plan was to measure and map local variants, then ascertain experimentally whether variations were heritable and could be artificially induced.[28]

With small grants from the Carnegie Institution, Crampton made a series of extensive field trips to the South Pacific islands, four in 1906–9 and more in later years. Within a dozen years he had collected some 80,000 shells from over two hundred valleys in the Society Islands alone and planned similar excursions to other island groups. It was clear—all too clear to his patrons—that Crampton had acquired a serious liking for expeditions and travel in the South Seas—among "the beautiful islands [and] their delightful inhab-

28. Henry E. Crampton, *Studies on the Variation, Distribution, and Evolution of the Genus Partula,* Carnegie Institution Publication no. 228 (1917), p. 9. John T. Gulick, *Evolution, Racial and Habitudinal,* Carnegie Institution Publication no. 25 (1905). Crampton to Robert S. Woodward, 13 Oct. 1905; Alfred G. Mayer to Crampton, 6 Oct. 1905; Charles B. Davenport to Woodward, 17 Nov. 1905; all in CIW.

itants." "The experiences incidental to the active life necessitated by such work," he rhapsodized in one report, "were many, varied, and interesting. . . . [T]he days and nights of arduous and sometimes dangerous effort included hours of beauty, while the chiefs and their families offered abundant hospitalities which it was a privilege to enjoy."[29] The beauty and endless variety of the Pacific shells was equally addictive, as many collectors had discovered, and opportunities to collect were endless.

Meanwhile, Crampton kept finding excuses to put off his intended experiments. Part of the problem was that *Partula* proved to be a most unpromising experimental animal. Producing only two offspring per mating and with only one parent knowable, a worse subject for genetic experiment could hardly be imagined.[30] Yet Crampton's patrons continually pressed him to give collecting a rest and do some experiments. They offered to pay for annual visits to tend experimental transplant colonies, but that was all: no more collecting. (Resourcefully, Crampton got funds from the Bishop Museum in Hawaii for another collecting expedition.) "I wish he or someone would . . . *stay* in the Islands long enough to do *breeding* experiments," Alfred Mayer sighed. "Crampton only nibbles around on the *outside* of the problem and can never solve it by his present methods of mere collecting." Francis Sumner recommended that someone else be enlisted to carry out experimental breeding if Crampton would not.[31] Field-workers were expected to move from an initial stage of collecting and observing to full-time lab work. But Crampton kept collecting until ill health and infirmity kept him at home. He did finally do some genetic experiments, with an unusual Hawaiian species of the genus *Limnaea* that had a good Mendelian character—right- and left-handedness—and tolerated life in the lab. But these experiments had nothing to do with the origin of species: it was transmission genetics in the Morgan style.[32]

Movement and instability were the common experience of border life, for several reasons. Evolutionists' attachment to the idea of determinate variation was bound to let them down (because it is false), and institutional support for mixed practices was scarce. But most important, I think, was the

29. Crampton, *Studies on the Variation* (cit. 28), pp. 35 (quote), 10–11.

30. Charles B. Davenport to Robert S. Woodward, 17 Nov. 1905; Henry E. Crampton to John C. Merriam, 25 Jan., 21 Oct. 1932; all in CIW. Crampton, *Studies on the Variation* (cit. n. 28), pp. 10–11, 94, 128–29, 305–11.

31. Henry E. Crampton to Robert S. Woodward, 24 Nov. 1919, 13 Nov. 1922; Walter M. Gilbert, "Memorandum for Dr. Merriam," 1 April 1921 (quoting Mayer); T. Wayland Vaughan to John C. Merriam, 24 May 1927 (quoting Sumner); all in CIW.

32. Henry E. Crampton to John C. Merriam, 1 Dec. 1930, 23 Oct. 1938, CIW.

idea of field experiment itself, which invited evolutionary biologists to extend laboratory methods rather mechanically to the field and gave them unrealistic expectations of their prospects. Given the prestige of experimental methods and biologists' inexperience in applying lab methods in the field, disappointment was in the cards. It took decades for biologists to evolve the productive mixed practices that would one day enable them to operate comfortably in the border region. Until then, those who attempted to combine experiment and fieldwork were likely to be drawn to one side of the border or the other, usually to the laboratory side, where practices were more dependably productive and creditable and institutional niches more secure.

Experiments in Nature: Ecology

Ecologists—the physiologists of the field—were also optimistic that the causes of community and succession would be revealed by experiments. How plants and animals altered their environments seemed an especially apt subject for experiment, Henry Cowles thought, because biological agencies were more amenable to experimental control than physiographic ones and acted fast enough to be "investigated with some exactness within the range of an ordinary lifetime."[33] In practice, however, early ecologists found it hard to take experiments afield, particularly since they had to deal with heterogeneous populations: one could not take individuals as surrogates for communities, as experimental morphologists did. That may explain why, despite their enthusiasm for experiment, first-generation ecologists attempted few experiments in nature, quite unlike the way they incorporated counting and measuring into their daily practice. Experimental projects seemed always to be stillborn or abandoned, or to revert to simpler laboratory or field modes.

Frederic Clements's first try at field experiment, for example, was a study of the forest-prairie ecotone in southeastern Nebraska. A high-school teacher, John Thornber, laid out quadrats and measured climatic factors, while George Hedgcock, a plant pathologist with the U.S. Department of Agriculture, investigated the physiology of prairie plants in the university's greenhouse. But the project was left unfinished and largely unpublished.[34] It was the

33. Henry Cowles, "The causes of vegetative cycles," *Botanical Gazette* 51 (1911): 161–83, on pp. 181–82, 171–73, quote on p. 172.

34. John J. Thornber, "The prairie-grass formation in region I," *Botanical Survey of Nebraska* 5 (1901): 29–143. George G. Hedcock, "The relation of the water content of the soil to certain plants, principally mesophytes," *Botanical Survey of Nebraska* 6 (1902): 5–79.

same with Forrest Shreve; he believed ardently in experiment, but his attempts to practice what he preached fell well short of the ideal. The experimental and ecological halves of his monograph on the Blue Mountains of Jamaica seem unconnected. He planned ambitious physiological experiments to confirm an observed correlation between winter temperature and plant distribution in the Santa Catalinas, but all he ever did (or at least published) was a few half-hearted experiments on the effects of freezing on cactus.[35]

And Henry Cowles: he preached that fieldwork was a prologue to experiment and chastised colleagues who did not work in the lab, like Roland Harper. But, as Harper observed, Cowles had never done an ecological experiment in his life that he knew of, though "if he doesn't do any experimenting himself he believes in it all right, just as I believe in chemical analyses of soils without ever having made any myself." (Cowles once referred to a forthcoming experimental paper on dune plants, but if he ever did the work, it was never published.)[36]

The problem was not a lack of ideas. Cowles easily imagined experiments: for example, on the ways that plant species competed for water and nutrients and how plants created soils, or on the biotic community of humus. Such experiments were doable and pertinent to his theory that plants were themselves the chief cause of succession. (The role of shading in competition seemed less amenable to experiment, because shading also altered humidity and soils.) Yet Cowles himself never pursued this experimental program but instead farmed pieces of it out to students.[37] William Ganong envisioned a similar program of experiments on soil physics, the reactions of marsh plants to soil and water conditions, and competition between plants in the "tension" zones where communities meet.[38] But like Cowles, he did nothing about any of these topics.

Not all ecologists were as keen on experiment as Cowles and Clements. Arthur Vestal, for example, felt no need for experiments on habitat selection in his work on the distribution of grasshoppers. Field experience alone

35. Forrest Shreve, *A Montane Rain-Forest: A Contribution to the Physiological Plant Geography of Jamaica,* Carnegie Institution Publication no. 199 (1914). Shreve, "The role of winter temperatures in determining the distribution of plants," *American Journal of Botany* 1 (1914): 194–202, on p. 201.

36. Roland M. Harper to Arthur G. Vestal, 8 Nov., 26 Nov. 1914, AGV box 2. Henry C. Cowles, "Ecological relations of vegetation on the sand dunes of Lake Michigan," *Botanical Gazette* 27 (1899): 361–91, on pp. 386, 361–63.

37. Cowles, "Causes of vegetative cycles" (cit. n. 33), pp. 173–79. Anna M. Starr, "Comparative anatomy of dune plants," *Botanical Gazette* 54 (1912): 265–305, on pp. 265–66.

38. William Ganong, "The vegetation of the Bay of Fundy salt and diked marshes: An ecological study," *Botanical Gazette* 36 (1903): 161–86, 280–302, 349–67, 429–55, on pp. 352, 360, 365, 447–53.

enabled him to predict merely from lists of local plants what species of grasshopper he would find in any place.[39] Observation was as good as experiment. Henry Gleason and Charles C. Adams held similar views; but it was a minority view. Most first-generation ecologists believed experiment would make field observation scientific, though few actually did experiments themselves.

But some did, and several modes of experimental field practice did take shape in the 1910s and 1920s, most notably the method of physiological life history. Though hardly mentioned in histories of ecology, this practice attracted some of the best early ecologists—Ganong, Shreve, and especially Victor Shelford. The method of physiological life history explained biotic communities in terms of the physiological needs and habits of their component species. (Ordinary life histories, in contrast, merely described essential activities such as feeding and nesting.) As Ganong put it, organisms live in a place because the needs imposed by their morphology and physiology coincide more or less with what a particular habitat affords; the worse the match, the shakier their occupancy. To get at the matching process, ecologists had, first, to measure the physical factors of a place, then ascertain by experiment the physiological requirements of the creatures living there. The first was relatively easy and widely done. The challenge, as ecologists like Shelford saw it, was to devise experiments on the physiology of germination, early growth and competition, habitat selection (in the case of animals), and reproduction—in short, all the activities that enabled creatures to settle and occupy suitable environments. That was the crux of "physiological life history."[40]

As Ganong imagined it (he did not practice it), physiological life history was laboratory physiology that dealt not with cellular functions and biochemical activities (a laboratory physiologist's bread and butter), but with the activities that spelled life and death to creatures in the wild. Like the genetics of species characters, physiological life history was performed in labs on individual animals—but for the purpose of explaining how natural communities worked. In principle it integrated laboratory experiment and field observation. But it turned out to be a prescription for perpetual movement across the lab-field border rather than for stable practices that integrated lab and field.

39. Arthur G. Vestal to Henry A. Gleason, 22 Dec. 1911; Vestal to Victor E. Shelford, 10 Sept. 1912, 13 April 1913; all in AGV box 2.

40. William F. Ganong, "The organization of the ecological investigation of the physiological life-histories of plants," *Botanical Gazette* 43 (1907): 341–44.

No ecologist did more to make physiological life history work than Victor Shelford, and no one's labors show more clearly the difficulty of keeping individual and population, lab and field in balance. Shelford's career oscillated from field to laboratory and back again. Trained in zoology "of the strictest morphological type" at Chicago, Shelford at first saw little value in ecology (he received grades of B- and C in courses in field zoology) and was drawn only gradually and grudgingly into the field. In his dissertation research he was surprised to discover that local ecology, not climate zones, explained the distribution of tiger beetles (they settled down in any place with open ground for hunting and moist clay slopes for laying eggs). Shelford began to talk ecology with Cowles and Adams and tagged along on excursions to the Indiana dunes. An assignment to assist in the field zoology course was an unwelcome "vacation" from laboratory work, but the experience was eye-opening.[41]

Shelford began to see how Cowles's physiographic method could be applied to animal communities and made more experimental. His idea was to create a method of classifying animal communities and successions based on the physiological reactions of animals to their habitats. Shelford envisioned a two-step sequence: first, an inventory of some animal community in the field; then lab experiments on the reactions of each species to environmental factors, to show how physiological requirements determined choice of habitat. Physiological life history, Shelford believed, would make field ecology experimental: "experimental study, conducted with due reference to the relations of the animals to natural environments, with conditions carefully controlled, and a single factor varied at a time."[42]

At Charles Adams's suggestion, Shelford chose pond and stream communities to test his method (fish are easier to see and catch than terrestrial animals). Two field sites within reach of the University of Chicago afforded well-defined communities and successions. One was a series of parallel, shallow ponds along the lakeshore just south of Chicago, which had formed at different times and constituted a successional series like Cowles's dunes. (The further from the shore the older the pond and the more successionally advanced its fauna.) Small creeks cutting into the lakeshore bluffs north of

41. Robert A. Croker, *Pioneer Ecologist: The Life and Work of Victor Ernest Shelford, 1877–1968* (Washington: Smithsonian Institution Press, 1991), pp. 12–21, 29. Victor E. Shelford, "Ecological succession, I: Stream fishes and the method of physiographic analysis," *Biological Bulletin* 21 (1911): 9–34, on pp. 9–10. Shelford, "Physiological animal geography," *Journal of Morphology* 22 (1911): 551–617, on pp. 552–56, 566–68, 586–91, 613.

42. Victor E. Shelford, *Animal Communities in Temperate America as Illustrated in the Chicago Region* (Chicago: University of Chicago Press, 1913), pp. vi (quote), 32–33.

the city afforded a similar set of stream habitats and communities arranged in a successional series, from the young fish communities of pebbly, fast headwaters to the older ones of silty lower reaches.

The initial fieldwork was a simple extension of Cowles's method of identifying and arranging communities in order of their place in a physiographic succession (see chapter 7). Shelford inventoried the species in his ponds and creeks, then tabulated them according to the physiographic age of their habitats. Definite clustering of species showed that each habitat was in fact occupied by a distinctive faunal community, which would in time be replaced by another as ponds filled in and streams eroded inland. Shelford likened the physiographic method to a laboratory instrument. It was to ecology, he wrote, what the microtome was to experimental anatomy, the electric needle to experimental morphology, or chemical reagents to physiology. Though not itself experimental, it was analytical and a guide to experiments in labs. Knowing a fish's preferred habitat, Shelford could predict its physiological requirements and reactions and design controlled experiments to prove the correlation.[43]

It followed that ecologists would have to leave the field for the lab after the initial field survey, since individual habits and physiology could be studied only in climate-controlled greenhouses with instruments that allowed environmental factors to be varied one at a time.[44] In 1913 Shelford ceased fieldwork and took to the lab, where he worked for the next fifteen years. The Illinois vivarium became a laboratory for physiological ecology: a lab for wild, not standard animals, where experiments could be done on the physiology of animals' relations to their preferred natural environments. It seemed simple enough: measure the reactions of each species in turn to variations in pertinent physical factors. If all the animals of a community reacted in the same ways, it would prove that communities were real entities and not just a bunch of animals that happened to be living in the same place.

As a test case Shelford chose the rheotactic reaction of a species of riffle fish: that is, the reaction that enables fish to maintain their position in the fast current of a shoal. Current and response were easily measured in a straight-current apparatus, and experiments showed that rheotaxis was the common trait of riffle fish communities, as Shelford expected. He also tested

43. Shelford, "Ecological succession, I" (cit. n. 41), pp. 31–33. Victor E. Shelford, "Ecological succession, II: Pond fishes," *Biological Bulletin* 21 (1911): 127–51.

44. Shelford, "Ecological succession, I" (cit. n. 41), pp. 31–33. Shelford, "Ecological succession, III: A reconnaissance of its causes in ponds with particular reference to fish," *Biological Bulletin* 22 (1911): 1–38, on pp. 2, 30–33. Shelford, "Ecological succession, IV: Vegetation and the control of land communities," *Biological Bulletin* 23 (1912): 59–99, on p. 98.

5.6 Victor Shelford (second from left) and other graduate students, University of Chicago, 1906. Courtesy of Lois Shelford Bennett, Urbana, Illinois.

reactions of fish to degrees of illumination and various kinds of simulated stream beds (sand, pebble, mud). He devised an apparatus to maintain gradients of oxygen in water or moisture in air and measured the movements of animals toward favorable environments or away from unfavorable ones. Simple experiments in habitat selection, Shelford believed, would create a physiological animal geography—a field discipline founded on laboratory experiments, a science of aggregates built up from knowledge of individuals' behaviors.

Although the test cases worked as expected, complications arose when Shelford applied his method systematically to other environmental factors and to several species of a community. One problem was that creatures of one community inhabited different vertical layers: riffle animals, for example, lived under, among, or over stones, and reacted in different ways to flow, light, and touch. Seasonality was another complication: in nature (but not in labs) animals reacted more vigorously to their surroundings in breeding seasons. And many species proved unable to accommodate to life in tanks and boxes and to the constraints of measurement. Slugs and snails were too slow to measure, and spiders too fast and aggressive (too given to eating their fellow material). Grasshoppers required too much vertical space, and tiger beetles were too fast for their movements to be recorded. Complete experimental data on whole natural communities was just not to be had, because few natural behaviors were amenable to the exacting requirements of experimental procedure. Not all of nature could be taken indoors.

The task was just too vast: no single experimenter could realistically hope to analyze all the animals of a complex community. Shelford had hoped to get much of his data from the literature of experimental zoology, but he found little of what he needed there. Experimental zoologists worked primarily with standard animals that were easy to collect and were preadapted to life in a cage. Unfortunately, such animals tended to be of the weedy sort, not members of any community that Shelford would want to study.[45] Furthermore, lab experimenters did not study the natural behaviors that underlay community relations, but only those that lent themselves to exact measurement. Shelford had assumed a communality of interest among physiologists and ecologists that did not exist. Physiologists of the field and of the lab worked on different problems to different ends.

The practice of physiological life history had one additional problem, and it was the fatal one for Shelford's project. Physiological reactions were measurable only in the narrow range of frequency or intensity that was convenient to human observers, and this range was different for every species and every behavior. Yet data on different species and variables could be compared only if they were produced in identical conditions—a cardinal rule of experimental method. So every time Shelford tested a new species or variable, he had to repeat previous experiments on other species and variables in the new and usually unsuitable conditions. It was an experimenter's worst nightmare: the more you work, the more work you have to do and the further you are from closure.

Shelford thus discovered that he could not apply experiments on individuals to the phenomena of mixed communities, because the stringent requirements of experimental method are incongruent with complex aggregates of different creatures. Shelford continued to experiment on simpler and more amenable subjects, such as reactions of marine fish to pollution and the causes of insect plagues, but his project of an ecology based on physiological life histories was finished.[46] Experimental methods could not be

45. Victor E. Shelford, "Basic principles of the classification of communities and habitats and the use of terms," *Ecology* 13 (1932): 105–20, on pp. 105, 114–15.

46. Victor E. Shelford, "An experimental study of the behavior agreement among the animals of an animal community," *Biological Bulletin* 26 (1914): 294–315, on pp. 294–98, 312–15. Shelford, "The significance of evaporation in animal geography," *Annals of the Association of American Geographers* 3 (1914): 29–42, on pp. 30–31. Shelford and Warder C. Allee, "The reaction of fishes to gradients and dissolved atmospheric gases," *Journal of Experimental Zoology* 14 (1913): 207–

applied to objects for which they were not designed, at least, not in the straightforward way that Shelford had imagined.

A similar fate befell Shelford's experiments on irruptions of codling moth and chinch bugs, to which he devoted himself between 1914 and 1928. As in his earlier work on fish communities, the idea was to determine experimentally the physiological requirements of these insect pests and correlate key elements of their behavior, such as hibernation, emergence, and survival in the larval stage, with environmental factors. If he could do that experimentally in the controlled conditions of the vivarium, Shelford thought, he ought then to be able to predict the seemingly random outbreaks of these pests that had plagued midwestern farmers for over a century. It was the practice of physiological life history, but now with single species as subjects rather than complex communities. But even single species proved intractable to experiment, and fifteen years of continuous labor proved no more fruitful than Shelford's experiments on pond and stream fauna.

As it turned out, the solution to the irruptions of codling moths and chinch bugs was found in a mere two or three seasons of part-time work in the field. Observing the variability of a natural phenomenon as it was playing out in nature achieved what fifteen years of laboratory experiment could not and with a tenth the effort. When there were heavy autumn rains, a high proportion of larvae survived the winter and became a plague in that area in the spring—it was as simple as that. One had only to observe insects in local habitats and take note of the particularities of local weather.[47] In nature the methods of natural history work best, and the methods and expectations of experiments and labs are out of place.

These experiences drove Shelford once again to the field, and for good. In 1928 he closed the vivarium door behind him and launched a comprehensive ecological survey of North America. It was not a sudden decision. The project had been on his mind since his ecological survey of the Chicago area in 1913, and his work as editor of a field guide for naturalists sharpened

66. Shelford, "The reactions of certain animals to gradients of evaporating power of air," *Biological Bulletin* 25 (1913): 79–120. His work on marine communities also soured Shelford on experiment: see Keith R. Benson, "Experimental ecology on the Pacific coast: Victor Shelford and his search for appropriate methods," *History and Philosophy of the Life Sciences* 14 (1992): 73–91.

47. Victor E. Shelford, "An experimental investigation of the relation of the codling moth to weather and climate," *Bulletin of the Illinois State Natural History Survey* 16 (1927): 311–440. Shelford, "An experimental and observational study of the chinch bug in relation to climate and weather," *Bulletin of the Illinois State Natural History Survey* 19 (1932): 487–547. Shelford, "Faith in the results of controlled laboratory experiments as applied in nature," *Ecological Monographs* 4 (1933): 491–94, on p. 492. Shelford to Warder C. Allee, 14 Jan. 1932, WCA box 22 f. 2. See also Stephen A. Forbes, "The humanizing of ecology," *Ecology* 3 (1922): 89–92, on pp. 90–91.

his desire to go afield. From 1921 on he made a point of spending part of every summer touring North America and seeing for himself what others had described.[48] No mixture of experiment and observation here: it was descriptive field ecology all the way down.

Fifteen years of laboratory experience gave Shelford a vivid sense of the limitations of laboratories and experiments and a renewed appreciation of the value of old-fashioned fieldcraft. Ecologists of his generation, he ruefully reflected, had tried too hard to turn themselves into standard-issue lab physiologists. They had taken too literally the ideal of ecology as a physiology of the field and had deferred too readily to the superior standing of experiment. It was a heartfelt confession:

> The experimental ecologist . . . sometimes wishes that he could adhere to the methods of the mechanistic physiologist. There is a great satisfaction in very carefully setting up an experiment with its parallel check, in tying down all the easily recognized physical and chemical factors with mechanical controls so that only one factor is varied, and in taking careful readings over a period of 20 minutes, two hours, or longer as the case may be. The writer once believed that the problems of outdoor nature could be solved in that way. . . .
>
> The faith in these carefully controlled experiments taken alone has been and still is too great. The difficulties of working with nature, with all the multiple factors involved, are really very great, the solution of problems comes slowly, and they have to be approached from a viewpoint different from that of the mechanistic physiologist. It is true that conclusions drawn from the observation of organisms in their natural habitat are often shown to be incorrect or only partially correct when subjected to experimental tests. The reverse is also equally true. The conclusions drawn from careful laboratory experiments are found to be erroneous when put to a test in the natural habitat.[49]

Fieldwork on communities was not a mere preliminary to lab work, as Shelford had once believed. "Progress . . . ," he wrote a friend, "means working back and forth from the community and physical environment on the one hand to the laboratory on the other."[50] Shelford himself never real-

48. Victor E. Shelford, *The Naturalist's Guide to the Americas* (Baltimore: Williams and Wilkins, 1926). Shelford, *The Ecology of North America* (Urbana: University of Illinois Press, 1963), pp. vii–viii. Shelford to Warder C. Allee, 19 March 1931, 18 Dec. 1923, 25 Aug. 1924; all in WCA box 22 f. 2.

49. Shelford, "Faith in the results" (cit. n. 47), p. 491.

50. Victor E. Shelford to Warder C. Allee, 14 Jan. 1932, WCA box 22 f. 2.

ized this ideal, because he was never able to devise a method of experiment that met laboratory standards and was appropriate to field problems. The method of physiological life history, like the genetics of complex species characters, was a standard laboratory practice applied to objects for which it was ill suited. Hence Shelford's abrupt flights from field to lab and back.

Experimental Taxonomy: Harvey M. Hall and Jens Clausen

As it happens, experiments can be adapted to field objects, and they have been. The trick is to pick methods that are appropriate to field conditions and not to insist too strenuously on standards of procedural rigor that can only be met in laboratories. A good example of successful field experiment is experimental taxonomy. This was a movement in the 1920s to 1940s to apply experimental methods, especially cytogenetics, to plant taxonomy. Among their varied accomplishments, experimental taxonomists demonstrated that species in nature are composed of distinct breeding populations or "ecotypes," which differ genetically and are adapted to their particular local environments. Unlike subspecies, ecotypes differ in such small and subtle ways that they cannot easily be distinguished by inspection in situ. Nonheritable variations in plant form obscure discontinuities between genetic ecotypes. Experiment makes ecotypes visible; not genetic experiment but simple reciprocal transplantation in standard gardens. Experiment eliminates the noise of phenotypic variation and makes the discontinuities between ecotypes visible. No stringent laboratory imports are required: transplantation experiments operate on populations and were in the 1920s already preadapted to field use.[51]

The technique of transplantation is simplicity itself: dig up plants (or collect seeds) from one place and plant (or sow) them in another, then measure precisely features of morphology or life function in the two populations. Variants of this technique are the reciprocal transplant, in which plants from two environments trade places; and the "variation transplant," in which plants from one place are relocated to a series of different places, to produce the full range of a species' variation in a few easily observed places. Typically, transplant experiments are performed in experimental gardens located along a transect across a mountain or valley, to take advantage of large differences of climate within short compass; it makes effects easier to

51. Joel B. Hagen, "Experimentalists and naturalists in twentieth-century botany: Experimental taxonomy, 1920–1950," *Journal of the History of Biology* 17 (1984): 249–70.

see and maximizes the number of stations that can be serviced with a minimum of travel. It is also usual to grow all the forms to be tested in a "standard" garden, as an experimental control. Digging, gathering seed, planting, and measuring are all that are needed. No vivarium is required, no special climate-controlled apparatus, no straining to relate lab data to field realities or the behaviors of individuals to aggregates.

Transplantation is like the quadrat: simple, flexible, multipurpose. It resembles the everyday practices of gardeners and horticulturists, as the quadrat does those of practical botanists and foresters. Transplantation was used mainly to measure the genetic variability of local races or "ecotypes," but could also reveal physiological effects of climate or soils, or simulate in fast-forward how vanguard or rearguard species of an ecological ecotone respond to climate change.

Transplantation and physiological life history make a revealing contrast. Where the method of physiological life history dealt with individual organisms, transplantation operates on natural populations. Where the one was conducted in labs, the other was (and is) carried out mainly in the field or in seminatural gardens in the field. Where physiological life history entailed movement between field and lab, transplantation is a true border practice, blending elements of field and laboratory practice and performed in a kind of place that has qualities of both lab and field. And whereas physiological life history proved to be a dead end, transplantation developed into a whole family of productive field practices. Physiological life history straddled a cultural border; transplantation opened up a settled border zone of mixed practice.

The places in which the first experimental taxonomists worked afford visible evidence of how experiment can be adapted to natural settings and to objects as recalcitrant as species and ecotypes. Transplant gardens are border places, both lablike and natural, where the different values of of experimental control and natural authenticity can be balanced. Let us visit these places and observe them.

In the United States experimental taxonomy began in the late 1910s with the work of Frederic Clements and Harvey M. Hall. Clements was keenly interested in species transformations, and the Alpine Laboratory at Pike's Peak was elaborately set up for transplant work, with a "neutral" or experimental garden at the lab's headquarters on Minnehaha Brook, and smaller "climate gardens" at intervals of two-thousand feet along the tourist cog railway from mixed prairie at six thousand feet to alpine tundra at fourteen thousand feet. Hall was a taxonomic botanist by training who became interested in using genetic and ecological methods in taxonomy. He worked closely with Clements in Colorado in 1918–22 but gradually became more independent as their interests diverged—Clements's toward neo-Lamarckian experiments, Hall's toward experimental taxonomy. In 1926 Hall set up on his own in the California Sierra Nevada with funding from the Carnegie Institution.[52] Hall died young in 1932, just as his project was getting fully organized, but his methods were continued and developed by the Danish-born botanist Jens Clausen and his regular coworkers, William Hiesey and David Keck. It is in Hall's pioneering work that we see most clearly the trade-offs that were an essential feature of adapting experimental techniques and values to the field.

Hall arranged his experimental gardens along a transect from his backyard garden in Berkeley, up the Sierra Nevada in the vicinity of Yosemite Valley, and down again to the arid alkali plain of the Great Basin. Initially he maintained eleven small gardens, but these proved impossible to keep up and were gradually winnowed down to four: one at Stanford in the experimental garden of the Carnegie Institution's main lab, one at Mather just outside Yosemite Park, and two further up toward the Sierra crest at Toulumne Meadows and Timberline. These field gardens were quite elaborate. At Mather, for example, there were two gardens, one on an irrigated flat and one on a dry slope, each subdivided into sun and shade gardens, and all fenced and cordoned off by firebreaks from the surrounding vegetation. A cabin doubled as laboratory and toolhouse, and there was a camp site with tent platforms and frames and cooking facilities for visiting scientists, at a safe distance from public areas. (At Toulumne Meadows, Hall's workers had to board at tourist lodges, which kept their plots dangerously close to tourists and livestock—notorious tramplers and browsers.)[53]

Hall's consolidation of small dispersed gardens into a few well-tended ones marked an important shift in his cultural location with respect to the lab-field border. Initially he had kept the conditions of his plots as natural as possible, doing his transplants in small patches of meadow or forest edge that were naturally sunny or shady, and in moist swales or on dry wooded slopes—found places and found objects, so to speak, for field experiments. Every effort was made to minimize human intervention in these natural ex-

52. Hagen, "Experimentalists and naturalists" (cit. n. 51). Hagen, "Clementsian ecologists: The internal dynamics of a research group," *Osiris* 8 (1992): 178–95. Ernest B. Babcock, "Harvey Monroe Hall," *University of California Publications in Botany* 17:12 (1934): 355–68.

53. Jens Clausen, David D. Keck, and William M. Hiesey, *Experimental Studies on the Nature of Species, I: Effect of Varied Environments on Western North American Plants,* Carnegie Institution Publication no. 520 (1940), pp. 4–15. Harvey M. Hall to John C. Merriam, 18 June, 1, 18 Sept., 20 Oct. 1926, CIW. Hall to Frederic E. Clements, 10 Aug., 2 Sept. 1923, FEC box 63.

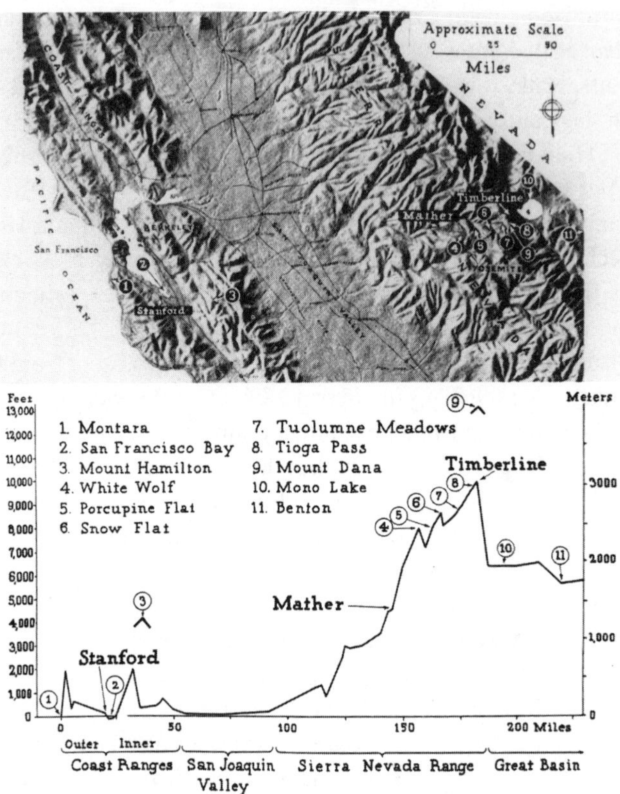

5.7 Relief map and profile transect of the California Sierra Nevada, showing locations of Harvey Hall's transplant gardens. Jens Clausen, David D. Keck, and William M. Hiesey, *Experimental Studies on the Nature of Species, I,* Carnegie Institution Publication no. 520 (1940), on p. 5.

perimental patches. In reciprocal transplants individuals were planted in the holes from which their pair had been removed. Hall gave his experimental patches minimal protection, allowing his transplants to take their chances with their wild neighbors and predators. In his selection of sites and his hands-off mode of experiment, Hall opted for natural authenticity over laboratory control. He set up the experiments but let nature do the rest.

However, these natural experiments turned out to be not very practicable. They were inconvenient: Hall's ad hoc experimental patches were scattered widely and were laborous to visit and work in rugged mountain terrain. It was difficult to provide emergency weeding, watering, and fertilizing, and impossible to keep out grazing animals and campers. Lack of care and tending meant that many test plants did not survive. Neglect is nature's way, to be sure, but neglect and low rates of survival are fatal flaws in human experiments, making conclusions uncertain and casting doubt on experimenters' diligence and competence. In labs, lack of care is carelessness, however natural. As Hall ruefully observed, his experimental plants did best in his backyard garden, "where attention brings nearly everything through."[54] But what did backyard results have to do with nature? That was the dilemma.

Even more fatal to experiment au naturel was the genetic contamination that occured in transplant populations growing amongst native forms. Native plants invade experimental plots physically by seeding themselves and genetically by cross-pollinating, producing hybrid offspring that then backcross or seed themselves in the experimental area, to the great confusion of experimenters. Hybrids are difficult to identify by sight and are all too easy to mistake for transplants that had been altered by growing in their new environments. Seeding could be controlled by chart quadrats and weeding, but to control genetic contamination one had to remove transplants to a place where their native cousins did not grow. That meant a less natural experimental environment, but genetic contamination was certain to occur within two or three years if plots were left unprotected and untended in the field.

The European pioneer of the transplantation method, the French botanist Gaston Bonnier, had made just that mistake, Hall knew. Bonnier had deliberately worked in strictly natural conditions. He did not weed, cultivate, fertilize, or protect his experimental plants from competing native vegetation. Transplants were kept in isolated spots on rocky hillsides that cows could not get to but that seeds and pollen could. The results of over thirty years of such experiments seemed to show that changes in environment did cause species to be transformed into others; however, it was becoming clear in the mid-1920s that Bonnier's results were the result of genetic contamination, not Lamarckian transformation. Hall was shocked when he heard leading French botanists pronounce as worthless the work of a man whom he had regarded as a hero of experimental field botany.[55] It was an embarrassment and a warning to all who attempted experiments in the field. A lifetime's work down the drain, because he was not a watchful and tidy gardener.

However, it did not pay to be too watchful and tidy either, as Hall became aware while visiting Goete Turesson, the Swedish ecologist who was at the time the leading practitioner of the transplant method and the man who famously proved the reality of local ecotypes. Turreson worked exclusively in

54. Harvey M. Hall to Frederic E. Clements, 10 Aug. (quote), 2 Sept. 1923, FEC box 63.

55. Clausen et al., *Experimental Studies* (cit. n. 53), pp. 396–401. Harvey M. Hall to Frederic E. Clements, 19 Feb., 12 April 1925, FEC box 65.

his well-tended garden at the University of Lund, not in field stations. This was because, unlike Hall, he had no interest in evolution (that was for philosophers, he said). For testing and classifying regional ecotypes it was necessary only to collect local variants from all over Europe and grow them in a "standard" garden, observing whether or not they maintained their local peculiarities. (Those that did were ecotypes.) He had no reason to mimic nature in his experiments. His garden was essentially an open-air laboratory.

On his visit to to Lund, Hall was impressed by Turesson's "beautiful and extensive series of experiments" and his unambiguous experimental demonstration of ecotypes, but felt that Turesson's vision was too narrow. Hall was interested in the evolutionary relations between ecotypes and the mechanisms of their formation. For that he had to forego the order and security of a lablike garden and standard experiments and create places and practices that were at once experimental and natural. As much as he admired what Turreson had achieved in his garden, Hall was not tempted to abandon his scruffy gardens in the field.[56]

As Hall and Clements remarked, balancing natural authenticity and experimental control seemed to be a zero-sum game. One could select sunny or shady places, wet or dry; but in the field environmental factors could not be rigorously controlled one by one as in laboratory experiments. That alone, as Jens Clausen observed, was enough to disqualify field experiments in the eyes of lab-trained biologists.[57] Even when they went afield to do experiments, field ecologists still operated in a cultural world where laboratory biologists decided what was good science and what was not, and where natural authenticity counted for less than experimental control.

Gradually Hall's scattered natural plots evolved from patches of untended nature into something more like standard gardens. In 1924 he reported to Clements that he was "coming more and more to believe in intensive work at a few stations, where the best attention and protection can be given."[58] He brought transplants together in large plots, which he weeded, watered in dry spells, and protected from browsers and genetic contamination with a cordon sanitaire. Though garden-tidy, these experimental plots were also natural, not located in a coastal suburb, where anything would grow, but in the harsher and less predictable environments of alpine and desert ar-

56. Harvey M. Hall to John C. Merriam, 1 July 1925, CIW.

57. Harvey M. Hall and Frederic E. Clements, *The Phylogenetic Method in Taxonomy,* Carnegie Institution Publication no. 226 (1923), pp. 20–22. Clausen, et al, *Experimental Studies* (cit. n. 53), pp. 23–24.

58. Harvey M. Hall to Frederic E. Clements, 14 March 1924, also 24 March 1924, both in FEC box 64.

5.8 Transplant gardens at Timberline station. Harvey Hall, "Heredity and environment as illustrated by transplant experiments," *Scientific Monthly* 35 (1932): 289–302, on p. 290.

eas where ecotypes had actually evolved. Hall's field gardens were hybrid places, physically and culturally, with elements of both lab and nature. Photographs of these hybrid workplaces reveal the compromises: we see squared-up, well-tended experimental plots that might be mistaken for campus gardens, were it not for the wild mountain scenery in which they are set. These hybrid places embody the accommodation that Hall sought between the values of lab rigor and authentic nature.

Jens Clausen and his coworkers took Hall's tendency to standard gardens and lablike field experiment still further. Experimental taxonomy as they developed it became more than a diagnostic test for ecotypes; it was a method of measuring genetic variation in species and mapping their fine structures. It was systematics done experimentally with live plants in nature rather than by sorting dried specimens in an herbarium. It was essentially a population genetics of wild species.[59]

Typically Clausen's field parties would collect the seeds of a species from several dozen locales along a transect across the Sierra Nevada, to get a full range of local environments. The seeds from each locale were mixed, and a sample was grown out in uniform conditions in the experimental gardens at Stanford or Mather. Then came precise measuring of physical and behavioral characters (dimensions, growth rates, seasonal timing of reproduction)

59. Jens Clausen, *Stages in the Evolution of Plant Species* (Ithaca: Cornell University Press, 1951).

and the construction of composite biometric representations of the variation of local forms. This sophisticated version of earlier "place modes" avoided the problem caused by tracking individual characters that varied in different directions, blurring gaps between different forms. Composite indices of variation made visible the discontinuities between the ranges of variation of local ecotypes. In this way the genetic and biogeographic structure of species could be mapped. It was slow work: typically, mapping a single species took from four to ten or even sixteen years. But it achieved by experiments on populations what taxonomy, biogeography, and genetics alone could not.

Clausen, Keck, and Heisey gradually added other kinds of field and laboratory experiments to their kit. Transplanting ecotypes in several climate zones displayed their full range of variation and made it easier for collectors to spot them visually in the field. Responses of individual plants to single environmental factors were studied in controlled setups in lab and garden. Most important, Clausen employed interspecies crosses, a standard genetic technique, in a novel way to map phylogentic relations in species groups by degrees of hybrid sterility.

Geneticists had long used interspecies crosses to test homologies between chromosomes, but they avoided crosses that produced sterile offspring, because sterility made further genetic analysis impossible. Clausen was not interested in genetic analysis, however, and realized that hybrid sterility, if measured quantitatively, was an indicator of evolutionary distance. The higher the proportion of sterile seeds from a cross of two forms, the further these forms had evolved toward complete reproductive isolation. Clausen was not interested in working out the genetics of wild species, nor in revising their taxonomies. For him taxonomy and genetics were tools for discovering how variation and adaptation operated to create new species. The products of experimental taxonomy were neither genetic maps nor taxonomic revisions, but graphic representations of the evolutionary fine structure of species and species groups.[60] What for lab geneticists was a failed experiment, was for Clausen and his group a valuable new mode of field experiment: a genetics appropriate to species in nature.

Experimental taxonomy and physiological life history make an instructive contrast. Hall and Clausen did not relocate laboratory procedures to

60. Clausen et al., *Experimental Studies* (cit. n. 53), pp. 1–3, 408–29. Clausen, *Stages in the Evolution* (cit. n. 60), pp. 11–12, 23–26, 29–31, 90–91, 94–95. Jens Clausen, David D. Keck, and William M. Hiesey, "The concept of species based on experiment," *American Journal of Botany* 26 (1939): 103–6, on p. 104. Clausen, Keck, and Hiesey, "Regional differentiation in plant species," *American Naturalist* 75 (1941): 231–50, on pp. 240–41.

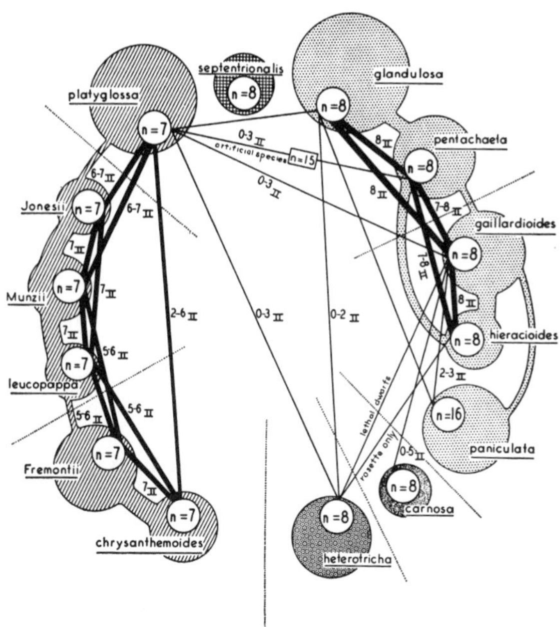

5.9 Phylogenetic relations between species of the genus *Layia,* as determined by the methods of experimental taxonomy. Black lines indicate successful hybridizations; shaded connections indicate degrees of possible gene exchange. Reproduced from Jens Clausen, *Stages in the Evolution of Plant Species* (Ithaca: Cornell University Press, 1951), p. 126, by permission of the Carnegie Institution of Washington.

the field, as Shelford tried to do. Nor did they take objects from nature into the lab for experimental analysis, or attempt to understand aggregates by experimenting on individuals. Rather, they used a simple field practice, transplantation, to do experiments that were appropriate to field objects and conditions. They used cytogenetic techniques not to map chromosomes but to measure evolutionary distance. Laboratory purists could object that Hall and Clausen's field gardens were nothing like a real laboratory. But these served as well as labs precisely because they were not slavish recreations of laboratories in nature. Where Shelford swung from field to lab for a period of years and then to field again, Hall and Clausen moved continually and easily between the one and the other. Where physiological life history was a discouraging failure that drove Shelford out of the laboratory to the field, experimental taxonomy blended laboratory and field methods and ends—a true border practice in a stable border zone.

The workplaces of experimental taxonomists make this evolving cultural geography visible. Follow Hall or Clausen from Berkeley to the high Sierra and back, and we perform a transect of their cultural ecotone. At the one extreme pure nature—the California outback, where Clausen's team explored for ecotypes; at the other extreme pure lab, in suburban Stanford; and in between the field gardens at Mather, Toulumne, and Timberline. These border places were joint products of laboratory and field culture, and practices were created there that were neither lab nor field but a distinctive blend of both—a new species of scientific place and practice.

Conclusion

Failed experiments and disappointed hopes are the common lot of most scientists, but they were especially familiar to the first biologists who attempted to take experimental methods into the field, like Sumner, Tower, Crampton, Shreve, or Shelford. The reason was partly inexperience and partly the fact that experimental logic and protocols were just not designed to work outside labs. (We think of experiment as universal, but in fact it is powerful only in its special places.) Trained in laboratory or field but seldom in both equally, biologists who took experimental methods to the field could hardly anticipate the difficulties of translation. It was a cultural problem as much as a practical one. Biologists applied experimental methods naively to field problems because they deferred too readily to the superior standing of laboratory science and lacked confidence in traditional field practices—habits formed when the new naturalists first occupied the lab-field border.

As we saw in chapter 3, memories of aimless measuring and Brush Creek ecology made ecologists unduly wary of natural-history methods and deprived evolutionists of techniques that were better suited to work on natural populations than laboratory experiment. Field biologists' predilection for lab-style experiment was encouraged by prevalent conceptions of natural groups as quasi individuals, of which Frederic Clements's idea of the association as a superorganism was the most notorious. This habit of mind was deeply rooted in the experiences of first-generation border dwellers, most of whom were morphologists who had worked on individual development in labs and who would naturally expect to see natural populations develop in the same orderly way. The cultural geography of the lab-field border and its initial settlement encouraged unselective borrowing at the same time that it discouraged more appropriate and fieldworthy approaches to field objects.

The first border biologists expected that their future as a community would be an experimental one, that field observation and survey were a pre-

Timberline, elevation 3050 m.

Mather, elevation 1400 m.

Stanford, elevation 30 m.

5.10 Jens Clausen's transplant gardens at Timberline, Mather, and Stanford stations. Note the large size and tidiness of these alpine gardens compared with Harvey Hall's in figure 5.8. Jens Clausen, David D. Keck, and William M. Hiesey, *Experimental Studies on the Nature of Species, I*," Carnegie Institution Publication no. 520 (1940), on p. 8.

liminary (though necessary) stage in the evolution of their disciplines that would be superseded by precision measurement and experiment. Even the supplementary use of description would disappear from ecology, Frederic Clements predicted, as a growing number of ecologists divided natural workplaces into smaller, more intensely worked territories. William Tower had a similar vision of historical progress from observation to experiment, and so did many others at the time. Biologists who tended to the field side paid a price, as Henry Crampton did. So did those who, like Arthur Banta, set aside modest but potentially rewarding field methods for experimental ones that promised more than they could ever deliver. A laboratory future was a feature of ecologists' "imaginative infrastructure" that powerfully shaped their choices of where to go and what to do.[61] In field biologists' imagined future, the movement of objects and practices and the course of careers were expected to progress from field to lab, because that was Progress.

The imagined future of an experimental evolution and a physiology of the field encouraged field biologists to apply laboratory methods to nature as though natural objects and places were like labs and experimental setups, as when Tower attempted a standard genetics with nonstandard animals, and Shelford pursued an experimental physiology of complex natural communities. To recall the analogy with Roman imperial frontiers, field biologists imagined themselves living in villas and eating wheaten loaves, as they continued to live in wattle-and-daub and eat rye and rabbit.

It was not that the cultures of lab and field are inherently incompatible; they are not. The cultural geography and dynamics of border life made them so. The cultural baggage that the first border practitioners brought with them caused them to trust too much in laboratory methods and too little in practices of the field. That is why border biologists resorted too much to laboratory practices and too little to those of the field. It is possible to devise methods that combine the ideals of lab and field, as we will see, but field biologists had first to become more relaxed about laboratory standards and more confident of their distinctive methods and achievements.

61. Frederic E. Clements, "The development and structure of vegetation," *Botanical Survey of Nebraska* 7 (1904): 5–175, on pp. 84–85. Clements, *Research Methods in Ecology* (Lincoln, Neb.: University Publishing Co. 1905), pp. 161–62. Tower, *Mechanism of Evolution* (cit. n. 18), pp. 304–7. Martin J. Rudwick, "Encounters with Adam, or at least, the hyaenas: Nineteenth-century representations of the deep past," in *History, Humanity, and Evolution,* ed. James R. Moore (New York: Cambridge University Press, 1989), 231–52.

CHAPTER 6

Troubled Lives

Material culture and practices are powerful agencies of change in science, and we do well to follow them into the lab-field border region. But they have this power only when wielded by individuals, whose choices of tools and practices are based on calculations more complicated than a simple one of scientific production. Practices are not just ways of making knowledge but also ways of life, and making a career has a social logic that may or may not coincide with professional credibility and output. The existential, affective dimension of careers and lives also plays into the choice of instruments and practices.[1] Career imperatives enter into choices of whether or not to count, measure, do experiments, or observe. And such choices, by shaping individual careers, also shape the cultural geography of developing border regions. We need to follow people, as well as tools and practices, into the lab-field border zone.

An exemplary study of this sort is Martin Rudwick's tracing of Charles Darwin's early career across the cultural map of London geology, out of the sphere of amateurs into the sphere of accredited

1. Lorraine Daston, "The moral economy of science," *Osiris* 10 (1995): 3–24. Daston, "Fear and loathing of the imagination in science," *Daedalus* 127 (1998): 73–85.

experts and thence into the elite and out again.[2] Our map of the terrain of laboratory and field biology is less well defined, but the experiences and career patterns of the early border biologists provide evidence of the developing topography of the border zone. The trajectories of careers transect cultural ecotones.

What was it like to live in the border zone? That is the question we need to ask. And how did the experiences of one generation shape the prospects for new generations of border crossers or border dwellers? How, in short, does experience shape cultural geography? Sources for a history of border lives are more scattered and fragmentary than sources for material culture and practices, more like hunting and gathering than agriculture—hit or miss. But there is sufficient evidence in published work and private correspondence to reveal what border life was like, though this chapter is more a reconnaissance than a systematic exploration.

Overall, it is fair to say, the experience of first-generation border biologists was unusually uncertain and frustrating. They operated in the field, but by the rules of the lab. They believed that the future was with laboratory methods of measurement and experiment, but they lacked doable mixed modes of practice, and the places in which they worked made it easier to preach laboratory ideals than to practice them. This was distinctly the problem of their generation. Nineteenth-century field naturalists had lived in their own social world and did not need to worry that their methods of observation and comparison did not live up to laboratory standards. Mid-twentieth-century biologists were to create more workable mixed modes of practice, which to a degree solved their problems of credibility. It was the first occupants of the border region whose lives were most perturbed by the conflicting imperatives of field and lab. "Pioneering don't pay," said the steelman Andrew Carnegie. No doubt; but it pays historians to watch those who try. Because, as the old robber baron knew perfectly well, it was the pioneers' mistakes that showed the next generation how to prosper.

Frustrated ambition is a leitmotiv of this chapter. There were the practical difficulties of taking laboratory methods into the field. Creatures well suited to biometric or biogeographical study were not so well suited to genetics or physiology; techniques that performed smoothly indoors did not perform well in nature; combining laboratory work and fieldwork was expensive and risky. Turning smart ideas into workable experiments is never easy, but it was especially difficult when laboratory and field imperatives

2. Martin J. S. Rudwick, "Charles Darwin in London: The integration of public and private science," *Isis* 73 (1982): 186–206.

both had to be met. The difficulties of operating simultaneously under two systems of cultural rules are what created the unusual gap between aspiration and reality.

It was relatively easy to imagine projects in experimental evolution and ecology. The ideas involved were not arcane or inaccessible, thanks in part to the agitation for a scientific natural history and the similarities of old and new field practices. But because the new naturalists expected too much of experimental methods, when things did not work out as expected the letdown was correspondingly great, and disappointment was most acute for the boldest and most ambitious. Moreover, since new naturalists were no longer content with traditional description, when more ambitious projects failed there was no fail-safe, no possibility of a modest but respectable fallback to reward the effort. Projects tended to be all or nothing, and as a result many first-generation new naturalists dropped out of research, became inactive in their middle years, or drifted away into pure laboratory work or systematics. Again, this tendency was most pronounced among those who took the ideals of scientific fieldwork most seriously. Discouragement was a disease to which the best and brightest were most susceptible.

In short, an unusual gap between ideals and real experience, and the absence of reliable fallbacks, resulted in a high mortality rate of attempted projects and a pattern of stunted, deflected, and wrecked careers. The evidence of this cultural dynamic is spotty, as it often is with things that do not happen, but sufficient. Dogs that do nothing in the night-time are difficult subjects for empirical investigation, but their evidence may nevertheless be vital for showing how things happen.

Midlife Crises: Ecology

It is remarkable how many first-generation American ecologists ceased to do research or left for other fields of science or other careers. Conway MacMillan is an extreme case: in 1906 he abruptly resigned from the University of Minnesota, took a job in a Philadelphia advertising firm, and never looked back. MacMillan was one of the best of the early ecologists, especially admired for his work on the ecology of boreal forests and bogs. (Henry Gleason advised a young protégé that if he did not know who MacMillan was, he certainly ought to look him up.)[3] Few went so far as to abandon biology, but a significant number gave up research for teaching or

3. Henry A. Gleason to Arthur G. Vestal, 1 Dec 1912, AGV box 2.

academic administration. William Ganong's pioneering and exemplary study of the ecology of the raised marshes of New Brunswick was his first and last significant contribution to ecology. As the professor of botany at Smith College, he devoted himself to teaching and tending the college's extensive botanical garden. Henry Cowles likewise never built on the foundation of his classical studies of dune vegetation. As professor of botany and ecology at the University of Chicago, he was a devoted and much loved teacher, but he essentially ceased research, or at least publication.

There were others. Thomas Kearney's promising early study of Atlantic coastal swamps remained his only contribution to ecology. Edgar Transeau, who did pioneering work on the relation of environmental factors to plant distribution, virtually ceased research in ecology when he became a professor of botany (at East Illinois Normal School, then Ohio State University). His subsequent publications were almost entirely on systematic botany. This was a common pattern. Henry Gleason, whose studies of Illinois sandhill and prairie-grove communities were models of their kind, gave up ecology for systematic botany when he became the assistant director of the New York Botanical Garden in 1919. His later celebrated papers attacking Clements's ideas of community and succession were purely theoretical and based on convictions gained during his early fieldwork. Frank Gates, one of the best of the early graduates of the Michigan school of ecology, likewise drifted toward systematic botany in his later career. Homer Shantz, one of Clements's most interesting disciples, drifted into university administration and conservation work.

In animal ecology it was a similar story. Charles C. Adams never fulfilled the promise of his early and much-admired study of the biogeography and ecology of the river snail, *Io.* He essentially gave up ecological research when he left the University of Illinois in 1914 for a post in the New York State College of Forestry; he then became head of the New York State Museum (in 1926).[4] Arthur Vestal, after a promising start studying with Adams and Gleason in the early 1910s, found his interest in ecology waning in midcareer and published only sporadically. Alexander Ruthven, Adams's most talented protégé, turned to systematics and then to academic administration as head of the University of Michigan's Museum in 1913 and as its president in 1929.

The roster of ecologists with stunted or diverted careers is a stellar one. Only a handful kept up continuous research and publication—Frederic

4. Juan Ilerbaig, "Allied sciences and fundamental problems: C. C. Adams and the search for method in early American ecology," *Journal for the History of Biology* 32 (1999): 439–63.

Clements, Victor Shelford, Forrest Shreve, and Frank Gates most notably, as well as such ecologically minded plant geographers as Roland Harper and George Nichols. It is no exaggeration that a majority of the most accomplished early ecologists either did not live up to their promise or achieved it in some other line of biological work.

A degree of churning and attrition is normal in scientific careers, especially in a new discipline that is not well defined and has not acquired an infrastructure of jobs, facilities, and patronage. It is no surprise that many undergraduate and graduate students who enlisted in ecological surveys dropped ecology upon graduating. There is bound to be a high attrition rate at that juncture in a discipline that relied heavily on student labor for fieldwork but offered few regular careers. What needs to be explained is the attrition among capable and accomplished ecologists in midcareer, those who were well employed and had opportunities for research.

Individuals drop or change lines of research for many reasons—the pressures and pleasures of teaching or practical work, lack of facilities, the higher salaries of business or administrative careers, or simple burnout. Money was apparently a factor in Conway MacMillan's decision to leave biology, or so he said. The frustration of continual battles with university administrators for salary and research support finally became too much. He was also heartily disliked by his zoologist colleagues, but MacMillan had a pugnacious personality and relished fights, so that was probably not a factor. For Thomas Kearney it was lack of opportunity: as an employee of the U.S. Department of Agriculture, he had to work on assigned tasks, and these were seldom ecological. His study of the ecology of the coastal swamps was a by-product of work assessing their potential for agriculture, and such opportunities were hit or miss. Likewise, Homer Shantz's duties as head of the Department of Agriculture's research on alkali- and drought-resistant plants kept his nose to the physiological grindstone, though he longed to get back to field ecology.[5]

But when so many individual career choices show a similar tendency, personal factors begin to seem an inadequate explanation, and one suspects underlying social structures that can shape the careers of entire generations. Although the evidence is circumstantial, I think the difficulties of first-

5. Sheri L. Bartlett, "The history of the Department of Botany, 1889–1989, University of Minnesota," n.d., on pp. 1–7; "Memorandum concerning the Conway MacMillan Memorial Research Fellowship in Botany," n.d.; both in UM f. MacMillan. Daniel T. MacDougal to Henry F. Nachtrieb, 8 July, 14 Dec. 1899, 4 Nov. 1901, 27 April 1902; and Nachtrieb to MacDougal, 1 May 1902; all in HFN box 1. C. Otto Rosendahl to William Trelease, 7 Jan. 1918, UM-Bot box 1 f. 6. Homer Shantz to A.G. Vestal, 5 Dec. 1914, AGV box 2.

6.1 Conway MacMillan, throwing a punch. University of Minnesota Archives (photograph files).

generation ecologists derive from the double bind that was their distinctive creation and burden. Defining ecology as a branch of physiology when its daily practice was natural history created a roadblock in the field's development and in careers. Ecologists were so eager to disavow their natural-history past for an experimental future that they found themselves with little ground in the present for incremental improvement. A natural-history present was simply an embarrassment, and the way to a physiological future seemed to lie in abandoning present modes of work before there were better ones to replace them. This disjunction was felt most deeply by ecologists who took laboratory ideals most seriously, yet took existential pleasure in fieldwork. (Brush Creek ecologists would have no existential qualms about not doing experiments.) So some just quit, while others did fieldwork under some other rubric, one that did not land them in a double bind: for example, amateur natural history, field teaching, or wildlife biology.

Of course, scientists accept and even thrive upon the sense of disparity between an imperfect present and a better future—their belief in progress is deeply rooted and a powerful incentive to work. But for the early ecologists that logic was complicated by a divided cultural terrain. Progress in ecology, unlike other disciplines, meant crossing the line between field and lab and moving from a cultural domain of lower standing and embarrassing connections to one of higher standing and honored neighbors. Improvement required ecologists to abandon their past and their places of work—to declare them a foreign country, so to speak. It was an act of abnegation that few laboratory scientists were required to perform.

No early ecologist was more deeply committed to ecology's physiological future than William Ganong, or took more pleasure in doing natural-history fieldwork. He concluded his superb study of the Bay of Fundy salt marshes—a purely descriptive work—with a vision of a program of experiment on the physiology of plant competition. He himself was particularly interested in the adaptive anatomical and physiological features of marsh plants (e.g., salt resistance) and did a few crude experiments testing roots of plants in saline solutions. But he made it clear that he did not expect to do much more than that himself, for lack of laboratory facilities. He hoped that one day someone would establish a field laboratory in the marshes, but he seemed certain he would not be the one, alas. "Fortunate will he be," he wistfully concluded, "who first has the proper opportunity to attack [these problems]."[6] To be sure, a proper laboratory was not in the cards for a teacher in a small women's college. But why did he not continue to do work of some kind, improvised experiments or more intensive field observations, in what he himself described as a nearly ideal place for ecological work? Why did he just abandon his beloved salt marshes?

The answer, I surmise, is that Ganong was unable to square his personal love of descriptive field ecology with his professional ideal of an experimental discipline. He chastised American ecologists for being too descriptive and offering "vague generalizations and nimble guessing" about plants' adaptations to physical environments, when what was needed was experiments. Ganong criticized his own work for failing in these very ways.[7] He published a textbook of plant physiology and a few items on laboratory instruments, but his real forte was natural history, and that he could not do without transgressing his injunction that ecology must abandon description and become experimental. Ganong raised the physiological ante so high that he was obliged to fold his ecological hand.

However, Ganong did not quit the field. For the rest of his life he spent part of each summer in New Brunswick—not as an ecologist but as a backpacking outdoorsman and naturalist. He expanded his long-standing interest in the natural history, physiography, and historical geography of New

6. William Ganong, "The vegetation of the Bay of Fundy salt and diked marshes: An ecological study," *Botanical Gazette* 36 (1903): 161–86, 280–302, 349–67, 429–55, on p. 453.

7. Ibid., pp. 450–453.

6.2 William Ganong, resting in the field, 1901. New Brunswick Museum, Saint John, New Brunswick (William Francis Ganong Collection).

Brunswick and wrote a series of popular short items for the local historical society. At his death he left behind a large unfinished manuscript on the natural and environmental history of the region and a vast collection of historical material.[8] Prevented by his professional ideals from doing fieldwork as ecology, Ganong did it as recreation and amateur natural history and geography. Much of what he did in the field was in effect ecology, but because he laid no claims to its being science, it had a meaning that did not conflict with his professional ideals. It was a highly creative use of a cultural boundary: assuming the social identity of a old-time natural historian enabled Ganong to continue doing what he liked without compromising his scientific ideals or setting a bad example to other ecologists.

Henry Cowles's career suggests a similar psychosocial dynamic. Like Ganong, Cowles was a vocal and uncompromising advocate of the view that ecology, to become a respected science, would have to leave its descriptive past behind for an experimental future. But he had a flair for descriptive fieldwork and loved it. His study of the Lake Michigan dunes, like Ganong's of the Bay of Fundy marshes, is descriptive and full of "nimble guessing" about the order and causes of succession: excellent guesses, as it turned out, but guesses nonetheless, not precise analysis. Cowles preached but did not practice measurement and experiment. He had no interest in field instruments and did no experiments that anyone knew of. After his first brilliant

8. J. C. Webster, ed., "William Francis Ganong Memorial" (St. John, N.B.: New Brunswick Museum, 1942). "William Francis Ganong," *New Brunswick Museum News* Special Supplement, Fall 1991. Shelf list for William Ganong Papers, New Brunswick Museum.

papers he published little else: a few interesting think pieces, but no more studies of plant associations and succession of the sort that had made him suddenly famous.

Cowles had no difficulty imagining things to do; that was not the problem. From titles of talks and hints of work in progress, quite a list can be compiled of projects he planned to carry out but never did. He spoke glowingly of the potential of statistics and mathematics in ecology but did nothing with it. He envisioned experiments on the bizarrely modified plants of the basswood dune formation, so different from their normal forms in river-bottom forests. (Would one form turn into the other in a single generation, Cowles wondered.) A second dune paper devoted to anatomical adaptations was planned but never published, and other experimental projects were farmed out to students.[9] The results of extensive fieldwork were never published: a follow-up study of the Indiana dune communities; a monograph with George Fuller on older stabilized dunes of the eastern shore of Lake Michigan; a comparative study of the dunes of Lake Michigan and Cape Cod; studies of the environmental causes of zonal vegetation on mountain slopes, and of physiographic successions in river valleys. He also planned but never wrote a general book on physiographic ecology based on his lectures, and a textbook of community ecology to match the one on the experimental side of ecology that he published in 1911.[10]

Like Ganong, Cowles did not abandon fieldwork. On the contrary, he pursued it regularly and with gusto—but as teaching, not research. His famous course in field ecology took classes on extended cross-country excursions to the varied plant formations of North America. He also guided visiting Europeans on tours of the famous sites of American ecology, especially the dunes. Cowles loved these field excursions. He was always in high spirits when he was afield and was remembered by his students as an im-

9. Henry C. Cowles, "Application of the quantitative method to the dynamical study of plant societies," *Botanical Gazette* 31 (1901): 72. Frank Lutz to Charles B. Davenport, 5 Nov. 1902, CBD-CSH-2. Cowles, "Ecological relations of vegetation on the sand dunes of Lake Michigan," *Botanical Gazette* 27 (1899): 361–91, on pp. 361–63, 386. Anna M. Starr, "Comparative anatomy of dune plants," *Botanical Gazette* 54 (1912): 265–305, on pp. 265–66. George D. Fuller, "Evaporation and plant succession," *Botanical Gazette* 52 (1911): 193–208. Earl E. Sherff, "The vegetation of Skokie Marsh, with special reference to subterranean organs and their interrelationships," *Botanical Gazette* 53 (1912): 415–35.

10. Henry C. Cowles, "The physiographic ecology of Chicago and vicinity," *Botanical Gazette* 31 (1901): 145–82, on p. 170. Titles of talks, *Proceedings of the American Association for the Advancement of Science* 52 (1902): 483–89. "Notes and comments," *Plant World* 20 (1917): 364. Cowles to Frederic C. Clements, 16 Nov. 1916, FEC box 79 f. 1916. Victor E. Shelford to Warder C. Allee, 19 March 1931, WCA box 22 f. 2.

6.3 Henry Cowles (third from left) and ecology students having fun on a field trip. University of Chicago Archives (negative no. AEP-ILP 246). Courtesy of the Department of Special Collections, The University of Chicago.

presario of enlightenment and fun. "He had the instincts of a scientific explorer, he was always a lover of wild nature," colleagues recalled. Although these excursions did not pause for intensive fieldwork, opportunities for research were not wholly lacking. Cowles told Roland Harper in the early 1910s that he had accumulated twenty-five thousand miles of "car-window" notes (that is, notes of observations from autos or trains), though as Harper noted "in his writing there is no hint of the existence of any such notes, apparently because that sort of exploration is not orthodox!" It was a shrewd observation. Perhaps car-window ecology, like the Brush Creek kind, was too much fun to seem properly scientific to a man who saw ecology as physiology of the field.[11]

We do not know why Cowles gave up research and publication. Perhaps he just preferred teaching, or was lazy when it came to writing up his work. But the conjunction of a visceral love for descriptive field ecology and a powerful commitment to experimental method suggests something beyond the personal. I fancy that Cowles could not work up much enthusiasm for experimental work but was unable on principle to pursue field observation as proper science, so turned to field *teaching,* which afforded the animal and

11. Charles C. Adams and George D. Fuller, "Henry Chandler Cowles, physiographic plant ecologist," *Annals of the Association of American Geographers* 30 (1940): 39–43. Roland M. Harper to Arthur G. Vestal, 26 Nov. 1914, AGV box 2.

intellectual satisfactions of fieldwork without compromising disciplinary ideals. I think that teaching was for Cowles what amateur natural history and geography were for Ganong: a way out of ecology's double bind, a social identity that served his conflicting personal and professional needs. Like natural history, teaching was a recognized and esteemed social activity, which could serve as useful epistemological cover and enable Cowles to do what he liked with a clear conscience. Student excursions did not result in publications. But that was the price he paid for doing what he liked in the field without threatening his profession's imagined laboratory future. If he himself could not participate in this future, well, his present at least was pleasing.

The same conflict may also have been a cause of Arthur Vestal's fading interest in field ecology. We get an enticing glimpse of it in a letter from Vestal to his wife, Wanda, from Colorado, where he was teaching summer courses in laboratory and field ecology. "I've been trying to present to the ecol[ogy] class a notion of what the vegetation here is like," he wrote, "and incidentally how veg[etation] may be arranged and analyzed [that is, classified]. It rekindles my enthusiasm in the subject. The darn stuff has got cold on me so often and stayed cold for such long intervals that anything which stimulates my interest is gratifying." It made him wish that he had taught the descriptive "synecology" course first, instead of "autecology," with its emphasis on individual plants and experiment.[12] But what cooled Vestal's interest in classification, despite the nostalgia that he felt when he gave it up?

We know that Vestal always had a taste for fieldwork and ecological travel. As a student in 1911 he traveled as much as he could on a miniscule budget, refusing a paying summer teaching job so that he could remain a "free-lance, scouting around the plains and different sand regions, collecting and seeing things all summer." In his project on the ecology of the Illinois sand regions he insisted on doing extensive first-hand comparisons with other sand areas in the region.[13] In 1923 and 1924 he toured the Southwest with his family to gather material on the ecology and biogeography of grasshoppers and to make "exploratory excursions" to the region's unusual plant associations, which he had read about and longed to see for himself. So fond was he of ecological touring that Clements felt obliged to warn him that fieldwork was not worth much unless combined with experiments. "I hope you will find it possible to make your work much more intensive than

12. Arthur G. Vestal to Wanda Vestal, 1 Aug. 1924, AGV box 5.

13. Arthur G. Vestal to Charles C. Adams, 2 Sept. 1912; Vestal to Henry A. Gleason, 22 Dec. 1911, AGV box 2.

extensive," he lectured his young friend. "You have done a great deal of the latter and you now need to carry the foundations deeper . . . with the fullest use of developmental and quantitative methods."[14]

Vestal agreed with the principle but found it hard to practice. Even as a student he had been reluctant to do laboratory work. Henry Gleason had had to push him to do a little if only for appearance's sake, and later, in his study of California grasslands and chaparral, he admitted to Clements that he was "rather at sea in planning what phases of experimental vegetation would be most promising."[15] He deferred to the authority and necessity of experiment in ecology, but for the kind of biogeography that he liked to do, instruments and experiments could be "more trouble than they were worth," he thought. "Observational methods are of greatest usefulness," he observed, "though not always given credit."[16] In short, Vestal knew that he ought to do more lab-style work because it was respectable, but he took more pleasure in traditional fieldwork, despite its lower standing. Individual and community values conflicted: no wonder that his interest in describing and classifying vegetation kindled and cooled. No wonder, too, that it was teaching that rekindled his enthusiasm for field observation. Teaching was a way of doing what he liked without a clash of values, as it may have been for Cowles.

Henry Gleason experienced the same conflict between experiment and field practice but resolved it differently. He had always been interested in the floristic side of ecology and the historical causes of plant distribution—what he called "developmental biogeography." He enjoyed reconstructing the vegetational history of places from close observation and recorded histories where available. (He later regretted missing the boat on fossil pollen analysis, or palynology; it might have kept him an ecologist.) He thought instruments impractical for use in the field, and the one time he tried them only showed how misleading quantitative measurement could be if field observations had not already revealed the answers. Gleason tolerated ecologists' deference to lab-style work as a fact of life that had to be lived with, but did not take it too seriously. He did not make speeches against experi-

14. Arthur G. Vestal to Henry A. Gleason, 22 Dec. 1911, AGV box 2. Vestal to Frederic C. Clements, 26 Sept. 1923, 13 April 1924; Clements to Vestal, 25 Oct. 1923 (quote); all in FEC boxes 63, 64.

15. Henry A. Gleason to Arthur G. Vestal, 6 Feb. 1912, AGV box 2. Vestal to Frederic C. Clements, 13 April 1924, 30 Nov. 1925, FEC boxes 64, 65.

16. Arthur G. Vestal, "A geophysical basis for plant geography," 25 Dec. 1917, AGV box 4 f. 1917.

ment but simply went about his business in his own way and did enough on the experimental side to maintain his credibility (more on that shortly).[17]

But the tension was there between what Gleason liked to do and what he regarded as good for ecology, and when the New York Botanical Garden offered him a job as a systematic botanist, he left ecology to take it. Gleason had been trained in systematics, liked it, and was good at it. (A higher salary and freedom from teaching were also attractions.) As the Botanical Garden's assistant director he was able to do some odd bits of ecological fieldwork, for example, as part of an ecological survey of Puerto Rico, but except for his theoretical papers, that was his last ecological work. Gleason missed ecology. In 1928 he asked that it be included in his official duties (three months in the summer with travel expenses) but was turned down. In 1940 he told a friend of his regret at giving up ecology: he got more kudos for his taxonomic work, but his early ecological work was the more important scientifically, he thought.[18]

We again get a sense of the conflict between what a field ecologist thought worth doing and what ecologists as a community valued. And again we see an individual escaping this double bind by crossing a cultural boundary and adopting a social identity for which that double bind was not a problem. A systematist could do pure fieldwork without being made to feel that he was doing something second-rate or harmful to his discipline. Experiments were not required of taxonomic botanists. As always, escape had its price: not, in Gleason's case, giving up field research or going amateur, but forgoing the credit that he should have had (and ultimately did have) for his heterodox ecological views of associations and successions. Was it the desire to escape the double bind that impelled Gleason and so many other leading ecologists to leave ecology for a more settled field? We do not know, but it seems a plausible inference from the scattered evidence.

A similar dynamic may have been what led Charles C. Adams to leave academic ecology for a career in wildlife biology and museum work. He never felt that field ecology was appreciated by the senior zoologists at Illi-

17. Henry A. Gleason, "The vegetation of the inland sand deposits of Illinois," *Bulletin of the Illinois State Laboratory of Natural History* 9 (1910): 23–174. Gleason and Frank C. Gates, "A comparison of the rates of evaporation in certain associations in central Illinois," *Botanical Gazette* 53 (1912): 478–91. Gleason, "The vegetational history of the Middle West," *Annals of the Association of American Geographers* 12 (1923): 39–85. Gleason to Arthur G. Vestal, 6 Feb. 1911, AGV box 2.

18. Malcolm Nicolson, "Henry Allan Gleason and the individualistic hypothesis: the structure of a botanist's career," *Botanical Review* 56 (1990): 91–161, on pp. 120, 130–31, 143–46.

6.4 Henry Gleason, circa 1904 or 1905. Henry Gleason Papers, box 5, LuEsther T. Mertz Library of the New York Botanical Garden, Bronx, New York.

nois, Stephen Forbes and Henry Ward, and even the advanced students in zoology knew nothing of ecology. Colleagues complained when he revamped the introductory course to include field and habitat studies, and it appears that there was little encouragement for teaching in the field. But in his new job at the New York State College of Forestry, Adams found a quite different attitude to field work. "Among foresters," he wrote a friend, "there is no prejudice against field work. So many of the young fellows here are as strongly interested in *out-doors* as they are in the money side of it. . . . I really believe that certain phases of forestry now furnish one of the best fields for a Naturalist." If being an academic zoologist meant giving up fieldwork, then he might as well give up research and teach forestry—that at least got him into the afield.[19] If the economic side of ecology seemed to offer more

19. Charles C. Adams to Arthur G. Vestal, 21 May 1915, AGV box 2. Adams to William M. Wheeler, 13 June, 21 Dec. 1909, WMW box 2. See also Adams to Alexander G. Ruthven, 16

opportunities for a field ecologist than the purely scientific side, perhaps it was because applied ecologists did not worry about the lower academic standing of fieldwork. The periphery was a more congenial place for someone like Adams than the disciplinary core, though it was not what he expected when he became an ecologist.

Ecologists' collective prejudice against Brush Creek methods could also deter field naturalists from becoming practicing ecologists. Some examples turn up in the correspondence of Frederic Clements, who liked to enlist local botanists in joint regional ecological surveys (they knew the local plants). But Clements also wanted them to live up to his high expectations, and that made recruitment difficult. For example, Bertram Wells, a botanist at North Carolina State University, felt obliged to explain to Clements why his study of his state's vegetation was purely descriptive and not up to Clements's standards. "I am well aware," he wrote, "that this paper comes under the condemnation laid down by you . . . that 'No study deserves to be called ecological that does not deal with the cause and effect relation of habitat and organism in a quantitative and objective manner.'" His only excuse was that there was no such description in the literature, and he hastened to add that he had already begun instrumental and phytometric work, so description would be a preliminary to the real thing.[20] Their joint project never materialized, and Wells went on to publish several substantial papers on the historical development of the region's distinctive plant communities—descriptive studies. He was in the end not deterred from doing the kind of ecology he liked, but did not become an ecologist, and one wonders how many other talented field botanists did not persist for fear of being seen by ecologists as less than scientific.

A similar case is Benjamin Tharp, a botanist at the University of Texas and an expert on the peculiar flora of the Texas coastal prairies. Clements was keenly interested for theoretical reasons and urged Tharp to extend his ecological analysis to other regions, since only by wider travel and comparison could he reveal how his community fit the larger picture of a climatic climax. Floristics were all very well, he wrote Tharp, but the "real story" was the response of plants to physical factors and the principles of development. Description was a mere preliminary. Tharp was aware of the slippery slope

July 1907, AGR box 51 f. 1. Adams to Ruthven, 3, 28 Feb., 26 March 1908, 1 March 1909, 1 May 1910, 10 July 1911, AGR box 51 f. 2. Adams to Charles B. Davenport, 16 July 1909, also 2 May, 19 Nov. 1908, 2 July 1910, CBD-Gen.

20. Bertram W. Wells to Frederic E. Clements, 15 April 1925, FEC box 65. Wells, "Major plant communities of North Carolina," *North Carolina Agricultural Experiment Station Technical Bulletin* 25 (1924): 1–20.

from floristics into ecological perdition: "I very much fear that I shall be accused by some ecologists of having floristic leanings," he wrote Clements, "and I might even go so far as to confess my guilt." But he doubted that he would ever have the time or the breadth of knowledge to work out the ecology even of the coastal prairie, much less the whole state of Texas. The best he could reasonably hope to do was to add his grain of local description.[21] Tharp would have made a good ecologist: certainly his instinct was sound that the coastal prairie was not a true climax, as Clements believed, but the result of frequent fires. However, he never became an ecologist, no doubt in part because he lacked institutional opportunity, but perhaps also because Clements had set the bar of participation so high that it seemed hardly worth trying. How many others were equally put off?

No border biologists were more eager than ecologists to combine laboratory and field methods, or more dependent on field description, and for that reason they were also the most likely to experience a cultural double bind. No wonder, then, that so many drifted into other pursuits, in which community values and daily practice were more congruent. Systematic botany and zoology, field teaching, and wildlife and conservation work were activities that field biologists could pursue without feeling that they were impeding progress toward an experimental future. These activities were further from the laboratory center of life science, but often that seemed an acceptable price to pay. The result was the pattern we see, of promising careers disrupted in midcareer, especially of those most ambitious to create a scientific field ecology, a physiology of the field. It was a characteristic feature of border life.

Ends and Means: Experimental Evolution

The careers of first-generation experimental evolutionists display a similar instability, but for reasons more mundane than ideological. Unlike ecologists, who were acquiring a distinct identity, evolutionists were zoologists or botanists pursuing novel practices within fields of settled identity. Evolutionary biologists did not use laboratory methods to keep tyros out, as ecologists did, and generally seemed less sensitive to the intellectual politics of methodologies. It was the logistic and financial hurdles of doing both labo-

21. Frederic E. Clements to Benjamin Tharp, 15, 29 April 1924 (see also 23 July 1924, 27 July 1925); Tharp to Clements, 29 March, 18 April 1924 (see also 16 Sept. 1925); all in FEC box 64.

ratory and fieldwork that kept would-be practitioners out of experimental evolution and stalled projects and careers in midcourse.

One difficulty was finding creatures that were suited to both biogeographical and experimental work; most, it seemed, were good for one or the other. For example, Roswell Johnson thought he had a promising subject in a little red copepod that had adapted to life in the fluctuating alkaline lakes of Eastern Washington, where Johnson was then employed. However, when his "little beast" refused to breed reliably in the lab, he abandoned it and began to collect lady beetles, which were biogeographically varied and seemed more amenable to life in a lab. Victor Shelford abandoned his tiger beetles when they failed to breed under glass. The zoologist Alfred Kinsey found gall wasps an ideal subject for biogeographical study in the field, but it would be hard to imagine a worse creature for experimental work than these highly specialized little creatures that are so fragile that they can hardly reproduce without killing themselves. Kinsey left zoology for a second career in the natural history of a creature with more robust and varied sexual habits.[22]

Lack of facilities also impeded mixed field-lab practice. Surveys of variations in the field were expensive, and experimental work on wild creatures required special climate-controlled facilities; the two together were too much even for those rare patrons who were sympathetic to the combination. For example, Frank Lutz had big plans for a study of the natural variation and genetics of crickets, which seemed an ideal creature (highly variable, especially in tropical species, with a dimorphic wing character that Mendelized). But he could never get Charles Davenport, his boss, to underwrite either the special laboratory or the field expeditions that would be required. Lutz, who liked fieldwork and fell in love with the tropics during a reconnaissance in Cuba, had to content himself with a standard genetic project, and soon left Cold Spring Harbor for a curatorship at the American Museum of Natural History.[23] Similarly, Carl Eigenmann nagged Davenport for years to support a study of blind cave fishes, which would have combined biodistributional work with experiments in the field laboratory he wanted to build in Mayfield's Cave. But Eigenmann was too much the tax-

22. Roswell Johnson to Charles B. Davenport, 29 Nov. 1903 21 Feb., 3 July 1904, n.d. [Feb. or March 1905], 15 May [1905]; Victor E. Shelford to Davenport, 14 Dec. 1912, 8 March 1917; all in CBD-Gen. Alfred C. Kinsey, *The Origin of Higher Categories in* Cynips, Indiana University Publication no. 4 (1936).

23. Frank E. Lutz to Charles B. Davenport, 8 Jan., 5 April, 19, 26 Dec. 1908; Lutz, "Report concerning a trip to Mexico and Cuba, May 8 to June 18, 1908"; Lutz, "Plans for cricket work," n.d.; all in CBD-CSH-2.

onomist ever to win Davenport's trust as an experimenter, and he finally gave up in disgust and devoted his career to the systematics of fish.[24]

For Roswell Johnson it was the field side that was difficult to arrange. He hoped, when he accepted a job at Cold Spring Harbor, that he would be allowed to combine his biogeographical work on ladybird beetles with experimental breeding. Once there, however, he was unable to keep up the fieldwork and soon found himself working on the genetics of guinea pigs. He finally did hire a beetle collector, but a few months later gave up biology to work in his family's oil-prospecting firm in Bartlesville, Indian Territory. (He did return to science after a few years, but as a petroleum geologist.) There are various good reasons to give up biology for oil, but one may have been the disparity between what he hoped to do in biology and what he was able to do. Johnson liked fieldwork and thought it essential for studies of evolution. He thought Davenport should maintain a full-time field collector to observe and collect for those who were stuck in the lab, as he was.[25] We see in Johnson's situation hints of the familiar conflict of values that scattered so many ambitious synthesizers out of border work.

For others, like Albert Bellamy, the impediment was the laboratory side. This young instructor in zoology and genetics at the University of Chicago was interested in the role of hybrids in the origin of species and conceived a project combining fieldwork on natural hybrids of fish with laboratory work on hybridization in simulated natural conditions. The field side seemed unproblematic; he could tag along on field trips with his friend Carl Hubbs. But for experiments he needed a large, compartmented experimental pond with controlled temperature and flow, and there was not a chance that the university authorities would build him such a thing. Bellamy was convinced, as he wrote Hubbs, "that a Taxonomist with a 'bent' toward experimental breeding and a Geneticist with an *active* and intelligent interest in taxonomic studies, working together, could really get somewhere with the problem of the origin of specific differences." But he could only hope that Hubbs, who had the facilities and wherewithal, might do the project himself. As it

24. Carl H. Eigenmann to Charles B. Davenport, 28 Sept., 6 Oct., 13 Nov. 1909; Davenport to Eigenmann, 2, 13 Oct. 1909; all in CBD-Gen. Davenport to Robert S. Woodward, 10 Aug. 1926, CBD-CSH-1. Eigenmann to Woodward, 27 Apr 1906; Eigenmann to "Dear Sir," 7 July 1906; both in CIW.

25. Roswell Johnson, "The role of imperfect segregation in evolution," sent to Charles B. Davenport 10 Oct. 1906; Johnson, "A proposed series of experiments," n.d. [1906–7]; "Budget for 1908," 13 Aug. 1908; Johnson to Davenport, 21 Feb. 1904, n.d. [Feb.–March 1905], 15 April 1905, 20, 26 Oct., 20 Dec. 1908; all in CBD-Gen. Roswell Johnson, *Determinate Evolution in the Color-Pattern of the Lady-Beetles,* Carnegie Institution Publication no. 122 (1910).

turned out, Bellamy never got his field trip either: his fellow zoologists, who were lab men and did not think much of the whole scheme, refused to give him leave from teaching.[26]

Taxonomists not infrequently imagined experiments that they might or might not carry out—but in fact never did. For example, Alexander Ruthven concluded his monograph on garter snakes with the observation that these hardy and prolific reptiles seemed excellent material for genetic experiments. Ruthven hoped to do some himself but to his chagrin kept drifting back to pure taxonomy; there always seemed so much to clear up in that line first. Alfred Kinsey considered experiments on gall wasps a poor investment. He had found a species that he thought might do for genetic experiments in an isolated valley in the Mexican Sierra. But getting to that remote place, glassing over a large stand of scrub oaks, spending years developing pure lines (the wasp had a two-year life cycle), and producing experimentally the hybrids that nature produced so abundantly—it seemed hardly worth the effort: "At the end of ten or twenty years, and an expenditure of perhaps a hundred thousand dollars, we would have a genetic analysis of one species of cynip."[27]

How common such experiences were we do not know, but there are structural reasons for supposing they might have been unusually common in the developing border zone. The earlier agitation against closet morphology had alerted experimentalists to the dangers of interpreting their results without some knowledge of natural variations. Academic training, including courses in basic laboratory disciplines, would have made younger taxonomists more amenable to experimental methods than the older, museum-trained generation. Thus on either side of the border zone it was more likely than ever before that practitioners might think of mixing biogeographical and experimental practices. It was relatively easy to see what might be done, and it was ideologically approved—but no less impracticable for seeming possible and desirable. The ideological imperative to combine laboratory and field practices, together with its practical difficulties, would have both raised expectations and impeded their realization. In the more settled laboratory and field disciplines, aspiration and infrastructure are usually more in balance.

26. Albert W. Bellamy to Carl L. Hubbs, 9 June (quote), 22 Nov. 1922; see also Ellinor H. Behre to Carl L. Hubbs, 16 May, 9 June 1921; all in CLH box 73 f. 55.

27. Alexander G. Ruthven, "Variations and genetic relations of the garter-snakes," *U.S. National Museum Bulletin* 61 (1908): 1–201. Ruthven to Charles C. Adams, 24 Sept. 1913, AGR box 51 f. 2. Kinsey, *Origin of Higher Categories in* Cynips (cit. n. 22), p. 65.

Identity

Unfixed or ambiguous identity was another hazard of border life. If ecologists saw themselves as physiologists of the field, those on the laboratory side might be forgiven for seeing them as field naturalists with a veneer of lab culture. If experimental evolutionists presented themselves as nonstandard geneticists, little prevented the real McCoys from taking them for biogeographers who did the odd experiment. Scientists everywhere operate across disciplinary boundaries untroubled by problems of identity—indeed, interdisciplinarity is commonly held to be stimulating and progressive. But it is one thing to combine two laboratory disciplines, like chemistry and experimental biology, and quite another to live comfortably while straddling the lab-field border.

Another reason why the identity of field biologists might be ambiguous is the social heterogeneity of the out-of-doors. Field biologists at work were easily mistaken for quite different sorts of people. Collecting chipmunks near Lake Champlain, C. Hart Merriam was mistaken for a bank robber on the loose, and in the mountains of Appalachia, for a revenue agent—potentially fatal mistakes. Collecting river snails along the Tennessee and Cumberland Rivers, Charles C. Adams was taken for a detective, a "dead-beat," and a crazy fellow. Aven Nelson, the leading expert on western plants, was multiply mistaken by townsfolk for a local lunatic and hermit; by Shoshone Indians as a medicine man; by tourists in Yellowstone for a mailman, a camper (his plant press resembled a folding bed), and a fisherman (flower specimens looked like bait); and by local sheriffs and suspicious citizens as an unlicensed peddler, canvasser, and tramp. (Nelson was less ambiguous to the foreign-born: railroad workers knew what he was up to, and a German cook engaged him in botanical conversation.)[28] In the homogeneous and regulated spaces of laboratories, such mistakes of identity could never happen.

Such confusions of identity are more than amusing and occasionally threatening episodes of field life. The resemblance of field biologists' activities and gear to those of sportsmen and vacationers affected their scientific standing. Was fieldwork play? To some it looked like it. Carl Welty, who did experiments on fish behavior, became aware of the ambiguity when he volunteered to help Paul Errington with a muskrat census in an Iowa marsh: following

28. Keir B. Sterling, *Last of the Naturalists: The Career of C. Hart Merriam* (New York: Arno, 1977). p. 39. Charles C. Adams to Charles B. Davenport, 2 Sept. [1900], CBD-Gen. Aven Nelson, "Popular ignorance concerning botany and botanists," *Plant World* 3 (1900): 33–36. Nelson was professor of botany at the University of Wyoming.

animals around seemed "boy-scoutish at first acquaintance," and Welty had to remind himself (and others) that it really was worthwhile science. In the field, work and play could feel much the same. The botanist Edgar Anderson wrote in his diary that "a guilty feeling always comes over me when I have to confess that I am walking for the fun of it—because I enjoy walking. I occasionally take refuge as today by saying that I am a scientist doing some fieldwork, but even at that I feel like a child caught in the jam pot." The physiologist Herman Spoehr told Forrest Shreve that some of the Carnegie Institution's trustees felt that "some of [his] work is not of a very serious nature, that much of the exploration and expedition work is a bit of glorified camping and vacationing."[29] Almost certainly that was Spoehr's view as well; he was relentlessly lab-oriented. Field biologists had too much fun to be accepted as entirely serious by laboratory biologists.

The association of fieldwork with outdoor recreation and play had real consequences. It was Herman Spoehr, for example, who finally closed down the Desert Lab and ended Shreve's career as a physiologist of the field. And the association of fieldwork with boy scouting could deter students from choosing careers in these lines. Henry Linville recalled Harvard undergraduates feeling that way: "I used to think some of the fellows in Zoology 1 looked upon part of the 'activity' work as play or something which they as men were above and of which they could not see the scientific importance." William Ritter opined that the reason field ornithologists did not get due respect from leading biologists was not just the fetish for laboratory methods, but also the "notion that out-of-door ornithology and its kindred branches are more or less boy-sciences."[30] Fieldwork was outdoor fun and games, kid stuff; microscopy and experiment were real work—a man's work. No wonder ambitious young biologists chose careers in the laboratory disciplines.

Field biologists were sometimes obliged to hide or camouflage their field identities. Frederic Clements, for example, did not like talking in specific detail about the popular course in fieldwork that he taught at the University of Minnesota, for fear it might seem infra dig to professional colleagues. Pressed by the English ecologist Arthur G. Tansley to give a concrete example of his teaching methods, Clements finally confessed why he was reluctant to go

29. Carl Welty to Warder C. Allee, 22 Dec. 1933, WCA box 24 f. 7. Edgar Anderson diary, 19 March 1921, quoted in John J. Finan, "Edgar Anderson 1897–1969," *Annals of the Missouri Botanical Garden* 59 (1972): 325–45, on p. 329. Herman A. Spoehr to Forrest Shreve, 13 July 1937, cited in Janice Bowers, *A Sense of Place: The Life and Work of Forrest Shreve* (Tucson: University of Arizona Press, 1988), p. 143.

30. Henry R. Linville to Charles B. Davenport, 17 Nov. 1897, CBD-Gen. William E. Ritter to Joseph Grinnell, 7 March 1908, JG-UC box 16.

6.5 Edith and Frederic Clements at work at the Alpine Laboratory, Pikes Peak circa 1900. American Heritage Center, University of Wyoming (FEC box 103, photograph album).

public: catching students' deepest interest "means beginning far below the level of the ordinary elementary text book and especially means the emphasis on field and garden work, which my opponents at Minnesota stigmatized as 'nature study.' To me this is a compliment, . . . but to my colleagues it was the last word of condemnation." He and his wife, Edith, had often considered publishing a description of the course but had always concluded that it would only damage its credibility.[31]

Sometimes field biologists censored their publications to hide the resemblance to natural history. Or were censored: this happened to Joseph Grinnell with his first publication as a professional biologist. Accustomed to writing in a popular style for a mixed audience of biologists and naturalists, Grinnell opened his paper with the customary itinerary and descriptions of places visited. But William Ritter, his academic mentor, strongly advised

31. Frederic E. Clements to A. G. Tansley, 25 June 1923 (see also 15 Oct., 11 Nov. 1923, and Tansley to Clements, 11 Nov. 1923); all in FEC box 63. Clements's mentor at Nebraska, Charles Bessey, thought ecology perilously close to nature study: Ronald C. Tobey, *Saving the Prairies* (Berkeley: University of California Press, 1981), pp. 37–38.

him to reduce the itinerary to a brief statement of where and when, lest the piece seem "too ornithological" for readers of academic monographs. (Ritter was no friend of laboratory zealots and knew about academic biases.) Grinnell protested that the itinerary was an essential part of his methodology and had to be described: "To any *field* zoologist this chapter will prove of especial interest," he wrote Ritter, "just as a summary of methods employed by the histologist has an important bearing in a histological paper. . . . Mine is a report of *field* work, and it seems to me an outline of methods is not out of place. Besides there is the *local* interest, itself of value." He finally relented and did as Ritter suggested.[32]

Frederic Clements likewise quietly edited manuscripts of associates who "dropped into the vernacular too frequently, and had an excess of local color."[33] Victor Shelford advised Arthur Vestal to omit a list of species and habitat descriptions from a paper he hoped to publish in a journal devoted to experimental work. Its readers, Shelford explained, "are more or less prejudiced against ecological work by the large amount of speculation on geography and mimicry, and they want merely the gist of the matter. They are interested if the work is new and concise but [if] they get the idea that it is a taxonomic list some of them do not care for it. I think you will find that it will pay to try to please them in this matter." Shelford had been told that people had stopped reading his papers on tiger beetles because he opened them with species lists and descriptions. He should have written the papers backward, he thought, with the laboratory work up front and the natural history discreetly in the rear, where readers might not notice it.[34]

The problem with a natural-history literary style was its similarity to the popular genre of hunting and fishing stories. For readers of such stories, the formulaic itinerary and description of participants and places were symbolic guarantees of authenticity, signs that the narrator was not an armchair writer but had actually been afield and experienced the places and events described. The narrative provided a kind of virtual participation, giving readers a sense of sharing the experience. Amateur naturalists knew the literary conventions and trusted a richly particular account. Laboratory scientists, however, were likely to read the same literary conventions in just the opposite way: as a sign that the work described was not serious science but a by-product of a casual outing—a fish story. For lab workers the neutral ab-

32. Joseph Grinnell to William E. Ritter, 29 Feb. 1908 (quote), also 31 March 1908, WER box 10.

33. Frederic E. Clements to Robert S. Woodward, 2 Feb. 1919, CIW f. Ecology projects proposed.

34. Victor E. Shelford to Arthur G. Vestal, 16 April 1913, AGV box 2.

straction and the passive voice of scientific papers were the signs of authenticity, because they mimicked the placelessness of labs and the convention that knowledge made there was made by no one in particular. Placeless prose was for laboratory workers a guarantee of trustworthy science, while the particularity of fish stories was a warning flag.

So when field biologists adopted the colorless style of lab reports, it was in part as protective coloration. Placelessness and lack of affect disguised the association of fieldwork with recreation and made it more credible to those who operated within a laboratory culture. When field biologists did occasionally revert to the you-are-there narrative mode, it was usually to report work that was a mixture of science and tourism, like trips to exotic tropical places, or ecological tours for foreign visitors.[35] Field biologists had to operate in disguise, always hiding their real identity. The men in white coats can revel unabashedly in their identity as "Scientists"; the fellows in checked shirts and galoshes have to be more circumspect in presenting themselves if they want to be taken seriously in a world of labs. That is border life.

Masquerade also figured in individuals' choice of work and careers. Ecologists, for example, might do the odd experiment to show the world that they were real scientists even if they did work mainly in the field. Henry Gleason engaged in this little subterfuge and advised Arthur Vestal to do the same for the sake of his career: "[I]t is necessary in these days of experimental science to throw down that peace offering to the experimenters," he wrote his young friend. "[Volney] Spalding did it, and I did it in my sand dune paper. . . . But we ecologists know that the experimental method has its limitations just as much as the observational method, and that either one when handled rightly leads to first class results." For Vestal's M.A. thesis work, a little morphology or physiology would do the trick: "Not a long problem, of course, but just something to keep you busy during the winter." But for Vestal's Ph.D. research, Gleason advised a physiological or morphological problem "of ecological application" done under the direction of the professor of plant physiology. "It will do you good; it will do me good; it will help you get a better position; it will smooth your path at Michigan; and it won't lose you a bit of your ecology and phytogeography."[36] Experiment as camouflage! How many ecologists adopted this disguise, and how many

35. E.g., Henry A. Gleason, "Botanical sketches from the Asiatic tropics," in *Torryea* 15 and 16 (1915–1916); Forrest Shreve, "Ecological aspects of the deserts of California," *Ecology* 6 (1925): 93–103.

36. Henry A. Gleason to Arthur G. Vestal, 6 Feb. 1912; Gleason to Vestal, 13 March, 1 Aug. 1913; all in AGV box 2.

were thus unintentionally diverted into experimental careers, we will never know, but I suspect it was not uncommon.

Border biologists also put their personal reputations at risk. For example, Charles Davenport described Arthur Banta as "somewhat of a slow and dogged plodder," quickly adding that his persistence in following a population of animals through 650 generations had given him a unique insight into "the very heart of evolution." Similarly, Thomas Hunt Morgan dismissed Henry Crampton as a mere "collector," which Davenport thought quite unfair, as Crampton was one of the few who actually tried to study mutations in nature.[37] That was precisely what troubled laboratory biologists like Morgan: would a first-rate scientist go to nature for mutations when they were so much more abundant and effectively studied in labs? In the laboratory world, mobility and cream skimming were the favored strategies of the best and brightest, not persistence in risky and inefficient work. It was how one recognized those on the fast track, and biologists who did not display these recognized signs of ambition could easily be taken for unimaginative "plodders" or mindless collectors. In the border region, careers were shaped by field conditions but judged by laboratory standards, and biologists who pursued mixed practices were at risk of being labeled second-raters whatever the quality of their work.

Genetics, True and False

The people most likely to be seen as non- or substandard were those who worked in the field but took laboratory ideals seriously. I doubt Morgan would have bothered to disparage Crampton had he really been just a collector; he would have been someone else's embarrassment. It was only because Crampton and others like him could be seen to pursue impure laboratory practices that they had to be noticed and, if necessary, declared out of bounds. The potential for boundary work, to use Gieryn's idea once more, was greatest in the hot spots of the lab-field border—where physiology and ecology, or genetics and evolution, met—because it was there that experimentalists would be most ready to see departures from their standards of practice. Francis Sumner and William Tower illustrate the point in different ways. Both sought to expand the purview of genetics to natural populations

37. Charles B. Davenport to John C. Merriam, 10 Aug. 1926, CBD-CSH-1 f. Merriam (on Banta). Davenport to Merriam, 30 April 1927, CIW f. Crampton.

of wild creatures, but were marginalized. Sumner fell into the role of dissident, railing against genetic orthodoxy, but was protected from outright banishment by his known respect for facts. Tower's ambition and work habits were less disciplined; accused of faking data, he became an outcast.

Sumner's relation with mainstream geneticists was strained but not hostile. Sumner regarded himself as a geneticist, though an unorthodox one. As cytogeneticists became more certain in the 1910s that "blending" inheritance was just recombination in multiple Mendelian factors, Sumner became a rear-guard defender of the older dualistic view. He peppered his publications and correspondence with potshots at the artificiality of standard genetics, its exclusive focus on individuals and neglect of environmental factors, its exaltation of the germ plasm—a "more or less imaginary entity" to a "position of well-nigh exclusive interest."[38]

Geneticists generally tolerated or just ignored Sumner's provocations, making no effort to bury his work or censor his rebukes. They respected Sumner for his courage and intellectual integrity and regarded his outbursts as the expressions of a unique and generally likable personality. A few were even sympathetic. His genetic work on *Peromyscus* was not up to laboratory standards, Charles Davenport admitted, but the difficulty of his material was an extenuating factor.[39]

Sumner, who was nothing if not an individualist, relished the role of dissident, but the experience of being ignored was a constant reminder of his second-class status as a geneticist. "Any studies which are not conducted according to the current methods of mendelian analysis are apt to be stigmatized as superficial or uncritical," he expostulated, "and their relevancy to true 'genetics' is questioned." In fact, he thought, geneticists' "so-called laws of heredity" had only the appearance of universality, created by avoiding complex or unfavorable material. (Edgar Anderson made a similar point.)[40] It was a shrewd insight into why experiment seems always to get easy credit, while fieldwork does not.

Significantly, Sumner's one real clash with the genetics establishment occurred when he tried to publish an orthodox genetic analysis of a Mendeliz-

38. Francis B. Sumner, "The organism and its environment," *Scientific Monthly* 145 (1922): 223–22, on p. 223.

39. Charles B. Davenport to Ross Harrison, 17 Sept. 1919, CBD-Gen. See also Charles Kofoid to William E. Ritter, 2 July 1913, WER box 13; Sumner to T. Wayland Vaughan, 7 May 1925, FEB-Fam f. 7.

40. Francis B. Sumner, "Results of experiments in hybridizing subspecies of *Peromyscus,*" *Journal of Experimental Zoology* 38 (1923): 245–92, on pp. 249–50. Edgar Anderson, "Hybridization in American *Tradescantias,*" *Annals of the Missouri Botanical Garden* 23 (1936): 511–25, on pp. 512–14.

ing mutant ("grizzled") that had accidentally turned up. His paper was rejected by *Genetics,* a first for Sumner, on the grounds that his analysis was "not sufficiently critical or convincing" (that is, numerically precise). Sumner protested that genetic work on wild animals ought not to be judged by the standards of work on laboratory animals. Naturally it was less precise, but it afforded insights into the genetics of speciation that work on lab animals never could. Sumner acknowledged that he had been wrong to think that the multifactor theory of inheritance did not apply to species characters. But in return he wanted mainstream geneticists to acknowledge that wild animals and species characters were respectable objects for genetic study, to be judged on its own terms.[41] Sumner wanted in, but on his terms. His "grizzled" paper was a kind of peace offering, a renegotiation of the lab-field boundary, with both sides giving some. The gambit was a mixed success. Sumner did get his paper published, but in a less prestigious journal. The episode suggests that the gatekeepers of disciplinary purity were not yet ready to relax their vigilance in this sector of the border. The idea of a genetics of the field remained, from the laboratory point of view, beyond the pale.

The same cultural dynamic also stymied Sumner's efforts to keep his *Peromyscus* project alive when Ritter's successor as head of the Scripps declared that everyone there would henceforth work on marine problems.[42] Sumner hoped the Carnegie Institution would take over the project, but on the advice of its genetics staff, it declined. The geneticist E. Carleton MacDowell complained that Sumner seemed unaware of the difference between genotype and phenotype, and though McDowell professed to have no bias against the study of subspecies, he doubted the value of any approach to evolution that was "essentially historical." Evolution was a genetic problem and was properly studied in labs using genetic methods and standard animals.[43] Sumner found himself on the wrong side of the distinction that to geneticists had made their discipline a respectable one in the early 1900s.

41. Francis B. Sumner to Donald F. Jones, 30 Jan. 1928, FBS-SIO. William B. Provine, "Francis B. Sumner and the evolutionary synthesis," *Studies in the History of Biology* 3 (1979): 211–40, on pp. 233–35. Sumner, "Observations on the inheritance of a multifactor color variation in white-footed mice (*Peromyscus*)," *American Naturalist* 62 (1928): 193–206. Sumner to William E. Ritter, 2 March 1924, WER box 20.

42. Francis B. Sumner to John C. Merriam, 20 Aug. 1929, CIW.

43. E. Carleton MacDowell to John C. Merriam, 25 May 1930; Francis B. Sumner to Merriam, 1 Jan., 23 April 1929, 23 March, 10 June 1930; Sumner, "Suggestions for continuation of genetic and ecological studies of *Peromyscus* or other small mammals"; Sumner, "Outline and objectives of *Peromyscus* program," n.d. [spring 1930]; all in CIW. Provine, "Francis B. Sumner" (cit. n. 41), pp. 211–12.

The Carnegie geneticists deployed the same argument when the mammalogist Lee Dice sought funding to continue Sumner's work at the University of Michigan. Dice, they observed, was a field biologist, adept in ecology and taxonomy, but the project required a "master of experimentation, trained in the use of analytical and critical methods." Genetics was the only proper way to get at the mechanism of evolution, and studying the effects of environment on physiological characters, which is what Sumner was then doing, was not genetics.[44] Because Dice did not see evolution as a problem in genetics, he was seen as "lack[ing] penetration in his analysis of the problem of speciation," in Davenport's words. That is how field genetics appeared from the lab, and when it came to jobs and grants it was that view that counted.

Tower's encounter with the genetic orthodoxy was rather more painful, resulting in accusations of fraud, a forced resignation from his professorship, and invisibility. Like Sumner, Tower regarded himself as a geneticist. He wanted desperately to make great discoveries in genetics but tried to do it with material—heterogeneous populations of wild beetles—that virtually guaranteed failure. It would have taken a man of strong character to handle that situation gracefully, and Tower was not that man. As Charles Davenport observed, Tower had many virtues: an eye for fundamental problems and the courage to tackle them, ingenuity in experimental methods, energy, and an extraordinary capacity for work. But he was also careless, dogmatic, opinionated, and obstinate, and his ambition to make brilliant discoveries was unchecked by critical judgment. His main difficulty, Charles C. Adams shrewdly observed, was "to follow his facts and not imaginary ones."[45]

From his earliest years Tower had astonished biologists with announcements of remarkable achievements, but there had also been doubts. Carnegie officials worried that his first monograph on *Leptinotarsa* contained "blunders," but Tower alleged that the inaccuracies were the result of oversight and came up with corrected data—a pattern that would become all too familiar. Charles C. Adams thought Tower's monograph "a fine piece of work,

44. Charles B. Davenport to John C. Merriam, 8 Feb. 1927, 30 March 1929; Albert F. Blakeslee to Davenport, 9 Feb. 1927 (quote); Lee R. Dice, "Studies of the factors determining speciation in mammals and reptiles," sent to Merriam, 18 Jan 1927; all in CBD-CSH-1. f. Merriam.

45. Charles B. Davenport to Robert S. Woodward, 2 Aug. 1917, CBD-CSH-1. Charles C. Adams to Alexander Ruthven, 9 Jan. 1909 (quote), AGR box 51 f. 2. See also William J. Moenkhaus to Davenport, 8 June [1899], CBD-Gen; and William E. Castle to Robert S. Woodward, 25 Feb. 1919, CIW f. Tower.

full of good things" but many questionable statements on theoretical points did make him wonder about his experiments.[46]

The geneticist William Castle was also impressed by Tower's claims and troubled by the absence of real evidence for them, so he decided to pay a visit to Tower's laboratory to see for himself. "I saw nothing but some beetles in some cages," he later recalled. "No explanations were offered which were at all illuminating. Most men on their own ground can usually tell you what they have and what it means. Tower either couldn't or wouldn't. I thought at the time 'wouldn't,' and so I withdrew. I now think 'couldn't'." Castle's doubts were strengthened in 1910, when Tower was challenged publicly on inconsistencies and again supplied corrective data. Castle concluded that Tower had simply faked data to rescue badly done experiments: He "got into a corner where he either had to admit himself in error, or else discover new marvels. The latter course was more to his taste."[47]

A second and more genetic monograph confirmed Castle's conviction that Tower was a faker, and he aired his suspicions in a letter to the head of the Carnegie Institution, Robert Woodward. Tower's genetics was entirely the product of his imagination, Castle thought. "He . . . works like mad on whatever subject happens to interest him . . . , [but h]is scientific work suffers from his monomaniac moods. He fixes attention on a doubtful question, arrives at a seeming solution of it, imagines a way of proving it and presto thinks he *has* proved it and writes down the proof in details which no man can either substantiate or disprove." Castle also pointed to a pattern of suspicious accidents and evasions that suggested deliberate cover-ups: a fire that destroyed a long-promised manuscript, a steam leak in his Chicago greenhouse that destroyed all his experiments and beetle stocks, failures to show up for scheduled talks, and so on. By 1917 Davenport reported a "nearly universal doubt about the reliability of Tower's reports," and as word of Castle's letter got around—as it obviously quickly did—what little remained of Tower's credibility evaporated. Under pressure Tower resigned and disappeared from sight, a nonperson.[48]

Did Tower fake data? Possibly, though my guess is that he simply recy-

46. Robert S. Woodward to Charles B. Davenport, 24 Nov. 1905, 18 Jan. 1906; Davenport to Woodward, 25 Nov. 1905; Tower to Woodward, 24 Dec. 1905; all in CIW f. Tower. Charles C. Adams to Alexander Ruthven, 15 Feb., 4 March 1907 (quote), AGR box 51 f. 1.

47. Castle to Woodward, 25 Feb. 1919 (cit. n. 47). See also Theodore Cockerell to Charles B. Davenport, 12 July 1910, CBD-Gen.

48. William E. Castle to Robert S. Woodward, 25 Feb. 1919, CIW f. Tower. Charles B. Davenport to Robert S. Woodward, 2 Aug. 1917, CBD-CSH-1.

cled data from experiments done earlier for some different purpose. That is what Gregor Mendel apparently did to get data that matched his theory with impossible accuracy, and I suspect it is fairly common in a cultural activity that changes fast and puts a high value on regular publication.[49]

But the real interest in Tower's downfall is what it reveals of the cultural dynamics of the lab-field border. It was Tower's ambition and self-deception that destroyed him, but such weaknesses are especially dangerous to their possessors in border areas where conflicting rules of practice blur the distinction between unorthodox and deviant behavior. As Davenport observed, "the field is so entirely Tower's own that there is no one man competent to detect any considerable proportion of [his errors]."[50] Davenport's point was that the Carnegie Institution was unlikely to be embarrassed by publishing his work; my point is that individuals who take wish for reality are less likely to be saved from themselves in border areas. In better-cultivated fields, social controls keep potential sinners on the straight and narrow.

The dual standards of border practices doubtless shelter unorthodox approaches against attacks by guardians of disciplinary purity, a good thing. But orthodoxies also serve as protection against self-deception. It is hard to imagine a drosophila geneticist or a systematic entomologist getting into Tower's fix. There would have been too many people working on related problems, too many knowledgeable gatekeepers and busybodies for anyone to stray too far from the straight and narrow. (It is no accident that others suspected of fraud have worked on unusual organisms in unorthodox ways.)[51] One of the real benefits of standard organisms and experimental protocols is that they offer some protection against the temptations that fast-moving activities like science offer to ambitious practitioners. In border sciences, with their mixed practices and dual standards, the line between unorthodoxy and fraud is harder to draw and easier to transgress inadvertently.

By laboratory standards Tower's work was disgraceful. Viewed from the field, however, Tower was perhaps guilty only of a failed experiment. As Carl Hubbs later observed, Tower's work in Mexico remained the most ex-

49. Federico di Trocchio, "Mendel's experiments: A reinterpretation," *Journal of the History of Biology* 24 (1991): 485–519. Floyd Managhan and Alain F. Corcos, "The real objective of Mendel's paper," *Biology and Philosophy* 5 (1990): 267–92.

50. Charles B. Davenport to Robert S. Woodward, 25 Nov. 1905, CIW f. Tower.

51. Jan Sapp, *Where the Truth Lies: Franz Moewus and the Origins of Molecular Biology* (New York: Cambridge University Press, 1990). Arthur Koestler, *The Case of the Midwife Toad* (New York: Random House, 1971). Sander J. Gliboff, "The case of Paul Kammerer," unpublished talk, History of Science Society, November 2000.

tensive study of variation in a natural population. "Perhaps," he reflected, "if all the energy which has been spent in condemning Tower's work had been combined in a repetition of some of his experiments, biology in general and experimental systematics in particular would be farther advanced." Similarly, it was field biologists and taxonomists who perceived most unambiguously the value of Sumner's work on the genetics of *Peromyscus.*[52]

Episodes such as these reveal how dangerous the border could be, a place where blinkered ambition or a failure of critical judgment could be fatal. Sumner was led merely into apostasy and was protected by his strength of character from the pitfalls of a treacherous cultural terrain. His integrity was widely known and admired. One colleague recalled Sumner at work: "I have known him, when some animal was acting exactly contrary to the way he should according to Dr. Sumner's carefully thought out theory, to groan audibly while making test after test to prove that his theory had been completely wrong." For many he was an exemplary biologist. "[N]ot that I saw much of him," wrote one biologist, "but he was always a man who was my idea of a scientist. . . . It always was a comfort to know that he was living and available for advice when needed."[53] Sumner was a man for border life. Tower was not. Personality and character always matter in science, but nowhere more than in border science, where social controls and clear rules of practice are no substitutes for virtue.

Physiologists and the Field

The ambiguities of border life are also revealed in Forrest Shreve's last work on desert ecology, which concluded in 1940 with his unwilling retirement and the closing of the Desert Lab. Of all ecologists' institutions, the Desert Lab was probably the one in which laboratory and field cultures mixed most freely. Shreve was devoted to the ideal of ecology as physiology of the field, but the Desert Lab's location at the northern edge of the great Sonoran desert afforded rich opportunities for fieldwork. Shreve's work shifted continually between field and lab, and though he saw ecology moving from natural his-

52. Carl L. Hubbs, "The importance of race investigations on Pacific fishes," *Proceedings of the IVth Pacific Science Congress* (1929) vol. 3, 13–23, on p. 19. David Starr Jordan to Francis B. Sumner, 16 July 1923, FBS-Fam f. 7. Gerrit S. Miller, Jr., to John C. Merriam, 15 May 1930, CIW f. Sumner.

53. Marston C. Sargent to Denis L. Fox, 28 July 1944, FBS-Fam f. 17. Waldo Schmidt to Carl L. Hubbs, 15 Oct. 1945, CLH box 24 f. 76.

6.6 Francis B. Sumner, circa 1930s. Scripps Institution of Oceanography Archives (FBS-Fam. f. 38).

tory to experiment, he himself moved slowly the other way, his eyes opened by trips with Daniel MacDougal to the Mohave and Mexican deserts in 1913–14 and 1923–24.[54]

However, in the 1920s the Carnegie Institution's policy shifted decisively toward the laboratory side. The principal mover was Herman Spoehr, a hard-line plant physiologist who looked to physics and chemistry as models and had no interest in field ecology. In 1926 Spoehr moved from Arizona to a new laboratory on the California coast, at Carmel, and in 1926–28 engi-

54. Forrest Shreve, "Investigations at the Desert Laboratory," 5 March 1927; Shreve, "Statement regarding the work and program of the Desert Laboratory," 23 Oct. 1928; both in CIW-DL f. 1927–31. Shreve, "Statement regarding the aims and work of the Desert Laboratory and the biological problems of the desert," 5 Nov. 1931, on pp. 6–13, CIW-DL f. 1931–72. Ray Bowers, *Mr. Carnegie's Plant Biologists* (Washington: Carnegie Institution, 1992).

neered a reorganization that demoted the ecological operations of Clements and Shreve to subsidiaries of a division headed by himself. The construction of a Carnegie laboratory for plant physiology at Stanford in 1929 completed the transformation: all work in the new division would be or would serve laboratory physiology. Harvey M. Hall's work on experimental taxonomy remained in good standing, but Shreve and Clements were isolated and obliged to adapt as best they could to Spoehr's unsympathetic and imperious rule. At the Desert Lab only Shreve, his wife, Edith, and a former student remained.[55]

In the next few years Shreve devised several programs of work that he hoped would enable the Desert Lab to survive in a harsher cultural environment (as his beloved desert species had in a drying climate). At first he emphasized the physiological side; not lab physiology, but a physiology of the field, which grew out of his interest in the environmental determinants of plant distribution. A previous study of the distribution and habitats of desert species had left him eager to ascertain by physiological experiments whether distribution was the result of present environmental conditions or of historical accidents that occurred when plants migrated into and settled a place that was becoming ever more arid.

Such experiments, Shreve argued, could not be done in controlled laboratory environments, but only in the desert, on plants living in their variable and changing local habitats. And they could not be limited to the biophysics and biochemistry of water exchange and metabolism, as laboratory experiments normally were, but must include the functions that most mattered in desert plants' survival: dispersal, germination, establishment, growth, reproduction. The environment was as much an object of experiments, in Shreve's view, as the plants themselves. It was precisely the variability of desert plants and their environments, observed and measured in the field, that Shreve hoped would reveal the physiological and historical causes of distribution.[56]

Shreve's proposal for a physiology of the field seemed appropriate to the Desert Lab's tradition and would, he hoped, be acceptable to Spoehr. It built on Shreve's fieldwork in the desert and moved it toward experimentation. It took advantage of the Desert Lab's unique natural endowment. (Spoehr had declared off-limits any research that could just as well be done

55. Bowers, *Mr. Carnegie's Plant Biologists,* pp. 13–15; Shreve, "Statement regarding the aims and work," pp. 15–18 (both cit. n. 54).

56. Shreve, "Investigations at the Desert Laboratory"; Shreve, "Statement regarding the work and program" (both cit. n. 54). Shreve, "Memo for Dr. Merriam," 7 April 1930, on pp. 4–5, CIW-DL f. 1927–31. Shreve, "Statement regarding the aims and work" (cit. n. 54).

6.7 Forrest Shreve measuring a cactus, Desert Laboratory, circa 1920s. Carnegie Institution of Washington Archives.

somewhere else.) Shreve would need a modest increase in staff (a plant physiologist and a soil physicist), but no new investment in labs, since the experiments would be done outdoors in the lab's desert preserve. It was the mix of field and lab that most appealed to Shreve, the prospect "of having unlimited supplies of material under natural conditions, of taking the problems of the field into the laboratory, and of testing the findings of the laboratory in the field."[57] The scheme looked like a winner.

However, Spoehr took quite a different view of Shreve's physiology of the field. From his perspective, proper experiments were not done on the life histories of wild species and were not performed out of doors. Variability of subject and environment were fatal flaws of experimental design, to be avoided at all costs. Spoehr quashed the idea of hiring a soil physicist, and declared

57. Shreve, "Statement regarding the work and program" (cit. n. 54), quote on p. 2. Shreve, "Memo for Dr. Merriam" (cit. n. 56). Shreve to Merriam, 17 April 1931, CIW-DL f. 1927–31.

that the physiology of water relations was best done indoors in a proper laboratory. Harvey Hall also worried that Shreve's project was unfocused and open-ended, and he urged Shreve to concentrate on the physiology of water in plant cells and tissues and go into life-history work only where needed. Where Shreve envisioned experiment supporting fieldwork, Spoehr and Hall saw it the other way around.[58] (Hall shared Shreve's ideal of field experiment but agreed with Spoehr that proper physiology was of the lab.)

The problem was that keeping experiments indoors made it impossible for Shreve to carry out his program. It would require a physiologist who was lab-trained but also experienced in desert work, and such a person did not exist. Furthermore, the laboratory at Tucson was hopelessly out of date for modern biophysics, and upgrading it to current standards would in the end only duplicate Spoehr's new lab at Stanford.[59] I do not think there was a conspiracy to make Shreve's field project unworkable (though Spoehr would have been glad to have both Shreve and Clements off his budget and the Desert Lab closed down). Rather it was the logic of border practice. Viewed from the lab, Shreve's vision of a physiology of the field was an impure form of physiology and a second-best investment. But that view made it impossible for Shreve to do any physiology at all: lab physiology was too expensive, and physiology in the field was unworthy. Imposing laboratory standards left Shreve no middle ground for a mixed mode of practice.

Thus prevented from pursuing his lifelong ideal of a physiological field ecology, Shreve pursued his growing interest in the natural history of deserts. In 1932, after a brief reconnaissance in Mexico, he proposed a full survey of the Sonoran desert in collaboration with LeRoy Abrams of Stanford's herbarium. The plan was unabashed natural history: Abrams would do the floristics, Shreve the biogeography and ecology, and others the physiographic history and paleontology. Shreve's aim was to map this distinctive desert type and reconstruct its history. Laboratory work on habitat relations was mentioned only as an adjunct to intensive observation and survey. The point was not to do experiments in the desert but to understand the desert as one of nature's grand experiments.[60]

58. Forrest Shreve, "Report of conference on work of the Desert Laboratory of the Carnegie Institution, 13 Dec. 1930"; Herman A. Spoehr to John C. Merriam, 31 Jan. 1928; both in CIW-DL f. 1927–30. Harvey M. Hall to Walter M. Gilbert, enclosing "A summary of the Desert Laboratory situation," 7 Feb. 1931, CIW-DL f. 1931–72.

59. Hall, "Summary of the Desert Laboratory situation" (cit. n. 58), p. 3; Hall to John C. Merriam, 17 April 1931, CIW-DL f. 1931–72.

60. Forrest Shreve, "Projected botanical investigations of the Southwestern Desert," 30 Nov. 1932; Shreve to John C. Merriam, 6, 30 April 1932, 30 Jan. 1933; Merriam to Herman A. Spoehr, 23 Dec. 1932; all in CIW-DL f. 1931–72.

Spoehr approved Shreve's desert survey, hinting darkly of a "radical readjustment" if Shreve strayed from the Desert Lab's original scope. For the rest of his career Shreve was a full-time field naturalist. His report on the Sonoran desert is a masterpiece of natural history, and his theory that the four types of North American desert are caused by different seasonal regimes of rain and drought remains the foundation of desert ecology to this day.[61] His explication of nature's experiments was probably a more valuable contribution to ecology than anything he might have achieved had he taken Spoehr's advice and stuck to the laboratory.

But Spoehr had no interest in the historical evolution of deserts, so when Vannevar Bush took charge of the Carnegie Institution in 1939 with a mandate to clean house, Shreve's project was an easy target. Spoehr provided a rationale for the prompt closing of the Desert Lab, declaring that it had accomplished its original objectives of desert exploration and biogeography. Its only unfinished business was physiology—of the biophysical and biochemical sort—for which the Desert Lab, isolated on the edge of a great desert, was quite the wrong place. Shreve's Sonoran monograph and his vision of a comparative study of North American deserts was off the map of Spoehr's scientific world, of interest to range managers or conservationists but not real scientists. Spoehr did not expect Shreve to understand; it was, he wrote Bush, "like urging a man to peep through a key hole when you know he has a glass eye."[62] Maybe so. But the windows of a physiology laboratory may not be all that transparent either, more a mirror than a window on nature.

Conclusion

Having to operate by someone else's rules (or to be seen to) is one of the defining experiences of border life, and it is a problem. Meeting two contradictory standards of practice is likely to provoke criticism from both the lab and field sides, and criticism tends to be taken as a sign that something is wrong—on the principle that where there is smoke, there is fire. And

61. John C. Merriam to Hermann A. Spoehr, 23 Dec. 1932, CIW-DL f. 1931–72. Forrest Shreve, *Vegetation of the Sonoran Desert,* Carnegie Institution Publication no. 591 (1951). Shreve, "The desert vegetation of North America," *Botanical Review* 8 (1942): 195–246. Bowers, *A Sense of Place* (cit. n. 29), pp. 110–15, 138–48.

62. Herman A. Spoehr, "Memorandum on the Desert Laboratory of the Carnegie Institution of Washington," 31 May 1939, pp. 6–12, CIW-DL f. 1931–72. Spoehr to Forrest Shreve, 20 June 1939; Vannevar Bush to Shreve, 23 June 1939; both in CIW-DL f. 1933–40.

since laboratory methods are accorded higher standing, field-workers are commonly held up to criteria of quality that were devised for laboratory work but are often not appropriate for work in nature. Mixed modes of practice, because they are likely to be read as imperfect examples of standard laboratory practice, are a less-certain basis for careers than practices that are securely either laboratory or field. Thus borderland inhabitants are more likely to have volatile careers than either naturalists or experimenters, who operate securely in their own domains.

Ambiguous identity is another defining experience of border life. Scientists take their identities from their work and workplaces—as is the case with most middle-class occupations. Laboratories guarantee secure identities for experimental biologists, but field biologists do not enjoy that security. Ecologists were liable to be mistaken for out-of-place physiologists who had wandered from the straight and narrow or did not know the rules. That, I think, is how Herman Spoehr regarded Shreve, though he would not have put it like that. Field biologists who held lab ideals and projected an experimental future for their discipline also ran this risk. They were also liable to be mistaken for the other sorts of people who operate in the field. As I suggested earlier, the resemblance of field biologists to vacationers or other nonscientific sorts is one reason why the field sciences have had a generally lower standing than laboratory disciplines. For the same reason, field biologists who strive to emulate laboratory standards invite being seen not as a distinctive kind of biologist operating by their own distinctive but appropriate rules, but as pretenders, or as lapsed or misguided or not very serious lab scientists, or as parvenus or masqueraders, or even as out-and-out frauds.

This problem of identity was a particular problem of first-generation border biologists, because they held experimental ideals in high esteem yet had not yet devised modes of practice that were appropriate to field conditions and that really worked. Projecting an image of experimental workers and an experimental future, they set themselves up to be misunderstood by guardians of experimental standards. This incongruity between aspiration and achievement was greatest in the 1910s and diminished in the 1920s and 1930s as concepts and methods were devised—most of them exact but not experimental—that dealt effectively with the complex problems of life in nature.

CHAPTER 7

Nature's Experiments

Adapting laboratory instruments and experimental practices for use in the field is not the only way to create border practices—or the best. The record of such attempts, we have seen, was decidedly mixed. And how could practices designed for places with no variation and no weather function reliably in the endlessly particular and ever-changing world of nature? A more promising strategy was to give traditional field practices the force of laboratory practices but without importing the paraphernalia or protocols of laboratory culture. In this and the next chapter I will argue that it was largely in this way that distinctive border cultures of field biology were shaped in the 1930s and 1940s: by intensifying or amplifying ordinary field practices. Field practices of observing and comparing were refashioned into instruments of causal analysis, some of which proved to be as effective as any experiment and as well suited to nature's particular places as experiment and precision measurement are to laboratories.

In one way or another these amplified practices all involve manipulating place—I call them "practices of place." In the field it is often the arrangement of spatial elements that provides critical evidence of relations between creatures and their environments, or the dynamics of succession and the origin of species. Places are to the field what experimental setups are to laboratories. Recognizing the places where observation and comparison can reveal how nature works, and learning to read the spatial evidence of these processes, are the crucial skills of modern field practice, just as inventing the right experimental tools is the crucial skill of laboratory work. Practices of place define field biology, as experiments define lab science; they are what field-workers do that no one else does.

By definition there are no practices of place in laboratory biology. Laboratories are designed to avoid the necessity of taking place into account, and their arrangements do not intrude in the analysis of experiments, unless something has gone badly wrong. (Imagine what laboratory science would be like if the first step of experiments was always to discover the right place, season, and weather conditions—as though it were a form of geomancy.) But in the field spatial elements are essential for measuring and interpreting natural processes. As Victor Shelford observed of ecological fieldwork, "The selection and analysis of the place of study is the most important step in the whole investigation."[1] To understand the nature of border biology we need to understand its practices of place.

Practices of place often involve reconstructing the history of places and events. Natural events are inherently historical, as laboratory events are timeless and without history. In labs repeatability puts occurrences into a kind of eternal present; natural events happen just once. Laboratory experiments have observable beginnings and ends; nature's are performed in deep time, unobserved. So to understand the causes of biotic communities and species one must usually know something of their histories and of the places where their histories unfolded, if only to distinguish historical contingencies from causes still operating in the present.

I will not be referring here to the actual reconstruction of ancient landscapes or lineages. Whole disciplines are devoted to that end—phylogeny, paleontology, paleoecology, historical biogeography, and others—and they deserve more attention than they have had from historians; but they are not my subject here. Nor are the special practices that have been developed to reconstruct ancient environments, like stratigraphy and physiography, coring, palynology, and paleoclimatology. My subject here is how field biologists reconstruct the processes that have made present places and biota what they are. It is not historical reconstruction as an end in itself, but as a means to understanding natural processes. It is the practices that use histories of places for causal analysis, as if in experiments.

1. Victor E. Shelford, "Ecological succession, II: Pond fishes," *Biological Bulletin* 21 (1911): 127–51.

Practices of place are varied. Some are simply intensified versions of normal practices of observing and comparing—these are the subject of the following chapter. In this chapter I describe practices that are conceptualized as experiments but do not rely on importing laboratory tools or protocols. These practices depend on the idea that natural processes can be interpreted and understood as experiments—not as our experiments, but as "Nature's experiments."

Nature's Experiments

The idea of "Nature's experiments" has an obscure history. It probably derives from natural theologians' trope of "Nature's book" and from secular versions that evoke nature as craftsman or artificer. In the late 1880s, for example, John Muir described the Yellowstone basin variously as a grand geological library and as a landscape crafted by a volcanic blacksmith, by a cook or chemist, and by a farmer (clearing ground with glaciers and sowing broadcast). "These valleys . . . may be regarded as laboratories and kitchens," he wrote, "in which, amid a thousand retorts and pots, we may see Nature at work as chemist or cook, cunningly compounding an infinite variety of mineral messes; cooking whole mountains. . . . In these natural laboratories . . . the awful subterranean thunder shakes one's mind as the ground is shaken."[2] In this maelstrom of metaphor we catch a glimpse of how the idea of nature's experiments might have been fashioned in the mid-nineteenth century, when laboratories began to dominate the scientific landscape. The expression occurs not infrequently in field biologists' public and private writings (as we shall see), but as an informal trope rather than a developed concept. Yet the casual way in which biologists make use of the idea suggests that it was an important part of their "imaginative infrastructure."[3] That is, it was an unexamined but powerful framework for thinking about how human experimenters can know nature.

The idea of nature's or natural experiments is also a framework for thinking about the relations between the practices of laboratory and field biology. It encapsulates in a vivid phrase field biologists' conviction (or hope) that nature offers them opportunities for causal analysis that are every bit as

2. John Muir, "The Yellowstone National Park," *Atlantic Monthly* 81 (1898): 509–22, on pp. 511–12 (quote) and 522.

3. Martin J. S. Rudwick, "Encounters with Adam, or at least the hyaenas: Nineteenth-century representations of the deep past," in *History, Humanity and Evolution,* ed. James R. Moore (New York: Cambridge University Press, 1989), 231–52, on p. 239.

good as anything that experimental biologists enjoy in labs—or better, because nature is an experimenter on the grand scale.

Like the conception of a universal experimental method, the idea of nature's experiments is an instrument of boundary work, in Thomas Gieryn's sense.[4] As the idea of a scientific method demarcates the boundary between science and not-science, so nature's experiment may have served to make the boundary between laboratory and field less absolute, more permeable. The notion that nature does experiments seems designed to reduce the disparity of credibility and standing that was created between lab and field science. It suggests an effort by field biologists to capture some of the authority of experiment without giving up the authenticity of nature. The term implies a contrast with the simplified and artificial character of laboratory phenomena. (Are lab experiments mere simulacra of nature's?) It implies that the methods of the field are just as scientific as laboratory methods and more authentically natural. If nature's experiments fall short of lab experiments in control and separation of variables, they are superior in their scale and variety, and in their adherence to life as it is really lived. These implications were never spelled out as a coherent doctrine but seem present in field biologists' common usage.

Thus the English zoologist Carl F. A. Pantin used the idea to undercut the commonplace and invidious distinction between active experiment and merely passive observation. Observing is anything but passive, he asserted. Field biologists in effect set up experiments when they choose the time and place to observe. "There are indeed 'natural experiments,'" Pantin insisted. "Instead of fixing conditions in the laboratory, we may wait for the appropriate time, or journey to the appropriate place, where we will find the necessary conditions to be fixed by natural events to satisfy the conditions for drawing an absolute conclusion." Selecting the right locale and the right moment to observe what nature does is analogous to what experimenters do when they frame a hypothesis and devise an appropriate test, and the results of "natural experiments" are as robust as anything done in a lab.[5]

Francis Sumner also invoked the idea of natural experiments to assert the equal dignity of fieldwork. He pointed to the "rigid limitations" of animal experiments, which were seldom conclusive and unequivocal and required more time to perfect than any individual had in his whole life (as he

4. Thomas F. Gieryn, *Cultural Boundaries of Science: Credibility on the Line* (Chicago: University of Chicago Press, 1999).

5. Carl F. A. Pantin, *The Relations Between the Sciences* (Cambridge: Cambridge University Press, 1968), pp. 16–17. For similar arguments in cultural ecology see Morris Freilich, "The natural experiment, ecology, and culture," *Southwestern Journal of Anthropology* 19 (1963): 21–39.

well knew). In his own work on *Peromyscus* he was thus "disposed to attach considerable importance to what have been called 'Nature's experiments.'" They, too, were inconclusive, he admitted, but no more than those done in labs, and they had the advantage of having been in progress for a very long time and in natural conditions.[6]

For Frederic Clements it was the sheer abundance and variety of "Nature's experiments" that made them superior to anything that humans could arrange in labs. Every plant association was for him an experiment on the effects of causal agencies ranging from climate cycles to human disturbance: "[T]he interaction of all these climatic and human processes impart to the biome a quasi-experimental value that is often not to be matched by actual experiments," he wrote. And if nature's experiments lacked human control, well, experiments in labs and greenhouses at best only approximated the conditions in which plants and animals actually lived. The paleontologist William Gregory remarked that experimentalists were wrong to dismiss natural experiments for their lack of controls. Were not fossils and archaic relicts in effect the controls for nature's experiments in evolution? Gregory thought so.[7]

In principle just about any thing or place can be regarded as one of nature's experiments, because all are the result of natural agency. "Every landscape . . . becomes a definite scientific proposition for analysis and explanation," Conway MacMillan observed. Frederic Clements regarded every plant community as an experiment in competition, and successions as "the best of all natural experiments" because they involved the widest range of natural processes—aggregation, migration, settlement, competition. Climatic cycles in the ecotone between prairie and semiarid plains were likewise "natural experiments," the results of which Clements thought could hardly be improved upon by laboratory experiment. For Charles C. Adams any variable and widely distributed genus was "one of nature's vast experiments" in evolution and a good subject for study.[8] What mere laboratory

6. Francis B. Sumner, "Genetic studies of several geographic races of California deer-mice," *American Naturalist* 49 (1915): 688–701, on pp. 696–97.

7. Frederic E. Clements, "The relict method in dynamic ecology," *Journal of Ecology* 22 (1934): 39–68, quote on pp. 41–42, 46. Harvey M. Hall and Clements, *The Phylogenetic Method in Taxonomy,* Carnegie Institution Publication no. 326 (1923), pp. 20–22. William K. Gregory, "Genetics versus paleontology," *American Naturalist* 51 (1917): 622–35.

8. Conway MacMillan, "Observations on the distribution of plants along shore at Lake of the Woods," *Minnesota Botanical Studies* 1 (1897): 949–1023, quote on p. 951. Frederic E. Clements, *Research Methods in Ecology* (Lincoln, Neb.: University Publishing, 1905), pp. 306–7, also 149–51. Clements, "Relict method" (cit. n. 7), p. 46. Charles C. Adams, "The variations and ecological distribution of the snails of the genus *Io,*" *Memoirs of the National Academy of Sciences* 12, no. 2 (1915): 1–184, on pp. 7–8.

could possibly match such an abundance of experiments? The social logic of the idea is clear enough.

Nature does not really do experiments, of course. Experiment is a uniquely human activity, a cultural category that does not apply to natural phenomena because nature is not purposeful and self-conscious. It is precisely the illogical mixing of categories in the expression "Nature's experiment" that jars us into rethinking invidious distinctions between laboratory and field practices. The idea describes not nature itself but our relations with nature. It is our search for cause and effect in nature's works that turns them into experiments. Nature does no experiments without us; our connivance is required. Trees fall in the forest whether we are there or not, but they do so experimentally only when a biologist is present with quadrat tapes and notebook. A natural phenomenon is an experiment only when we read it as one.

Legibility is the big problem for those who make use of nature's experiments. How do we "read" nature's grand-scale experiments when we do not design and set them up or watch them proceed, as laboratory experimenters do theirs? Another problem is that nature may be an avid and prolific "experimenter" but is a lousy record keeper. As species evolve, they are transformed or go extinct; stages of ecological succession are wiped out as one community replaces another. Nor can nature's "experiments" be rerun, as experiments can with standard instruments and protocols in unvarying environments. So how can short-lived and shortsighted human observers turn nature's abundance to their advantage?

The answer is, by the systematic use of observation and comparison, and by the clever use of the particular places where nature's processes seem (but only seem) to be experiments. We read nature's experiments, in other words, by employing practices of place. Sometimes the sheer abundance of those "experiments" is sufficient: there are so many going on all the time that evidence can be observed somewhere of any intermediate stage in the process of, say, succession or speciation. Comparison of variations in places sometimes enables observers to infer causes and effects. Such practices are the subject of chapter 8. But perhaps the most intriguing practices are those that make use of places whose elements are fortuitously arranged as if for an experiment, places that a human observer familiar with experimental setups can read as an experimental setup. In such places field biologists can reconstruct or relive "experiments" on the grand scale that they might fondly imagine doing but can never do. This practice of place—the tricks of selecting and reading the places of nature's actions as if they were experiments—is our present concern.

Reading natural places as experiments is an odd border practice. It reveals cause-and-effect relations, not by experiment but by methods of natural history. Its mental operations are like those of an experiment, but the actual skills and techniques are not the same as or even derived from laboratory practice. Traditional field methods are heightened and amplified to do in the field what experiments do in labs; yet implicitly the superiority of experimental method is acknowledged in the very idea of "Nature's experiments." What are we to make of a practice whose techniques are of the field but whose rules of knowing are of the lab? Is it an insidious insinuation of laboratory values into field biology—a kind of cultural body snatching? Or is the idea of natural experiments a sign that field biologists were reclaiming laboratory values and refashioning them for their own purposes? Both, I would suggest: we see in these practices of place a mixed culture beginning to take shape in the lab-field border zone, a culture that resembled those of both laboratory or field but was distinctly different from either one.

Places in Process

But why bother with reconstructing nature's experiments, one might ask, when they are going on all the time all around us. Why not just watch and record them as an eyewitness in real time? Speciation is too slow a process to observe, of course. Not even Peter and Rosemary Grant's twenty-year study of Darwin's finches in the Galapagos Islands or studies of the rapidly speciating cichlid fishes of Lake Victoria have witnessed the origin of new species.[9] However, plant succession moves fast enough to be caught in the act by ecologists who have but one brief lifetime to see it happen.

There are many places in the world where nature keeps resetting the clock of succession back to start. Erosion and deposition, volcanic eruptions, glaciers, floods, landslides, tides, and fire—all create new places where nature's experiments in succession can be observed. The scree slopes of young mountain ranges attracted Frederic Clements to the Rocky Mountains' Front Range. Henry Cowles found similar opportunities in Lake Michigan's dunes, where wave, ice, and wind keep the successional clock at zero. Charles C. Adams was delighted with the actively eroding ravines along the Ohio River valley at Cincinnati. "Another big flood is on," he wrote a friend in 1907, "and the

9. Jonathan Weiner, *The Beak of the Finch: A Story of Evolution in Our Time* (New York: Knopf, 1994). Tijs Goldschmidt, *Darwin's Dreampond: Drama in Lake Victoria* (Cambridge: MIT Press, 1996).

washing in the ravines and landslides are beyond anything I ever saw. . . . [T]he whole bottom of one ravine was cleared away. . . . [T]his region is simply superb . . . like Africa and simply full of problems."[10]

River valleys are rich in newly cleared ground, from the eroding headwaters to the expanding delta mouths—huge conveyer belts for shifting sediments downhill. Victor Shelford thought river floodplains were as good as dunes for studying succession. Cowles was impressed by how the islands formed by prairie rivers moved slowly downstream, wearing away at one end and growing at the other. No American ecologist worked the great river deltas of the Gulf and east coasts, as Charles Flahault did the delta of the Rhône, in France, but William Ganong glimpsed a similar opportunity in the raised marshes of the Bay of Fundy in New Brunswick, which were built up not from silt washed down from hills but of sediments scoured from the bay's soft sandstone bed and washed up to its northern end by its famously huge tidal bores—a river delta in reverse, as Ganong put it.[11]

Glaciers are another large-scale clearer of ground, mere remnants now of the great Pleistocene ice sheets, but windows on some of nature's most stupendous ecological and biogeographic experiments. Smaller but equally impressive are volcanic ash falls and lava flows. Volcanic islands like Krakatau, St. Vincent, and Hawaii also attracted scientific attention in the late nineteenth and early twentieth centuries.[12]

And there is us. The destruction of North American forests and prairies in the past two hundred years was an ecological feat matched only by continental ice sheets, and it has the advantage over ice sheets of having an observed and recorded history. Some of the driest and wettest regions were still being cleared, ploughed, and drained when the first ecologists came on the scene, and in eastern regions forests were beginning to return to vast stretches of abandoned agricultural land. Here were natural experiments (or unnatural, if you prefer) on a scale that no lab or garden experiment could come close to mimicking.

Ecologists were divided over whether these human experiments were

10. Frederic E. Clements, "The development and structure of vegetation," *Botanical Survey of Nebraska* 7 (1904): 5–175, on p. 141. Henry C. Cowles, "The physiographic ecology of Chicago and vicinity," *Botanical Gazette* 31 (1901): 73–108, 145–82, on pp. 96–108. Charles C. Adams to Alexander Ruthven, 19 March (quote), 12 April 1907 (see also 11 Dec. 1905), AGR box 51 f. 1.

11. Victor E. Shelford, "Basic principles of the classification of communities and habitats and the use of terms," *Ecology* 13 (1932): 105–20, on p. 108. Cowles, "Physiographic ecology of Chicago" (cit. n. 10), pp. 96–100. William Ganong, "The vegetation of the Bay of Fundy salt and diked marshes: an ecological study," *Botanical Gazette* 36 (1903): 161–86, 280–302, 349–67, 429–55.

12. Daniel T. MacDougal, *The Salton Sea,* Carnegie Institution Publication no. 193 (1914), on pp. 115–16.

proper objects of scientific study. Some had no qualms. Joseph Grinnell, for example, saw the ecological transformation of North America as a great natural experiment and an opportunity. "If we now had the accurate records of fauna in the Atlantic states a century ago," he mused, "how much might we not be able to adduce from a study of the changes which have taken place!" There were no such records, but Grinnell made it his business to preserve the record of similar changes that were transforming the hills and valleys of California even as he worked to stay one step ahead of ploughmen and loggers. Grinnell expected to see significant changes not just in the distribution of animals, but in their habits and physical characters: it was an experiment in evolution on a grand scale.[13]

Other naturalists were less sanguine about the uses of human-assisted "natural" experiments. Victor Shelford, for example, thought agricultural clearances were worthless as ecological experiments because the initial conditions were unrecorded. Charles C. Adams likened the study of humanized environments to the laboratory study of diseased organisms: neither meant much without prior knowledge of normal healthy organisms or natural places undisturbed by human activity.[14] This view was almost certainly the prevailing one. An unsystematic survey of ecological journals suggests that ecologists mainly avoided humanized environments until the mid-1930s, when for a variety of reasons, such as New Deal conservation projects, topics like succession on abandoned farmland or cutover became legitimate subjects for ecological fieldwork.

There were a few sustained efforts in the 1910s and 1920s to observe succession in real time, but only a few, and their results were surprisingly meager. One such project was William Cooper's study of succession on ground vacated by retreating glaciers. A small glacier on Mount Robson, in the Canadian Rockies, first caught Cooper's eye. The glacier had paused five times during its recent retreat, leaving distinct moraines, four of which were accessible and datable from tree rings, and shingle outwash flats that were not yet eroded away and forested. "[A]n ecologist," Cooper exclaimed, "will seize with eager delight upon an opportunity to study the successional processes directly" as year by year vegetation moved in to occupy the new

13. Joseph Grinnell, "Museum of Vertebrate Zoology," *Report of the President of the University of California 1908–10* (1910), pp. 117–24, quote on p. 122. Grinnell to Annie Alexander, 2 Nov. 1907, AA box 1. Grinnell to William E. Ritter, 29 Feb. 1908, WER box 10.

14. Victor Shelford, "Ecological succession, V: Aspects of physiological classification," *Biological Bulletin* 23 (1912): 331–70, on p. 350. Charles C. Adams, *Guide to the Study of Animal Ecology* (New York: Macmillan 1913), pp. 25–28.

ground.[15] Standing there one could easily conjure up vanished Pleistocene landscapes of retreating ice sheets and advancing vegetation.

Unfortunately it was logistically almost impossible to work in a high alpine valley on the edge of an uninhabited wilderness, but two years later Cooper found just the right place for his project, at Glacier Bay, on the Alaskan panhandle. Glacier Bay, a large fiord left behind by a large, rapidly-retreating tongue of a mountain glacier, had been the object of scientific attention since John Muir visited in 1879 and wrote a book about it. Geologists had mapped the glacier's retreat in 1890, 1907, and 1919, giving ecologists an exact physiographic history. Moreover, the glacier's retreat was uncovering datable fossil forests destroyed during earlier advances, so that a history of the changing terrain and vegetation could be constructed going back several centuries. And though the place was remote and uninhabited, it was at sea level, and ecologists could tend their research sites around the fiord and ice fronts quickly and easily by boat.[16] Cooper established permanent quadrats in 1916 in the expectation of returning at regular intervals to record the changes in vegetation as the glacier retreated and vegetation moved in. He would make three more trips to Glacier Bay in the next twenty years.

Daniel T. MacDougal saw a similar opportunity in the accidental flooding and recession of the Salton Sea in 1904–13. The Salton Sea lies in the Salton Sink, the northern end of the Gulf of California rift valley and the site of the recently reclaimed Imperial Valley. Below sea level and isolated from the gulf by a low watershed built up by the Colorado River's vast depositing of silt, the Salton Sink has over the centuries been periodically flooded when the Colorado shifted course and flowed northward into it, forming a large lake that, when the land barrier again formed, once more dried out.

In 1904 an ill-advised enlargement of the canal diverting water from the Colorado to Imperial Valley helped this natural cycle along, causing potentially catastrophic flooding. The canal paralleled the Colorado but had a slightly steeper grade, so when a combination of human meddling and unusual seasonal floods enlarged the canal, the Colorado abandoned its channel to the gulf and flowed entirely into the sink, threatening to drown Im-

15. William S. Cooper, "Plant successions in the Mount Robson region, British Columbia," *Plant World* 19 (1916): 211–38. Cooper, "The recent ecological history of Glacier Bay, Alaska," *Ecology* 4 (1923): 93–128, 223–46, 355–65, quote on p. 93.

16. Cooper, "Recent ecological history" (cit. n. 15).

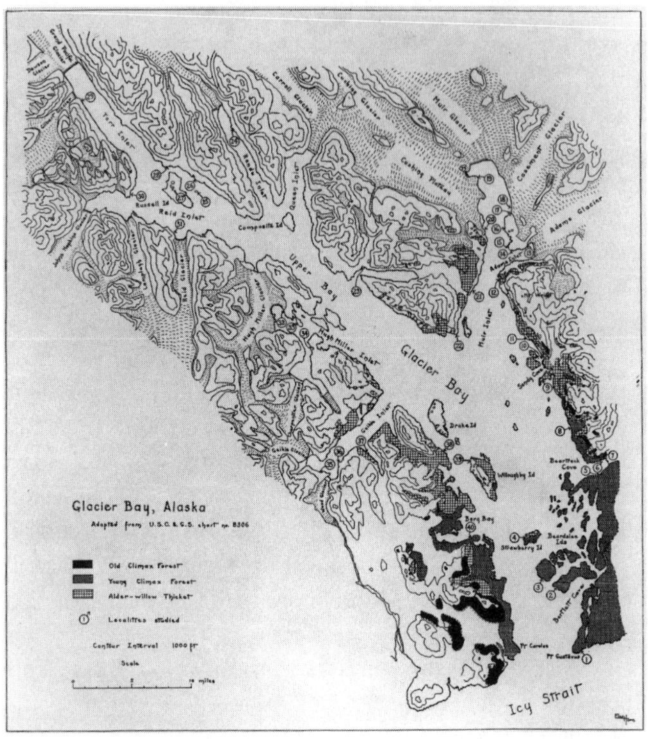

7.1 Glacier Bay, Alaska, showing field stations worked by William Cooper in his study of succession. William S. Cooper, "The recent ecological history of Glacier Bay, Alaska, I," *Ecology* 4 (1923): 93–128, facing p. 93.

perial Valley in a recreated "Lake Chauilla"—a few hundred years ahead of schedule. Massive earth moving by engineers of the Southern Pacific Railroad narrowly prevented that disaster, but not before seven cubic miles of water had swelled the Salton Sea to 250 square miles.[17]

What was a near disaster to California's agribusiness was for ecologists a once-in-a-millennium opportunity to see how desert plants and animals reoccupied the lifeless mud flats that emerged as the flood receded. It was a vast natural experiment in succession, without the usual complications of established vegetation. MacDougal was especially interested to see how plant dispersal worked as the Colorado washed seeds and propagules into the Salton Sink from as far away as Wyoming. It was a natural experiment

17. MacDougal, *Salton Sea* (cit. n. 12), pp. 1–2, 115–20. William DeBuys, *Salt Dreams: Land and Water in Low-Down California* (Albuquerque: University of New Mexico Press, 2000).

7.2 The Salton Sea at the height of its unplanned flooding in February 1907 and during its recession a year later. Daniel T. MacDougal, *The Salton Sea,* Carnegie Institution Publication no. 193 (1914), plate 17, at p. 124.

that could never be simulated so naturally or on such a scale. MacDougal also hoped that the drying of the Salton basin would throw light on the origin of deserts, recapitulating in fast-forward their ancient evolution as the Pleistocene climate turned slowly arid and vast glacial lake beds became dry salt flats. A botanical survey in 1904 had already established a baseline, and expeditions from 1907 to 1913 recorded the details of revegetation. MacDougal foresaw no difficulty in reading the evidence: the receding water was "writing its story very plainly on the shores."[18]

However, the results of these projects were disappointingly meager. In the Salton Sink, invading nondesert plants quickly died as mud flats dried,

18. MacDougal, *Salton Sea* (cit. n. 12), pp. 142–53, 159–65, 180–82. MacDougal to Robert S. Woodward, 9 Feb. 1906, CIW-DL f. Founding. MacDougal to Woodward, 14 May 1908, CIW-DL f. MacDougal.

and MacDougal was unable to follow the fate of individual plants. The vast, soft mud flats were difficult to work, and their ecology so complex and uncontrolled that MacDougal could learn little about the process of invasion. (He had somewhat better luck with laboratory simulations, but not much.)[19] The yield from William Cooper's expeditions to Glacier Bay was not much greater; his reports suggest that he learned more about the physiography of glacier retreat than about plant succession.[20] Places that seemed so well suited to observing nature's experiments were in fact not like experiments at all. Why?

One reason was the enormous difference in the time scales of nature's experiments and ecologists' careers. Nature changes slowly, and scientific careers depend on a steady flow of discoveries. Human observers can seldom stay with a project long enough to see significant changes. Laetitia Snow, a disciple of Henry Cowles, began a study of physiographic succession in the dunes and barrier islands of the Delaware coast in 1902, but on revisiting her study sites a decade later, she found that little had changed. William Cooper had a similar experience revisiting Isle Royale sixteen years after his first study there.[21]

Another reason is that nature's experiments are not equally interesting as experiments. The causes of natural processes are readable only in places where some special spatial arrangement or history makes them legible as experiments. In most places various natural agencies work in concert and cannot be disentangled, and the Salton Sink, Glacier Bay, and the Delaware dunes were such places.

Reading Places

For reading places, naturalists have over the years developed ingenious tricks that were not inspired by laboratory models but taken from traditional biogeographical field craft. For example, contiguity has long been

19. MacDougal, *Salton Sea* (cit. n. 12), pp. 159–65, 171–76.

20. William S. Cooper, "A third expedition to Glacier Bay, Alaska," *Ecology* 12 (1931): 61–95. Cooper, "A fourth expedition to Glacier Bay, Alaska," *Ecology* 20 (1939): 130–55.

21. Laetitia M. Snow, "Some notes on the Delaware coast," *Botanical Gazette* 34 (1902): 284–320. Snow, "Progressive and retrogressive changes in the plant associations of the Delaware coast," *Botanical Gazette* 55 (1913): 45–55. Snow was professor of botany at Wellesley College. William S. Cooper, "Seventeen years of successional change upon Isle Royale, Lake Superior," *Ecology* 9 (1928): 1–5.

used as an indicator of genetic relationship, and things conspicuously in or out of place as clues to the direction of moving ecotones. Biogeographers since the time of Alexander Humboldt have used the rule of thumb that spatial proximity is evidence of phylogenetic closeness. From chains of related species one adjacent to the next, they "read" the stages of an evolving genus as it migrated from a center of distribution into new environments. It was a way of translating spatial arrangements into temporal sequences, reading the past from the present. True, such evidence was circumstantial, and many supposed migrations, land bridges, and phylogenetic schemes ended up on the scrap heap of speculation. But some proved robust. Biogeographers also invented rules of thumb for interpreting spatial patterns of variation. For example, the larger bulk and shorter extremities of animals living in colder environments (Bergmann's rule and Allen's law), and the darker color of animals in moister climates (Gloger's rule) were read as evidence of adaptive evolution driven by environment.[22]

Early ecologists inherited similar practices of place: for example, the rule of thumb that the striking vegetation zones of mountain slopes are a compacted vertical equivalent of the great globe-girdling vegetation belts. Ecologists like to say that a few miles' hike up a mountain peak is the ecological equivalent of a journey of hundreds or thousands of miles along a line of longitude.[23] Vertical becomes horizontal space and elevation becomes latitude, making patterns too vast to perceive visible at a glance. (In practice life zones are often not that regular. Roland Harper, who visited the very spot in the San Francisco Mountains where Merriam had done his fieldwork, found it "rather obscure, to say the least," and thought that "an accurate map . . . would show things arranged in patches rather than zones.")[24]

With the same trick one could also transform space into time, if one believed, as Frederic Clements did, that vegetation evolves in long cycles toward climatic climaxes. For Clements, following vegetation zones up mountain slopes was not just a virtual trip along a line of longitude but a journey in a time machine through the phases of climate cycles. It was, as he put it,

22. Mary P. Winsor, *Reading the Shape of Nature: Comparative Zoology at the Agassiz Museum* (Chicago: University of Chicago Press, 1991), pp. 76–80. Keir B. Sterling, "Clinton Hart Merriam and the Development of American Mammalogy, 1872–1921," Ph.D. diss., Columbia University, 1973, pp. 254–59. Janet Browne, *The Secular Ark: Studies in the History of Biogeography* (New Haven: Yale University Press, 1983).

23. Sterling, "Clinton Hart Merriam" (cit. n. 22), chap. 6. Charles S. Kendeigh, "History and evaluation of various concepts of plant and animal communities in North America," *Ecology* 35 (1954): 152–71.

24. Roland M. Harper to Arthur G. Vestal, 31 Aug. 1924, AGV box 5.

"a long look backward into the past and forward into the future of the biome."[25] Few ecologists shared Clements's theoretical vision, so most found it hard to share his readings, but the habit of reading vertical into horizontal space, and space into time, remains deeply rooted in ecologists' field practice.

Other practices of place enabled ecologists to read biogeographical boundaries, like the long shifting ecotone between prairie grasses and riverine woodland in the central Mississippi Valley. Such boundaries appear static but are in fact always in motion, as climate changes and species compete in zones in which neither has an established advantage. But they move so very slowly and haltingly that it is hard to say in what direction. Are grasses advancing on the forest, or trees on the prairie? Are the isolated prairie groves and forest barrens (grassy woodland glades) of the prairie "peninsula" of Illinois and Indiana the vanguards of expanding communities or the rear guards of their retreat? And what was the meaning of the patches of boreal sphagnum and tamarack that ecologists found in swamps of northern Indiana, far south of the boreal forest belt, or the patches of coastal pine barren that live anomalously in the mountain valleys of Tennessee?[26] Things out of place suggested movement, but in what direction, and impelled by what forces?

Often it was ecological and spatial evidence that made such places legible. For example, Henry Gleason's "relict" method used knowledge of habitat preferences to read motion into vegetation boundaries. If an isolated patch of vegetation occupies an environment to which its various species are well suited, then it is likely to be a pioneer of an expanding community getting a toehold where favorable habitat gives it a competitive edge over species already established there. But if an isolated patch occupies a habitat to which the surrounding vegetation is not well suited, then it is likely to be a rear guard relict, holding on in places from which the better adapted invaders find it harder to evict them. (Vanguards expand first into the most suitable environments, and rear guards hang on in places that are not ideal for them but even less suited to invading species.) Gleason hit upon this simple but ingenious method during his work in the prairie-forest ecotone in central

25. Clements, "Relict method" (cit. n. 7), p. 41.

26. Edgar Transeau, "On the geographic distribution and ecological relations of the bog plant societies of northern North America," *Botanical Gazette* 36 (1903): 401–20. Thomas H. Kearney, Jr., "The pine-barren flora in the east Tennessee mountains," *Plant World* 1 (1897): 33–35. Henry A. Gleason, "An isolated prairie grove and its phytogeographical significance," *Botanical Gazette* 53 (1912): 38–49. Transeau, "The prairie peninsula," *Ecology* 16 (1935): 423–37.

Illinois in 1908 and later showed that it also applied to the boundary between boreal and temperate forests.[27]

Comparison was another simple but effective practice of place. As laboratory biologists repeat experiments with variations to sort out causal relations, so field biologists use nature's prolific and varied experiments to the same end. Mere repetition sometimes suffices to give circumstantial evidence the force of experimental proof. That is how Henry Gleason explained the striking oak groves that dot the prairie of central Illinois. He suspected that they were remnants of a once-continuous forest, most of which had been destroyed by prairie fires set by Native Americans and Euro-Americans. (Forest edges were favorite sites for Indian camps and early European settlers.) It was suggestive, he thought, that Burr Oak Grove, his main study area, occupied the higher ground in an area of sinuous low ridges and swampy sloughs (an old moraine). Probably this grove had been protected from fire by the wet sloughs, but Gleason could not be sure from a single case. Suspicion became conviction when he visited twenty groves and mapped their topography and surroundings: all but two, he found, were located downwind from sloughs or lakes, protected from prairie fire.[28] No experiment could have been more persuasive.

Bertram Wells used a similar device to solve the much-disputed puzzle of the grassy "balds" at the tops of southern Appalachian mountains. Comparing the topography of twenty-two such balds, Wells found that fifteen occupied protected southern slopes near old Indian trails and upland springs. There could be little doubt that the balds were the result not of ecological differences of soil or climate but of human cutting and burning.[29] These cases illustrate how simply observing and comparing could render legible nature's endlessly varied experiments.

The thousands of pothole ponds and bogs of the glaciated upper Middle West, the "solution ponds" (plugged sink holes) of Indiana's limestone belt, the spring-fed pools that dot the arid west, tiny relicts of once-great river systems, the clusters of offshore islands isolated from the Atlantic mainland by rising sea-levels—all these multiple places have offered naturalists a natural

27. Henry A. Gleason, "The vegetation of the inland sand deposits of Illinois," *Bulletin of the Illinois State Laboratory of Natural History* 9:3 (1910): 23–174, on pp. 44–45. Gleason to Arthur G. Vestal, 27 Oct. 1911, AGV box 2. Gleason, "Twenty-five years of ecology, 1910–1935," *Memoirs of the Brooklyn Botanical Garden* 4 (1936): 41–49, on p. 47.

28. Gleason, "An isolated prairie grove" (cit. n. 26).

29. Bertram W. Wells, "Southern Appalachian grass balds," *Journal of the Elisha Mitchell Scientific Society* 53 (1937): 1–26.

equivalent of repeatable experiments. Such places are abundant in areas affected directly or indirectly by the cycles of continental glaciation, and because many formed more or less at the same time during the last headlong retreat, they are especially suited to comparison. Infinitely varied but synchronized in time, they are natural experiments in succession on a vast scale.[30]

But comparison alone may or may not be very revealing. Nature does experiments helter-skelter, and finding places where comparison reveals causes is hit or miss. The best places are those in which special features of topography or history resemble laboratory setups.

A good example is Frederic Clements's study of succession in the lodgepole-pine forests of Estes Park, Colorado in 1907–8. Clements was asked by the U.S. Forest Service to find out why burned areas grew back sometimes as lodgepole pine and sometimes as other mixed forests, and he seized the chance to investigate the intimate mechanics of succession. It is a complicated case. Occupying a belt between eight and ten thousand feet, between yellow pine below and alpine Englemann spruce above, lodgepole-pine forest is a patchy mosaic of lodgepole pine, Douglas fir, and aspen. It is also a zone of frequent forest fires, fueled by the species' dense growth and fired by violent late-summer storms. Because lodgepole pine, Douglas fir, and aspen have similar climatic requirements, burned areas may grow back as pure lodgepole, pure aspen (in wet areas), or mixtures of all three. Successions are thus both frequent and varied, depending on particular local circumstances, and it was that variation that made Clements think that the place might reveal how the process worked.

Fires were especially common in Estes Park, because its delightful scenery and climate made it a favorite summer camp for plains Indians and a magnet for Euro-American settlement. With human occupation came more frequent fires and a historical record of their occurrence. Clements was able to identify and map the outlines of thirteen burns between 1707 and 1905, which he dated by tree rings and historical records. He studied five in some detail, using tree rings, stumps, and analysis of burn scars to reconstruct the type and age of the forests that had burned, how each had been reseeded,

30. Edgar Transeau, "The bogs and bog flora of the Huron River valley," *Botanical Gazette* 40 (1905): 351–75, 418–48; 41 (1906): 17–42. Charles H. Shaw, "The development of vegetation in the morainal depressions of the vicinity of Woods Hole," *Botanical Gazette* 33 (1902): 437–50. Henry A. Gleason to Arthur G. Vestal, 6 Feb. 1912, AGV box 2. Forrest Shreve, "Conditions indirectly affecting vertical distribution on desert mountains," *Ecology* 3 (1922): 269–74. Carl L. Hubbs and Robert R. Miller, "The zoological evidence: Correlation between fish distribution and hydrographic history in the desert basins of western United States," *Bulletin of the University of Utah* 38:20 (1948): 17–166. Victor E. Shelford, ed., *Naturalist's Guide to the Americas* (Baltimore: Williams & Wilkins, 1926), p. 374.

7.3 The burn of 1901 in the lodgepole-pine forest in Estes Park, Colorado, where Frederic Clements studied succession. Frederic E. Clements, "The Life History of Lodgepole Burn Forests," *U.S. Forest Service Bulletin* 79 (1910), plate 1 (frontispiece).

and what kind of forest grew in its place. (It is amazing how much information Clements was able to extract from the details of scars.) Topography, aspect, soils, and moisture were painstakingly observed, measured, and mapped. Clements paid particular attention to the edges of the old burns, because it was from these edges that burned areas were reseeded, the critical variable in succession, Clements surmised.[31]

The most valuable places of all for Clements were those in which one burn happened to overlap or include earlier ones, because these overlaps were most like experiments repeated with variations. The burn of 1901 was exceptionally revealing. Because it burned an area that had regrown unevenly from the burn of 1878, it left a mosaic of burned, half-burned, and unburned patches. It had also occurred in an exceptionally mixed forest that included pure lodgepole, lodgepole mixed with aspen and yellow pine, pure aspen, and several transitional types, providing varied opportunities for reseeding. From this variety of evidence Clements could infer the precise features of topography, seed trees, environmental factors, and ground cover that enabled lodgepole pines to take over a burned area. (Seed trees uphill, well-timed rains, and lack of cover for seed-eating rodents proved to be the critical factors.)[32] It was not the multiplicity of nature's experiments that counted, but the singular places that by accident were legible as experiments by biologists accustomed to reading experimental setups.

31. Frederic E. Clements, "The life history of lodgepole burn forests," *U.S. Forest Service Bulletin* 79 (1910), on pp. 7–27, 46–47, 53–56.

32. Ibid., pp. 13–27.

Physiographic Ecology

Places where physiographic and historical circumstances conspire to preserve a record of nature's experiments have been especially dear to border biologists. Instead of an illegible palimpsest, with the record of one stage obscuring earlier ones, in these special places the whole process of succession or evolution is inscribed legibly upon the landscape, as the process of a laboratory experiment is inscribed on a scroll of graph paper by recording instruments. First-generation ecologists became adept in finding and reading such places.

The zoned bogs of glaciated regions are one such place. Arranged in a visually striking concentric pattern from open water in the center to dryland forest at the outer margin, these zones are a visual record of a process of plant succession that begins with the pond plants of open water and proceeds through stages of reeds, marsh, swamp, and grass, to forest. What enables ecologists to read this record is their understanding of the physiographic life cycle of pond bogs from their birth as barren glacial lakes to their final fading into the enveloping forest. As an accumulation of silt and organic debris destroys the habitat of one zone of vegetation, it recreates that same habitat in the adjacent zone of slightly deeper water. The whole array of zones moves slowly inward, recreating itself continually over a long period of time and preserving a continuous record of intermediate stages.

When ecologists read a place in this way they transform space into time, reading a vertical sequence in time from a horizontal sequence in space. Starting at the forest edge and moving toward the lake's center, ecologists can in effect move back through time from the present (or future) to the time when nature's experiment in succession began at the glacial front. Standing in one place in the sequence one could see in one direction into the past, and into the future in the other.[33] This trick of reading time from space was one of ecologists' first and most successful practices of place.

These spatial patterns are especially striking in the glaciated areas of the American Middle West. Occupying the kettle holes of old glacial moraines, these ponds are typically small (and thus short-lived), round, and shallow,

33. Conway MacMillan, "On the formation of circular muskeag in Tamarack swamps," *Bulletin of the Torrey Botanical Club* 23 (1896): 500–507. Cowles, "Physiographic ecology of Chicago" (cit. n. 10), pp. 79–80. Howard S. Reed, "A survey of the Huron River valley, I: The ecology of a glacial lake," *Botanical Gazette* 34 (1902): 125–39, on pp. 125–26, 134, 138–39.

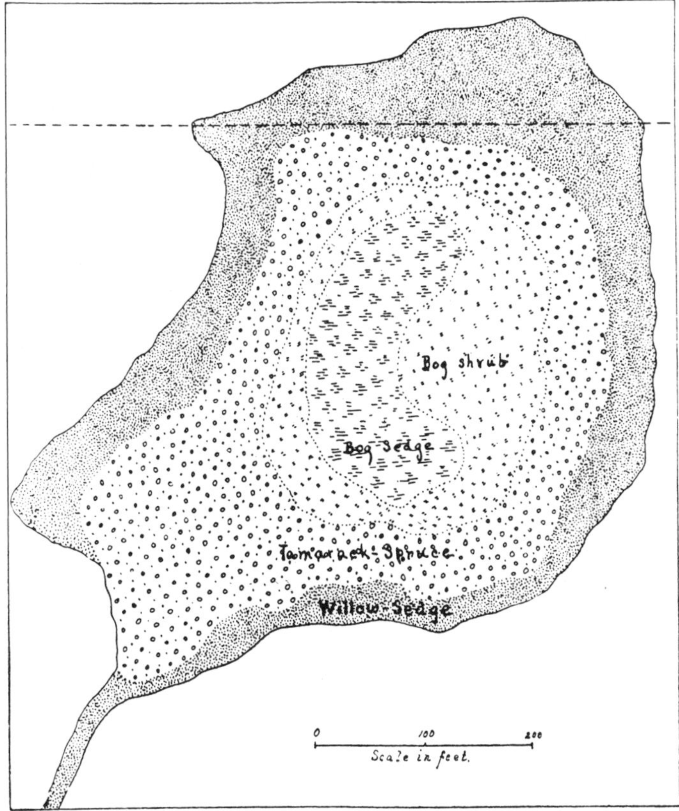

7.4 Diagram of a concentric bog, indicating succeeding zones of vegetation. Edgar Transeau, "The bogs and bog-flora of the Huron River valley," *Botanical Gazette* 40 (1905), on p. 439.

with gently sloping profiles, so that infilling produces well-defined, concentric vegetation zones. Created during the ice sheets' last, rapid retreat, these ponds filled in at a rate that obligingly left a full record for late-interglacial ecologists of nature's experiments in succession. (Much slower and the series would be incomplete; much faster it would have disappeared before ecologists came upon the scene.)

Scandinavian and German botanists were the first to study zoned bogs, so Conway MacMillan knew their value when he found some fine examples during botanical fieldwork in Minnesota's North Woods in the early 1890s. Foresters pointed him to the best, in an area of primordial forest: six water-plant zones clearly visible, plus four on the shore were "[e]xceedingly pretty examples" of the dynamic process of plant succession at work, MacMillan thought.[34]

Records of nature's experiments in succession were also preserved when the water level of large glacial lakes dropped, leaving lagoons, beaches, and dunes high and dry and open to invading plants. Where the water level dropped in stages, it left behind a series of progressively younger fossil beaches, like a series of experiments in succession started at different times, and as readable as the zones of glacial bogs. This physiographic scenario was played out in the history of the Great Lakes, which are remnants of larger ancestral lakes that formed as meltwater was impounded at the edges of retreating ice sheets. They took their present form in successive drops, as melting ice sheets uncovered new and lower drainage outlets. Most of the lakes' shorelines are too steep to record these changes, but where they are flat, a complete and legible record was laid down of the process that slowly changed bare sand to forest. The great dune belt at the southeastern end of Lake Michigan is one such place. For the first American ecologists the Indiana dunes were a special, almost sacred place, because it was there that Henry Cowles developed his celebrated method of physiographic ecology.

Cowles's idea was that ecological succession follows physiographic change, and that plant communities can be classified by the physiographic character of their habitats. Dunes, because their physiographic development is relatively rapid, are a good place to construct successional series, and the Indiana dunes are a remarkable example of their kind. Along the lake a belt of active, wandering dunes rise to over thirty meters (one is nearly sixty meters) and extend inland one to two kilometers. Three to eight kilometers from the lakeshore, belts of older stabilized dunes blend insensibly into the surrounding forest and prairie. These older dunes are fossil dunes and strands created when glacial Lake Chicago was sixty feet higher than Lake Michigan. As new outlets were uncovered by the ice sheet's retreat, Lake Chicago dropped in steps of about twenty feet, creating succeeding belts of strand and dunes, left high and dry and deprived of the sand that was transported by steady northwest winds to the dune region. What would ordinarily have been a thin belt of dunes thus became a large series dating back to the time of the continental ice sheets.[35]

34. MacMillan, "On the formation of circular muskeag" (cit. n. 33), p. 503. Conway MacMillan, "Notes for teachers on the geographical distribution of plants," *Journal of School Geography* 1 (1897): 97–102, on p. 101.

35. Henry C. Cowles, "Ecological relations of vegetation on the sand dunes of Lake Michigan," *Botanical Gazette* 27 (1899): 36–91, 95–117, 167–202, 281–308, 361–91. Rollin D. Salisbury and William C. Alden, "The Geography of Chicago and its Environs," *Bulletin of the Geographi-*

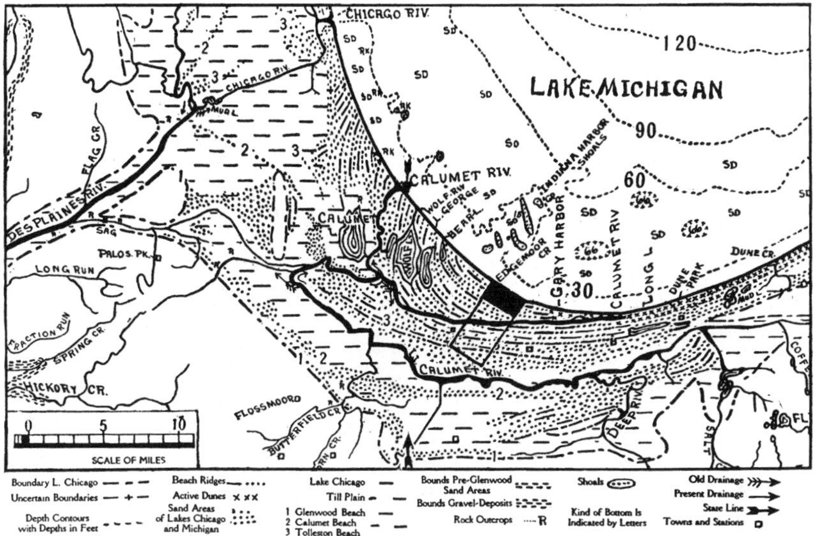

7.5 Map showing the glacial physiography of the south end of Lake Michigan. Deep dune belts are on the far right, and the black rectangle demarcates the fossil beaches and ponds where Victor Shelford worked. Various dashed lines numbered 1, 2, and 3 indicate the ancient shore lines of glacial Lake Chicago. Victor E. Shelford, "Ecological succession, II: Pond fishes," *Biological Bulletin* 21 (1911): 127–51, on p. 129.

Henry Cowles was not the first to be attracted to the Indiana dunes. Their unusually diverse flora (a mingling of southern and northern forms) made it a popular place for local botanizers. John Merle Coulter first visited the dunes in 1879 and subsequently used it for student field excursions. It was on one of these trips, in 1896, that Cowles first encountered the place that would bring it and him ecological celebrity. In the early 1900s the Indiana dunes were an island of wild nature in a rapidly developing region, of no use to farmers or lumberers, or (yet) to industrial developers, and just an hour or so away from the University of Chicago by interurban trolley. Frequented by picnickers, nature worshippers, and collectors, it became for ecologists a natural laboratory for studying nature's experiments in succession.[36]

---

cal Society of Chicago 1 (1899), pp. 60, 62. George B. Cressey, "The Indiana sand dunes and shore lines of the Lake Michigan basin," *Bulletin of the Geographical Society of Chicago* 8 (1928), on p. 72. Jerry S. Olson, "Rates of succession and soil changes on southern Lake Michigan sand dunes," *Botanical Gazette* 119 (1958): 125–70, on pp. 126–31.

36. J. Ronald Engel, *Sacred Sands: The Struggle for Community in the Indiana Dunes* (Middletown: Wesleyan University Press, 1983), on pp. 141–42.

7.6 A basswood and dogwood forest (left background) being slowly swallowed by moving dunes, Dune Park, Indiana. Henry C. Cowles, "The ecological relations of the vegetation on the sand dunes of Lake Michigan," *Botanical Gazette* 27 (1899): 281–308, on p. 283.

What made these dunes a special place for ecologists was their complete record of physiographic and ecological succession. In the active lakeshore dunes Cowles saw the first stages of succession, as sand-holding plants caused blowing sand to accumulate in ridges that moved slowly downwind; further inland he saw dunes stabilized by a succession of distinct plant communities as humus built up in sandy soils. The older the dune zone, the further advanced was the physiographic and ecological process of transforming sand to soil, beach grass to forest. Walking from the lakeshore inland, Cowles moved forward in time, as it were, from the bare till of the retreating glacial front 12,000 years ago to the present forest six miles and that many years inland.

These belts of dunes were like a series of experiments in succession, each of known starting point and duration, and with the results laid out for ecologists to read. Nature accomplished by starting a process over and over what laboratory experimenters do when they start a series of identical experiments and stop them at different times: for example, a series of developing embryos. Restarting the successional clock enabled Cowles to read vertical time from horizontal space.[37] The stages of succession are arrayed in their historical order in the Indiana dunes.

37. Cowles, "Physiographic ecology of Chicago" (cit. n. 10), pp. 79–80.

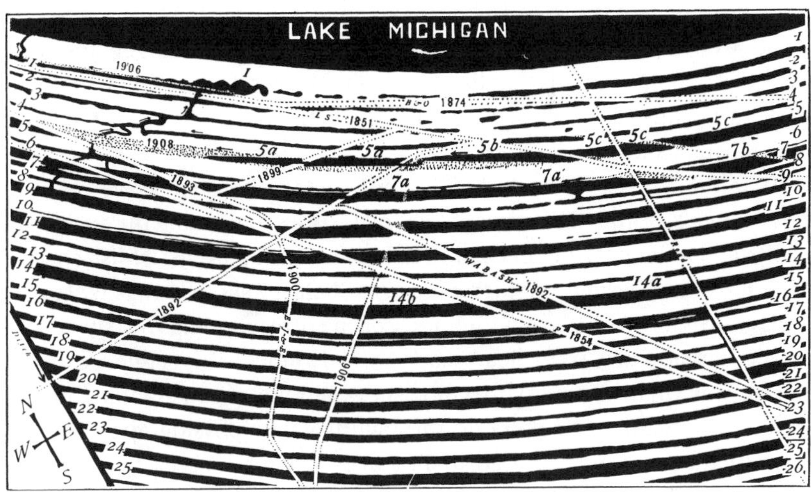

7.7 Fossil beach ridges (white) and ponds (black) on the south shore of Lake Michigan, used by Victor Shelford to study physiological succession. Victor E. Shelford, "Ecological succession, II: Pond fishes," *Biological Bulletin* 21 (1911): 127–51, on p. 131.

Cowles was not especially concerned to reconstruct a generic or a complete historical sequence for a particular place, that is clear from his publications. What he sought in the dunes were examples of plants' adaptations to environments that he could take into the laboratory to examine experimentally. "[N]owhere else could many of the living problems of ecology be solved more clearly . . . ," he wrote; "nowhere else could ecological principles be subjected to a more rigid test."[38] Knowing the history of a succession suggested its causes and the crucial points at which these might be studied experimentally—that was Cowles's vision. The physiographic method enabled ecologists to read the record of nature's experiments and to plan further experiments of their own in the lab.

The dunes were not the only place on Lake Michigan's shores where ecologists could read the record of nature's experiments. Victor Shelford found another such place, just a few miles west of the dunes, for a Cowlesian study of animal succession. Here some seventy-five to ninety low sand ridges alternated with long shallow ponds, each about one hundred feet wide and from a few inches to a few feet in depth. These were not stranded dunes but fossil beaches and offshore sandbars, which had been left high and dry by the last twenty-foot drop of Lake Chicago. These ponds, like dunes, made

38. Cowles, "Ecological relations" (cit. n. 35), pp. 96 (quote), 361–81. Cowles, "Physiographic ecology of Chicago" (cit. n. 10), pp. 173–77.

possible a trip into the past. The further a pond was from the lake, the longer it had been filling in with silt and organic debris, and the "younger" the communities of pond animals that lived in them. The whole series of ponds constituted a continuous record of a vast experiment in the evolution of a place and its fauna and flora.

With the help of his students, Shelford made a thorough survey of a series of ponds, noting their physical characteristics and listing the species of plants and animals living in each of them. Arranging his list of fish species in order of increasing physiographic age, Shelford showed that there was a sequence of distinctive communities, each corresponding to a successive physiographic stage, from clean sand to weedy muck. This sequence, he inferred, was what Methusalan ecologists would have seen standing in one spot and watching a single pond through its entire physiographic history. Because the ponds represented all the stages of animal succession, Shelford could peer into the deep past and reconstruct the changing pattern of life in a changing place.[39] It was a striking practice of place.

Shelford used a variant of this method and another special place to read the succession of fish communities in streams. To visit this place we take the interurban trolley from Chicago, not south this time but north, beyond the expanding suburban fringe to an area of lakeshore bluffs, in whose clay-banked ravines Shelford had once collected tiger beetles. Shelford knew that river fauna change along the length of a river from mouth to source; some species like slow water and muddy bottoms, others, fast water and stony rapids. It seemed evident that the fish communities characteristic of each habitat would migrate upstream as streams aged and eroded their way up their gradients, in the same way that bog plants move inward as their habitat fills in. If an ecologist could stand at one spot for a few thousand years, he would see a slow succession of creatures corresponding to the changing stream environment. Or, if that proved inconvenient, he could do the same thing virtually, by arranging in a single series a set of streams of different ages all connected to a common faunal pool—Cowles's physiographic method. The small streams cutting into the lakeshore bluffs formed just such a series.

Originating in the morainal upland some sixty feet above Lake Michigan, these streams differed in length from a few hundred yards to about two miles, and they had cut back into the bluff in stages as the level of Lake

39. Shelford, "Ecological succession, II" (cit. n. 1), pp. 128–35. Victor E. Shelford, "Ecological succession, III: A reconnaissance of its causes in ponds with particular reference to fish," *Biological Bulletin* 21 (1911): 1–38.

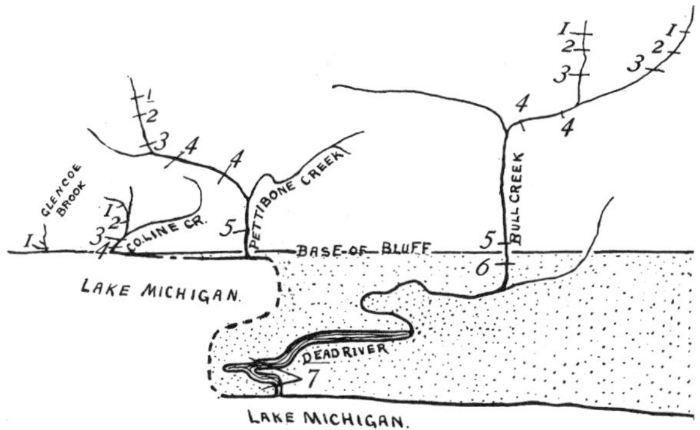

7.8 Lakeshore streams north of Chicago where Victor Shelford studied the physiographic succession of fish communities. Victor E. Shelford, "Ecological succession, I: Stream fishes and the method of physiographic analysis," *Biological Bulletin* 21 (1911): 9–34, on p. 17.

Chicago dropped and gave them a steeper gradient. (It was the sand eroded out of these ravines that piled up as dunes further south and east.) The longer the stream, the older it was, and together the series constituted the stages in the physiographic history of a generic stream. Likewise, the communities of fish inhabiting these streams represented stages in a biotic succession.

Shelford again began with a simple survey: list the species present from mouth to headwater and observe the physical environment in which each was found, then arrange the species in geographical sequence and look for clumps indicative of communities. In the longest stream Shelford found seven distinct communities of fish, and in successively smaller streams the same communities in the same order but truncated one by one, until in the shortest stream he found just the community that lived nearest the headwater of the longest stream.[40] Although the historical record of any one stream was destroyed, the lost past was observable in the ensemble. From place, read time.

Shelford's analysis derived its logical force from the fact that in streams of different ages the same fish communities turned up in the same order and in the same physical habitats. The consistent pattern showed that the sequence of fish communities was the result not of conditions peculiar to a particular stream but of a generic developmental process, the same in all. That is what gave this natural-history practice the force of experimental

40. Victor E. Shelford, "Ecological succession, I: Stream fishes and the method of physiographic analysis," *Biological Bulletin* 21 (1911): 9–34, on pp. 13–19, 26–33.

proof. Shelford even claimed to predict ecological futures: in the communities of the longer streams he could see those of the shorter creeks when in the distant future they had eroded further inland.[41] It was a provocative claim, since the power of prediction was by common consent the unique property of experiment and the reason for thinking it superior to field methods. In special places observation and comparison could have the power of experiment—there the idea of "Nature's experiments" was more than a rhetorical trope.

Panoramas

At the heart of Cowles's and Shelford's method was an act of visual conjuring, of imagining places moving fast-forward through time. Physiographers and ecologists "regard the flora of a pond or swamp or hillside not as a changeless landscape feature," as Cowles put it, "but rather as a panorama, never twice alike. The ecologist . . . endeavor[s] to discover the laws which govern the panoramic changes." "Panorama" is the key word—an imagined visual trip into deep time and back, from the moment when a place emerged from beneath the retreating continental ice sheet, through stages of revegetation, to its present state and beyond into a distant and different future. Places like the Indiana dunes encouraged this act of visual imagining because of the powerful impression of impermanence and change that the ever-changing dunes gave to human visitors. It was, in Cowles's words, a "restless landscape," never the same from one visit to the next.[42] Great rivers were another place that inspired Cowles to conjure up a panorama of physiographic and ecological change: "[B]y visiting the parts of a river from its source to its mouth," he wrote, "we can imagine what its history at a given point has been or is to be."[43] At its headwaters one sees what it was as a young stream hundreds of miles away and countless millennia in the deep past. This visual imagining was performed in the mind but was stimulated by the direct experience of particular places—a practice of place.

Scattered evidence suggests that this habit of virtual time travel was not an individual but a communal one. Charles C. Adams took an imaginary trip to the end of the ice age to observe a changing landscape in fast-forward.

41. Ibid., p. 29.

42. Cowles, "Ecological relations" (cit. n. 35), pp. 95 (quote), 194–96.

43. Henry C. Cowles, "The causes of vegetative cycles," *Botanical Gazette* 51 (1911): 161–83, on pp. 181–82.

"Let us imagine ourselves," he invited readers, "standing upon some vantage ground (say the forested region of southern Wisconsin), and watching the succession of forms slowly and gradually pass northward close to the base of the retreating ice." First a zone of bare ground fringing the ice, in the middle distance mosses and lichens, in the distance bogs, sedges, and willows. Then (next frame) in the foreground grasses and shrubs replacing mosses, in the distance conifers crowding in and breaking the grasses into islands. Then deciduous trees wedging into the conifer forests up sinuous stream valleys, and so on to the present. It is a vivid picture of a place in motion through time.[44] Conway MacMillan envisioned the panorama of vegetation moving back and forth across Minnesota with succeeding glacial advances and retreats. What botanists termed "northern" and "southern" species should be more properly called "south-bound" and "north-bound," he thought, so powerfully did the panoramic vision occupy his imagination. Frederic Clements had a similar but vertical vision of vegetation moving up and down the slope of Pike's Peak with the cycles of climate change, causing species to mingle and evolve in a profusion of "natural experiments."[45]

This visual practice was as important for the first ecologists as the ability to conceptualize deep time was for geologists and paleontologists.[46] It differentiated the new ecology, with its focus on development and cause, from both descriptive botany and historical geography. Ecologists see places not just as they are or once were, but as they come to be. It is not like a trip in a time machine, where time trippers squeeze into a black box and are whisked back to some time in the past. Ecologists experience in their imaginations the flow of time and change, reconstructing or reviewing the process of succession. This act of envisioning places moving through time helped make the theory of succession thinkable and plausible, as the ability to imagine deep time made the theories of Lyell and Darwin thinkable. A panoramic vision made it possible to imagine nature as experimenter and encouraged efforts to extract the principles of those experiments from imperfect records.

The panoramic vision of ecologists was one of a family of visual practices shared by all the nineteenth-century field sciences—geology, physiography,

44. Charles C. Adams, "Postglacial origin and migrations," *Journal of Geography* 1 (1902): 303–10, 352–57, on pp. 356–57.

45. Conway MacMillan, "Metaspermae of the Minnesota Valley," *Geological and Natural History Survey of Minnesota,* Botanical Series 1 (1892), p. 584. Frederic C. Clements, "Chapter 4. Alpine Laboratory and Transplantation Garden," n.d. [c. 1940?] p. 90, FEC box 84 f. Colorado.

46. Rudwick, "Encounters with Adam" (cit. n. 3), p. 239. Martin Rudwick, *Scenes from Deep Time: Early Pictorial Representations of the Prehistoric World* (Chicago: University of Chicago Press, 1992).

paleontology, biogeography, evolution. Henry Cowles was a geographer before he was a botanist, and we may speculate that he carried the visual practices of that science into ecology. The practice may also have been rooted in Victorian cultural practices. One thinks, for example, of the "panoramas," or huge scrolling paintings, that enraptured mid-Victorian audiences; or of a more modern form of visual entertainment—movies. Frederic Clements thought that the sequential stages of succession "give a picture of the life movement of a forest much as the individual films [i.e., frames] give motion in a moving picture." Movies also came to Henry Gleason's mind when he described how biogeographers constructed historical change from present scenes, each "one exposure in the long film of an activity which has been continued since the origin of the species." The experience of early railroad passengers may also have reinforced a panoramic habit.[47] Indeed, a good deal of ecology and biogeography was done from the window of a train or automobile.

Cowles and Shelford's physiographic practice was widely but not universally admired, and the few dissenters afford a useful insight into its peculiarities. Some biologists failed to grasp that its purpose was not to reconstruct actual histories of places but to use an ideal history to reveal the principles of succession. For example, some of Shelford's colleagues complained about the lack of actual fossil evidence for his reconstruction of pond succession.[48] It was a natural mistake for those whose only models were the historical field sciences. But physiographic ecology was different. At its heart was a model of experimental inference—a laboratory paradigm enacted in field practices of reading places. Cowles and Shelford read nature's experiments not to reconstruct their histories but to understand how they worked, in the same way that lab experimenters design setups and protocols not just to mimic natural phenomena but to reveal their principles.

A similar logic underlies Henry Gleason's dislike of the practice, which began when he was a student and grew stronger with time. He objected that Cowles's method did not enable ecologists to reconstruct the actual history of postglacial revegetation. The climates, landscapes, and biogeographies of the present and postglacial periods were just too different for anything to be inferred about how plants and animals had in fact reclaimed the glaciated

47. Frederic E. Clements, "Plant formations and forest types," *Proceedings of the Society of American Foresters* 4 (1909): 50–63, on p. 51. Henry A. Gleason, "The vegetational history of the Middle West," *Annals of the Association of American Geographers* 12 (1923): 39–85, on p. 41. Wolfgang Schivelbusch, *The Railway Journey: The Industrialization of Time and Space in the Nineteenth Century* (Berkeley: University of California Press, 1986).

48. Shelford, "Ecological succession, II" (cit. n. 1,), pp. 144–46.

regions. As a junior member of Cowles's and Adam's expedition to Isle Royale in 1904–5 Gleason had dutifully applied the Cowlesian paradigm, but reluctantly and with a public disclaimer. Provocatively, he likened it to the infamous "biogenetic law," which also projected present phenomena into the deep past and had produced nothing but spurious imaginings.[49]

Gleason saw more clearly than most the act of imagining at the heart of the physiographic method, but since he did not believe in communities and successions he had little use for an act designed to make these ideas plausible. But Gleason failed to see that physiographic ecology was not about reconstructing actual historical environments but about making ecology a causal, quasi-experimental science. It did not matter to Cowles how a place and its vegetation had actually developed. The important thing was to have some plausible generic picture of historical development to serve as a framework for classifying and studying vegetation. What mattered were the mechanisms of historical processes, the principles of nature's experiments, and those are what the panoramic vision revealed.

Evolution

Evolutionary biologists also had their practices of place. If the panorama of an evolving species was harder to conjure up than an ecological succession, biologists could hope that the present arrangement of the places in which speciation events had occurred might afford clues to how they happened. As with ecology the point was not to reconstruct historical faunas but to understand the principles of "Nature's experiments," and that was most easily accomplished in places that resembled laboratories or experimental setups. Evolutionists could read natural events as experiments in places that invited laboratory modes of inference.

No one was more persistent in searching out such places than Francis Sumner. He was aware from the start that special places might serve his purpose. Studies of variation in island races might reveal whether isolation played a role in the origin of new subspecies, for example; or animals living on unusually dark or light ground might show the effects of natural selection. Or the structure of mixed populations inhabiting the zones of in-

49. Henry A. Gleason, "The ecological relations of the invertebrate fauna of Isle Royale, Michigan," *Report of the Board of the [Michigan] Geological Survey* 1908, 57–78, on pp. 57–58. Gleason, "Further views on the succession concept," *Ecology* 8 (1927): 299–326, on pp. 309–10.

tergradation between subspecies' ranges might afford insights into the role of geographical barriers and hybrids in evolution.[50]

The question of protective coloration was an old Darwinian chestnut and an obvious place for Sumner to begin. The literature contained many reports by mammalogists and collectors of abnormally dark races of animals found living in places with dark lava soils. C. Hart Merriam, Wilfred Osgood, Vernon Bailey, and others had made such claims, but on the basis of a very few specimens and without careful study of their environments. The idea persisted as "a sort of perennial rumor," as Joseph Grinnell put it. The problem was to find a place that was so arranged that a proper test could be done. Sumner had investigated one lava area in 1917, but too few mice lived there. The place had to be just right: an area of black lava contiguous to an area of light-colored desert (the experimental control), both abundantly mousy. Grinnell assured him such a place could be found in an accessible part of the Mojave. So when Grinnell invited him to join an expedition to Death Valley and the Mojave Desert, Sumner leapt at the chance.[51] Sumner expected a null result but would keep an open mind: "I don't care much how the thing turns out," he wrote Grinnell, "so long as it does turn out—one way or the other. A small and dubious difference, making the whole result uncertain, is the thing I am dreading."[52] A lablike place could turn coincidence into cause and effect.

What Sumner found in the Mojave lava area pleased him. Its mice were normally colored, biometrically indistinguishable from mice of the nearby sandy desert; no sign of natural selection at work. Sumner was also pleased with the place as a quasi-experimental setup. The lava area was smallish (about three by five miles), isolated, and geologically new, mostly bare jagged rock. Trapping in the desert around the lava area revealed that lava mice did not venture far from home and so were effectively isolated—as in a laboratory experiment. If natural selection could produce local races it would do so in just such a biotic island. The arrangement of the place seemed to Sumner a guarantee that the null result was significant.[53]

50. Francis B. Sumner, "Report to staff," 19 April 1917, FBS-SIO box 5 f. 532. Sumner to Joseph Grinnell, 7 April 1916; Grinnell to Sumner, 10 April 1916; both in JG-MVZ. Francis B. Sumner, "Continuous and discontinuous variations and their inheritance in *Peromyscus*," *American Naturalist* 52 (1918): 177–208, on pp. 180–90.

51. Joseph Grinnell to Annie Alexander, 17 April 1920, AA box 2. Francis B. Sumner to Grinnell, 7 April 1916, 25 Feb. 1920, n.d. [received 11 March 1920]; Grinnell to Sumner, 10 April 1916; all in JG-MVZ.

52. Francis B. Sumner to Joseph Grinnell, 27 May 1920, JG-MVZ.

53. Francis B. Sumner, "Desert and lava-dwelling mice, and the problem of protective coloration in mammals," *Journal of Mammalogy* 2 (1921): 75–86, on pp. 79–83.

However, other biologists were not persuaded by Sumner's reading of this natural experiment. It was a negative experiment, for one thing, since the lava mice were not a dark variant. Sumner's critics also faulted his choice of place: it was too small, and too young (just five hundred to a thousand years) for natural selection to have produced a locally adapted race. If Sumner wanted a clear-cut test case, why not go to the place where Merriam had reported local dark varieties—the San Francisco Mountains of Arizona? So Sumner determined to redo Merriam's fieldwork with statistically proper samples and precision photometric measurement.[54]

Sumner's main collecting area at San Francisco Peak was a temperate plateau of lava rock at 6,500 feet on its north flank called Dead Man's Flat. Control sites were established 2,000 feet lower in the Painted Desert, hot, dry, and sandy. Sumner quickly confirmed Merriam's report of a local mouse distinctly darker than the desert form. Even better, Sumner discovered an area of Kaibab Limestone thirty miles along the plateau at the same elevation as Dead Man's Flat and with the same climate, vegetation, and mice. The two sites differed only in soil color—the crucial variable to be tested. The setup was as good as an experiment, "a 'natural experiment,' if you please," Sumner wrote Grinnell. Sumner was further pleased when the mice of the limestone site proved to be about as dark as the lava mice, proving as surely as he could wish that the dark lava race had not been created by selection. Some other agency was the cause, probably humidity, Sumner thought.[55] It was as if nature had run three experiments at San Francisco Peak, two differing just in the variable to be tested, and the third a control. It was the logical equivalent of a laboratory experiment, with the selection of a place achieving in nature what experimental setups do in labs.

Places of a different sort were required to test the role of isolation in the origin of species, and for those, too, Sumner kept a weather eye. Islands were the obvious places. For a time he had hopes for the Coronado Islands, a group of islets just offshore from San Diego where an unusual subspecies of *Peromyscus* was known to reside, but when and how it got there was not evident from the islands' topography or geological history.[56] A more prom-

54. Francis B. Sumner, "The supposed effects of the color tone of the background upon the coat color of mammals," *Journal of Mammalogy* 5 (1924): 81–113, on pp. 81–84. Sumner, "A proposed collecting trip in Northern Arizona," to Joseph Grinnell 13 Dec. 1921; Sumner to Grinnell, 13 Mar., 12 Nov. 1922; all in JG-MVZ.

55. Sumner, "Supposed effects of the color tone" (cit. n. 54), pp. 83–89, 95–97, 111–13, map on p. 113. Sumner to Joseph Grinnell, 23 Oct. 1922; Sumner to Harry Swarth, 24 Dec. 1922, 12, 31 Jan., 31 March 1923; all in JG-MVZ.

56. Francis B. Sumner, "Outline of proposed continuation of experimental and distributional studies of rodents," 12 Nov. 1923, pp. 3–4, FBS-SIO box 5 f. 534.

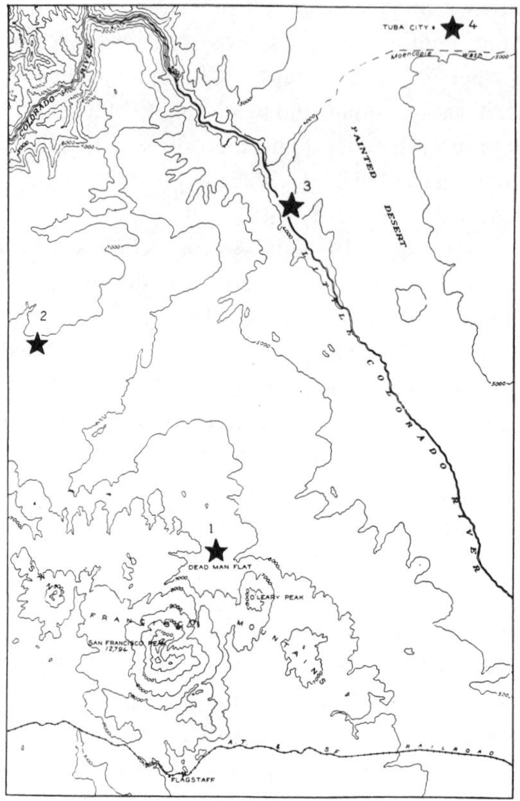

7.9 The north slope of the San Francisco Mountains, Arizona, showing the topography and test sites of Francis Sumner's work on the evolution of dark "lava" mice. Francis B. Sumner and Harry S. Swarth, "The supposed effects of the color-tone of the background upon the coat-color of mammals," *Journal of Mammalogy* 5 (1924): 81–113, on p. 113.

ising place was the narrow peninsula that enclosed Humboldt Bay, near Eureka, California. Wilfred Osgood had reported that an unusually pale local race of the darker coastal species of *Peromyscus* lived in this isolated strip of dunes—just six specimens, but enough to suggest an incipient subspecies in the making.

It was the particular topography of Humboldt Bay that attracted Sumner. A pale mouse living on open, light-colored sand suggested environmental cause and effect. So did its isolation, a marshy area preventing the mice of the outer peninsula from interbreeding with those of the coastal dunes. And just to the north at Mad River, the coastal dune mice met a darker subspecies of the coastal redwood forests in a boundary with no ob-

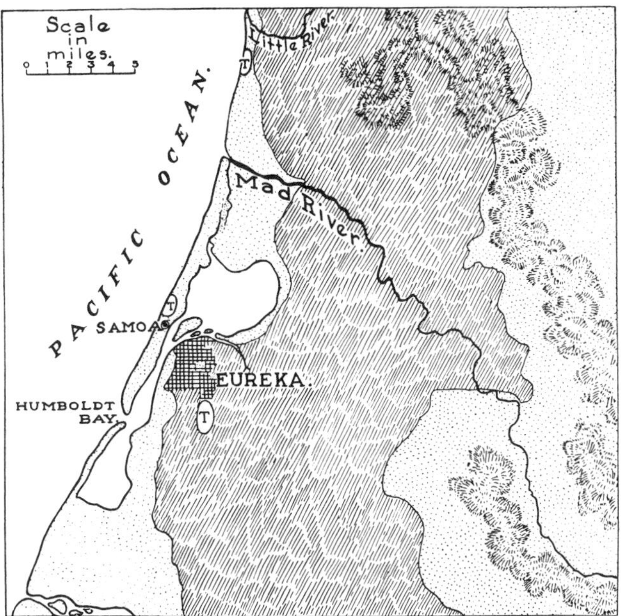

7.10 Humboldt Bay, California, showing the topographical arrangement of "nature's experiment" in the evolution of light-colored dune mice. The letter *T* indicates Sumner's test sites; shading indicates redwood forest. Francis B. Sumner, "The role of isolation in the formation of a narrowly localized race of deer-mice (*Peromyscus*)," *American Naturalist* 51 (1917): 173–85, on p. 178.

vious physical barrier. It was virtually an experimental set up: the isolated sandy peninsula and its very pale mice constituted the test case, the normally light mice of the northern dunes the critical comparison, and the dark forest mice the control. Isolation was the variable that changed, others were constant. Sumner's reading of this natural experiment seemed unambiguous: the northern dune mice had the pelage normal for the coastal subspecies, measured photometrically, so it appeared that the light peninsular race was indeed the result of isolation. Unfortunately, the biometrics of characters other than color varied in other directions, so the evidence for isolation as a cause of evolution remained inconclusive.[57] Eureka (unfittingly) was not the place where Sumner found nature experimenting.

But he kept his eyes open, and in 1920 a place came to his attention that seemed right in every way. It was another abnormally light-colored beach

57. Francis B. Sumner, "The role of isolation in the formation of a narrowly localized race of deer-mice (*Peromyscus*)," *American Naturalist* 51 (1917) 173–85, on pp. 181–82, map on p. 178.

mouse (*Peromyscus polionotus leucocephalus*), almost albino, and found only on the island of Santa Rosa, a small barrier island of open, brilliantly white sand off the coast of the Florida panhandle. The extreme modification of the island mouse and the matching extremity of its environment appeared to be a clear case of environmental causation, though whether the cause was selection or the effects of a very humid climate was unclear. Sumner's interest was further aroused by the chance appearance of a very pale mutant in one of his laboratory stocks that differed only in two genetic alleles from the wild type. If *leucocephalus* also differed from its coastal parent stock by just a few mutations—a hypothesis easily tested—the mechanics of speciation might finally be revealed. It meant another expedition, well beyond Sumner's usual range, but it seemed a unique chance to catch nature in the act of making a new species.[58]

The topography of Santa Rosa Island also seemed an ideal setup, even better than Humboldt Bay, and its geological history was known. Just a mile or so from the mainland, the island was completely isolated and always had been, having appeared as a sand reef out of the gulf when sea level dropped a few thousand years earlier. No need to speculate about what had happened: *leucocephalus* must have arisen when a few mice of the coastal subspecies (*P. p. albifrons*) were washed across the lagoon and marooned in an alien environment. Perfect isolation and rapid genetic change were historical facts, not guesses. It was if the place had been arranged as a kind of natural experiment. In 1924 Sumner traveled to the Gulf Coast to read the record.[59]

Principally Sumner worked Santa Rosa Island and the mainland across Choctawhatchee Bay, preparing large series of specimens for biometric and photometric analysis and trapping live mice for experimental breeding. He also took advantage of the topographical variety of the Florida coast, collecting on a semi-isolated sandy peninsula just to the west of Santa Rosa, on a stretch of mainland fronting directly on the Gulf just to the east, and on nearby Ono Island and St. George Island, 150 miles to the east. A farm in inland Alabama provided a series of *P. p. polionotus,* the dark ancestral stock of the somewhat lighter coastal *albifrons* of the sandy coast and the very pale *leucocephalus* of Santa Rosa.

What Sumner was after, evidently, was a set of places that offered a varied mix and match of variables—soil color, vegetation, degrees of island and

58. Sumner, "Outline of proposed continuation" (cit. n. 56). Sumner to Joseph Grinnell, 8 April 1924, JG-MVZ.

59. Francis B. Sumner, "An analysis of geographic variation in mice of the *Peromyscus polionotus* group from Florida and Alabama," *Journal of Mammalogy* 7 (1926): 149–84, on pp. 150–54.

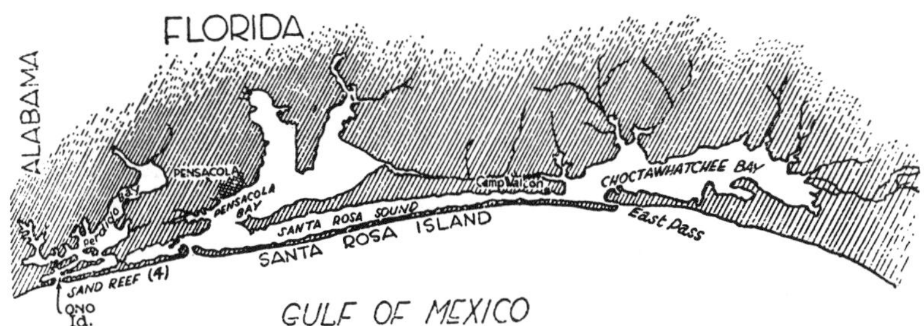

7.11 Santa Rosa Island and the coast of the Florida panhandle, where Francis Sumner studied variation and evolution in *Peromyscus*. Francis B. Sumner, "An analysis of geographic variation in mice of the *Peromyscys polionotus* group from Florida and Alabama," *Journal of Mammalogy* 7 (1926): 149–84, on p. 155.

peninsular isolation—as in a set of experiments. From this varied evidence of place a case for isolation might be unambiguously made, and perhaps for natural selection as well, though Sumner still doubted that.[60] Sumner's experimental habits are plainly evident in his Florida project, but so is his naturalist's eye for place. Selecting places in which variables were combined in different ways did in the field what experiments did in labs.

Lab-trained biologists like Sumner were not the only ones who pursued these practices of place. Field naturalists like Joseph Grinnell also took an interest in nature's experiments. Grinnell visited areas newly opened to irrigation, to see what happened when species once separated were forced into potentially breeding contact, and traveled to the scenes of local rodent irruptions to study the causes of these natural experiments, like the infamous mouse plague at Bakersfield in 1927.[61] Grinnell was delighted when a colony of English sparrows set up housekeeping at Furnace Creek Ranch, an isolated oasis in the middle of Death Valley. The isolation and extreme environment made a perfect setup, as good as a laboratory experiment, he thought. The more extreme the conditions, the faster changes should occur, and isolation would prevent local variants from being swamped. "How in-

60. Ibid., pp. 154–59. Francis B. Sumner to T. Wayland Vaughan, 7 May 1925, FBS-Fam f. 7. Sumner, "Memorandum to Dr. Vaughan," 5 Oct. 1925, FBS-SIO box 5 f. 535. Sumner to Joseph Grinnell, 23 Oct. 1927, JG-MVZ. Sumner, "Report of work accomplished . . . ," n.d. [20 Aug. 1928], CIW. Sumner, "The analysis of a concrete case of intergradation between two subspecies," *Proceedings of the National Academy of Sciences* 15 (1929): 110–20, 481–93.

61. Joseph Grinnell, "Geography and evolution in the pocket gophers of California," *Smithsonian Institution Annual Report* (1926): 343–55, on pp. 348–50. Francis B. Sumner to William E. Ritter, 7 April 1914, FBS-SIO box 5 f. 31. Grinnell to Annie Alexander, 2 Nov. 1907, 20 Jan. 1927, AA boxes 1 and 2. Grinnell to William E. Ritter, 29 Feb. 1908, WER box 10.

tensely interesting it will be," Grinnell mused, "to watch the course of this 'experiment' now under way . . . in Death Valley, with 'controls' vigorously maintaining themselves (against man's wish!) in San Diego, Berkeley and Boston." Critics, he acknowledged, might complain that it was not a true controlled experiment because environmental factors were not varied separately and because the animals were not under wire. But were animals confined in breeding cages not also subject to unknown factors?[62] Nature's experiments were in no way inferior to those of labs, in Grinnell's view.

Conclusion

The practices of place described in this chapter were the first to realize in action the ideal of a scientific natural history. They were not imports of material culture and protocol from labs. They used instruments and quantitative methods but were not dominated by them, and they did not attempt actual experiments in unsuitable places. These practices of place drew upon the methods and accumulated experience of the older field disciplines like biogeography and physiography, but in lablike ways. Practitioners of place borrowed the general forms of experimental method—isolating setups, amplifying conditions, separating variables—but achieved these ends by selecting or manipulating natural places. Practices of place operated not so much across a border between laboratory and field as within a border region in which the distinction was not a polarizing one. Cultural conflict was not entirely eliminated, but it was considerably less acute for practitioners of place than for those who tried to import laboratory culture wholesale to the field.

But it is also true that the results of reading nature's experiments were mixed. Studies of places in process—dunes, glacier retreats, floods, and eruptions—revealed little of the principles of succession. The physiographic approach, for all its appeal to early ecologists, was in the longer term not a very practicable program for empirical fieldwork, and its theoretical assumptions of community and development came under increasing attack in the 1930s. Likewise, experimental approaches to evolution mostly came to nought. Francis Sumner was persuaded that isolation and natural selection could produce new species but was no closer to proving how evolution worked. Practices of place never equaled laboratory experiments as engines of scientific production. Perhaps they were too dependent on a limited

62. Joseph Grinnell, "The English sparrow has arrived in Death Valley: An experiment in nature," *American Naturalist* 53 (1919): 468–72.

number of special places. And though we may read nature as experimenting, in fact the natural world of aggregates, variability, and chance is nothing like a lab.

Border biologists like Clements, Cowles, Shelford, and Sumner came to the field with unrealistic expectations of an orderly world. Mostly lab-trained, they expected to see determinate variation, an orderly tendency of change built into the fabric of individual organisms; they expected succession to be as rigidly programmed and predictable as the development of individual animals or plants. Doubtless these expectations were a legacy of nineteenth-century scientific culture, in which the study of individual organisms was the central activity. The early inhabitants of the border zone projected upon nature a vision, and practices, that were distinctly lablike. Understandably: populations that vary in an orderly way and aggregates that change predictably are objects that are likely to reward scientific activity. Importers of instruments, performers of out-of-doors experiments, and practitioners of place all in their different ways saw nature through the cultural aperture of the lab, because of their own experience and because they expected (rightly) to be judged by laboratory standards.

Perhaps because of their exaggerated expectations of order, first-generation border inhabitants often took from their experience of nature a profound sense of its restless variability, not disorder exactly, but an order beyond their comprehension.

The ever-shifting, unpredictable variations in his beetle colonies reminded William Tower of Darwin's image of nature as a tangled bank: "This production in the population of . . . [variations] first in one direction, then in another, withdrawing from the first but dropping behind . . . show precisely the condition that Darwin saw. . . ." The amoeboid figures that he devised to represent these shifting patterns give a visual sense of nature in flux.[63]

William Cooper was a loyal disciple of Cowles's physiographic vision, but two decades' of field experience left him doubting all theoretical systems. None, he thought, captured "the all-important fact of CHANGE." Cooper used a scene from nature—the unstable landscape of a prairie river—to express his feelings of a changing world:

> [T]he vegetation of the earth presents itself as a flowing stream, undergoing constant change. It is not a simple stream but a "braided" one, of enormous complexity, with its origin in the far distant past. Its more

63. William Tower, *The Mechanism of Evolution in* Leptinotarsa, Carnegie Institution Publication no. 263 (1918), pp. 304–307, 237–46, 270–340, 324–26, 335–37, quotes on pp. 304 and 306.

> or less separate and definite elements branch, interweave, anastomose, disappear, reappear. We ourselves watch its advancing front, just as one may watch the advancing front of a mountain torrent, born of a cloudburst, as it travels down its canyon bed. Vegetation as we see it today is thus a mere cross section of this complex stream.[64]

There was no progressive tendency in succession, he thought, no development and no stable climax—only change. It was time, Cooper thought, to "unsystem" ecology and deal simply with the particulars of change.[65] Cooper's image of a wandering prairie stream vividly projects the experience of a generation of biologists who took laboratory methods and ideals into the field, only to learn that nature is no lab.

Edgar Anderson also chose a meandering midland river to express his sense of the unpredictable flux of nature—not an imagined river like Cooper's but an actual one, as he wrote Francis Sumner:

> For my own education I have studied variation and natural selection in all the plants of a small gravel bar (where I went swimming once a day during the summer) on a river in the Ozarks for several years. I was . . . particularly impressed with the constant variability in the direction (or directions) of selection. The environment on a gravel bar, like Mr. Roosevelt, is forever jumping on horseback and riding off violently in two or three directions at once.

Anderson, a lab-trained geneticist and a convert to fieldwork, reported "very great difficulty in communicating to biologists, who have not had such an experience, anything like an adequate sense of the complexities of the problem."[66]

Henry Cowles was similarly impressed by the restlessness and variability of his beloved dunes:

> The dune-complex is a restless maze. . . . While there is a general advance of the complex as a whole in the direction of the prevailing winds, individual portions are advancing in all directions in which winds ever blow. . . . All stages of their life-history may be seen; the

64. William S. Cooper, "The fundamentals of vegetational change," *Ecology* 7 (1926): 391–41, on p. 397.
65. Ibid., pp. 392, 396–99, 402–9.
66. Edgar Anderson to Francis B. Sumner, 6 Jan. 1943, FBS–FAM f. 17.

> beginning, the climax, the destruction. . . . From a distance the complex seems always the same, a barren scene of monotony, but the details are never twice alike. . . . [It] is not only a maze but a restless maze. . . . [It] is like a river with its side currents and eddies at many points, but with the main current in one direction.[67]

The restlessness of nature frustrated border biologists' efforts to know it in the way that experimentalists knew their standardized and controlled "material." But their discovery that change was the essential meaning of nature's work, though a little shocking to a generation trained mostly in labs, was an important achievement, because it was the starting point for the invention of more effective—and unlablike—ways of grasping the orderly disorder of nature's restless mazes.

Practitioners of place were on the right track in thinking that traditional methods of natural history could be redirected to more analytic, lablike ends. They were just a little too impressed by the power of laboratory culture. But their disappointments were the foundations on which others built. The interpreters of nature's experiments were the first practitioners of a distinctive border culture that combined elements of laboratory and field but was different from either one. Their achievements and their failures are symptoms of this distinctive culture in the making.

67. Cowles, "Ecological relations" (cit. n. 35), pp. 194–96.

CHAPTER 8

Border Practices

Before about 1930 biologists in the lab-field border zone could import practices of counting and measuring, or they could read nature's work as experiment. In either case, biologists remained dependent on the material culture or protocols of laboratory science. However, in the next two decades practices evolved that were less dependent on laboratory methods and models. These practices were traditional field methods, but intensified and amplified in a lab-like way. They possessed the analytical power of laboratory experiment, but they did not imitate experimental logic and were appropriate to field problems and conditions. And they produced important conceptual advances in evolution and ecology. Some were the foundation of whole disciplines of field biology, for example, ecosystem and population ecology, population genetics, and other varieties of evolutionary "synthesis." They supported a general abandonment of ideas of equilibrium and orderly development in favor of stochastic views of nature permanently in flux.

I will support my thesis with a few well-chosen examples: Ernst Mayr's work on geographical speciation in birds, Edgar Anderson's on hybrids and introgression in plant evolution, Raymond Lindeman's invention of ecosystem ecology, and Robert Whittaker's gradient or continuum ecology. All are representative in some way of a larger family of field practices. I could as well have chosen others less well known; for example, Carl Hubbs's work on hybrids and speciation in fish, or Paul Errington's studies of population dynamics and predator-prey relations, or the ethological work of Margaret Nice and others. But a few of the more familiar episodes seemed the best vehicle for getting across a somewhat unconventional view of biology in this period.

Although they may seem an odd lot, these border practices—if I may so call them—are in principle quite similar. First, they all employ quantitative methods. It is well known that biometrics and statistics reentered evolution and ecology in the 1930s. However, historians have focused primarily on theoreticians like Sewall Wright, R. A. Fisher, and J. B. S. Haldane, not on practical field biologists like Edgar Anderson or Carl Hubbs, who pioneered mass-collecting methods and applied biometrics to the study of natural variation. Simply counting, measuring, and tabulating large numbers of plants and animals could and often did yield important new insights. Anderson counted and measured variation in species of plants over their entire ranges. Ernst Mayr measured tens of thousands of specimens of Polynesian birds. Raymond Lindeman counted the small creatures living in a eutrophic bog, season after season for over five years. Painstaking counting and tabulating proved to be the form of quantitative culture most appropriate to field phenomena. These methods generally did not require high theory, were easy to understand (though arduous to perform), and paid off in ways that laboratory methods never had. As Edgar Anderson observed, biology advances when appropriate qualitative units are identified and used with quantitative precision.[1]

A second feature of border practices is that they take populations as their fundamental units of investigation, not individuals. Field biologists have always dealt with aggregates—plant communities, taxonomic series—but without conceptualizing them as populations. Early ecologists saw associations as quasi individuals or superorganisms; taxonomists identified species with individual type specimens. It was not until the 1930s and 1940s that field biologists invented practical ways of conceptualizing and studying aggregates as populations. Ernst Mayr has long insisted that populational thinking was the key to the "evolutionary synthesis."[2] I think he is right and

1. Edgar Anderson, "Hybridization in American *Tradescantias*," *Annals of the Missouri Botanical Garden* 23 (1936): 511–25, on pp. 512–15.

2. Ernst Mayr, "Prologue: Some thoughts on the history of the evolutionary synthesis," in *The Evolutionary Synthesis: Perspectives on the Unification of Biology*, ed. Mayr and William B. Provine (Cambridge: Harvard University Press, 1980), 1–48. Mayr, "The role of systematics in the evolutionary synthesis," ibid., 123–136, on pp. 127–128.

would only add that populational thinking was also fundamental to the changes in ecology and other field disciplines that occurred at the time. The "individualistic" ecology of Henry Gleason and Robert Whittaker was, despite its name, a populational approach, which self-consciously rejected superorganism models of communities and succession.[3] Theodosius Dobzhansky made diverse populations of wild flies the objects of his study, replacing standard flies. Evolutionary taxonomists like Mayr, Anderson, Jens Clausen, and Carl Hubbs took local populations as their basic units of analysis.

The turn to populational thinking signified a fundamental change in field biologists' relation to experimental culture. Experiments operate on individual organisms—even genetics or population modeling, which deal with standard animals, or clones. Traditional experiment was not designed to deal with heterogeneous populations; it deals with variability by avoiding it. So when field biologists took populations as their objects of study, experimental methods became less obviously the preferred ones and thus lost some of their unquestioned authority. It could no longer be assumed that progress meant replacing inferior practices like taxonomy with superior laboratory practices like genetics, "because," as Carl Hubbs put it, "systematic groups are populations and must be investigated by population analysis." Ernst Mayr made the same point about evolution in almost identical words: if populations are the units of evolution, "it becomes clear that only a population analysis can lead to valid conclusions."[4] In the field it was experiment that was out of place and suspect. Populational concepts and practices gave fieldwork the force and standing of laboratory work without the dependency that had always been the price of cultural borrowing.

Comprehensiveness is a third common feature of these border practices. They involved exhaustive, total knowledge of places and biota, combining the intensive methods of local studies with the extensive reach of regional surveys. The combination was a powerful one. Ernst Mayr could make a strong case for the role of isolation in the origin of species because he had an essentially complete knowledge of the bird species of the South Pacific archipelagoes. Lindeman could calculate the energy dynamics of trophic levels because his knowledge of the species of Cedar Creek Bog and their seasonal variations was total—each exhaustively identified, counted, weighed,

3. Malcolm Nicolson, "Henry Allan Gleason and the individualistic hypothesis: The structure of a botanist's career," *Botanical Review* 56 (1990): 91–161.

4. Carl L. Hubbs, "Racial and individual variation in animals, especially fishes," *American Naturalist* 68 (1934): 115–28, on pp. 115–16. Ernst Mayr to Willard H. Camp, 1 Nov. 1943, EM-HU box 2.

and tabulated. It was this comprehensive knowledge of a whole ecosystem that enabled Lindeman to turn Charles Elton's qualitative concept of trophic structure into a quantitative picture of the flow of energy through an ecosystem—its "metabolism." Anderson's theory of introgressive hybridization likewise derived from a total knowledge of variation in plant species on a continental scale. And Paul Errington won novel insights into population dynamics from a total knowledge of the lives of predators and prey in small locales over a dozen years or more.

Total knowledge of natural places and communities gives field methods something like the power of laboratory experiment. It enables field biologists to analyze variables without manipulating them experimentally, simply by observing and comparing. With border practices field-workers can be as confident of causes and effects as any laboratory worker can be with artificially simplified setups. For example, Ernst Mayr remarked that ornithologists' total knowledge of bird species enabled them to do quasi experiments: "The ornithologist knows his material so well that he can do what the geneticist also does," he wrote, "that is to pick out one particular character and study its fate under the influence of geographical variation, and in the phylogenetic series."[5] What geneticists did by manipulating chromosomes experimentally, naturalists did by tabulating field data. Paul Errington made a similar point when he observed that "equivalents of excellent experiments have been furnished by some of the more comprehensive field or statistical researches [on natural populations]."[6] His own researches exemplify his point.

Finally, place was an important element of border practices. Selecting the right place was a crucial step. Not every place was suitable for work that aimed to be comprehensive and quantitative. A right place was not too big to count or too wild to work easily, for example. But the spatial requirements of border practices were on the whole less stringent than those of practices that were more dependent on imported experimental tools or protocols. For example, places for border practice did not need to be small enough to control experimentally, like an experimental garden. They did not even need to be suitable for deploying instruments. Border practitioners simply observed, sampled, counted, measured; they did not in general rearrange and manipulate. Nor were they restricted to those rare and spe-

5. Ernst Mayr to Edgar Anderson, 28 Jan. 1941, EM-HU box 2.

6. Paul L. Errington, "Predation and vertebrate populations," *Quarterly Review of Biology* 21 (1946): 144–77, 221–45, on p. 145.

cial places where evidence of nature's experiments had luckily been preserved. Border practices tended to work just about anywhere, not just in special lablike places but in humdrum, unremarkable places where creatures went about the daily business of eating and reproducing. Amplified forms of traditional field methods are more flexible and adaptable than methods borrowed from labs—hence their popularity with field biologists.

Border practices somewhat redressed the epistemological asymmetry between lab work and fieldwork. Carl Hubbs, in an ebullient mood, asserted that the new systematics was as superior to the old as T. H. Morgan's genetics was to Francis Galton's.[7] Edgar Anderson wondered if the basic units of genetics (Mendelian alleles) and of taxonomists (breeding populations) were really all that different. Both, he observed, were qualitative units that overlooked much individual variation. (A truly quantitative genetics in which traits like bristles or eye color are treated biometrically would be a mess.) Was genetics a more quantitative science than taxonomy? Were geneticists more sure of the reality of genes than taxonomists were of the reality of populations? Anderson doubted it: "[T]o hear some of them [geneticists] talk," he quipped, "you would suppose they had been inside a chromosome and walked around ticketing the genes."[8] In a way, that is precisely what field-workers *could* do: know natural systems by walking around inside them and observing, counting, and ticketing significant items. That practice was as good as an experiment—maybe even better, because it did not simplify complex natural objects.

The new border practices of the 1930s and 1940s were traditional field methods amplified by numerical treatment systematically applied to natural units of population. They were not imports from genetics or physiology, not adaptations or makeshift approximations of laboratory methods, but were homegrown. Even those that most resembled laboratory practices were distinctively of the field. Border practices evolved in the field as intensifications of traditional methods of observing and comparing. Most were invented by field naturalists, and they perpetuated the traditional aims and methods of natural history by giving it some of the analytical force of laboratory precision and causal analysis. They were of the field but met laboratory standards of evidence and inference.

7. Hubbs, "Racial and individual variation" (cit. n. 4), on p. 116. In a spoken version of his paper Hubbs hedged, saying he may have exaggerated. Hubbs, "An experimental attack on the species problem," 21 Nov. 1934, CLH box 75 f. 101.

8. Anderson, "Hybridization in American *Tradescantias*" (cit. n. 1), pp. 512–15. Anderson to Ernst Mayr, 9 Jan 1941 (quote); Mayr to Anderson, 28 Jan. 1941; both in EM-HU box 2.

Geographical Speciation: Ernst Mayr

Ernst Mayr did not invent "allopatric speciation," that is, the idea that isolation is an essential condition for the origin of species. That idea goes back to Darwin's time and was espoused by leading German ornithologists, especially those in the circle of Erwin Stresemann, Mayr's mentor. Mayr absorbed his belief in isolation from his teachers; his contribution was to produce solid empirical evidence for the theory, using simple field methods. He was also the most insistent among evolutionary systematists that biogeography and taxonomy—field disciplines—were as important in the evolutionary "synthesis" as genetics. The title of his landmark book *Systematics and the Origin of Species* (1942) makes this point in its echo of the other founding text of the "synthesis," Theodosius Dobzhansky's *Genetics and the Origin of Species* (1937). It is a point that Mayr has been pressing upon historians, who have tended to give most of the credit to theoreticians and experimenters.[9]

Mayr was a border person with a border career. He was a student of medical science and an ardent amateur birder when Erwin Stresemann recognized his "fabulous taxonomic instinct" and persuaded him to become a professional zoologist and begin an apprenticeship in systematics at the Berlin Museum. As Mayr recalled, Stresemann set him the task of sorting out a genus of creepers so difficult to tell apart that even experienced taxonomists took days to do it and still made mistakes; Mayr got it perfect in half an hour. Like his mentor, Mayr believed that the future of ornithology lay in physiology and anatomy, and he might well have taken up physiology in the early 1930s but for a crucial exposure to fieldwork.[10]

In 1928 Stresemann arranged for Mayr to join an expedition dispatched by the Rothschild Museum at Tring to New Guinea and the Solomon Islands. The three years he spent in the South Seas (his only major fieldwork) were some of the happiest years of his life, he recalled, realizing romantic dreams

9. Jürgen Haffer, "Erwin Stresemann (1889–1972)—life and work of a pioneer of scientific ornithology: A survey," *Acta Historica Leopoldina* 34 (2000): 399–427. Ernst Mayr, *Systematics and the Origin of Species From the Point of View of a Zoologist* (New York: Columbia University Press, 1942). Mayr, "Introduction, 1999," in *Systematics and the Origin of Species,* 2nd ed. (Cambridge: Harvard University Press, 1999), xiii–xxxv. Theodosius Dobzhansky, *Genetics and the Origin of Species* (New York: Columbia University Press, 1937).

10. On Mayr's student years and Stresemann's influence see Jürgen Haffer, ed., *Ornithologen-Briefe des 20. Jahrhunderts, Ökologie der Vögel* 19 (1997): 1–980, on pp. 64–68, 370–73, 804–10; and Ernst Mayr, "How I became a Darwinian," in *Evolutionary Synthesis,* ed. Mayr and Provine (cit. n. 2), 413–23, on pp. 413–17. The anecdote of sorting birds is in Mayr to Edgar Anderson, 18 Jan. 1941, EM-HU box 2. Haffer mentions Stresemann and Mayr's belief in anatomy and physiology, *Ornithologen-Briefe,* p. 381.

of exploration and exotic places inspired by the books of explorer-naturalists like William Beebe. In 1931 his familiarity with South Seas birds got him a job at the American Museum of Natural History in New York, working up the vast collections of the Museum's Whitney South Seas expeditions (1920–39). It was thus in the museum's specimen rooms that ideas about isolation and species taken on faith from Stresemann and other German ornithologists took a more concrete form, as Mayr slowly brought order to the tens of thousands of varied and often difficult specimens of South Seas birds.[11]

The basic idea of allopatric speciation is simple: a segment of a geographically variable species becomes isolated by some geographical barrier or by migrating beyond the margins of the species' normal range. Through genetic reshuffling and selection, the isolated population gradually evolves anatomical or physiological traits that isolate it reproductively from its parent stock. Should the two populations again come together, they would no longer be able to interbreed. A gap has appeared: two species where there was one. (Mayr believed that speciation cannot occur "sympatrically," in a single population without isolation, but later evidence has shown this type of speciation also does occur.)[12]

Mayr's conception of speciation was a spatial one, and it rested primarily on evidence of spatial distribution. Circles of subspecies constitute one striking example of such evidence. Each subspecies in the circle interbreeds with its neighbors, except for the two extreme forms that close the circle; after long separation they will meet as different species, no longer able to interbreed. The others are transitional forms, caught in the act of divergence. The circumpolar herring gulls were the best such case, and Mayr learned of others in New Guinea. Offshore islands provided other striking evidence, when one group of immigrant mainland birds remained isolated long enough to develop reproductive isolation before the next wave arrived.[13] Such cases were like natural experiments: one could simply see from the spatial arrangements what had happened and how.

11. On Mayr's participation in the Whitney expedition see Haffer, *Ornithologen-Briefe* (cit. n. 10), on pp. 68–70, 810–11; Mayr to Jossylen Van Tyne (for William Beebe), 21 Sept. 1949, EM-HU box 6; also Mayr to Julian Huxley, 9 Dec. 1953, EM-JH-HU box 1. On Mayr's years at the American Museum see Haffer, *Ornithologen-Briefe*, pp. 71–78; and Mayr, "How I became a Darwinian" (cit. n. 10), pp. 417–19. I am grateful to Professor Mayr for permission to use his correspondence and for useful pointers and references; also to Professor Haffer for references.

12. Mayr, *Systematics and the Origin* (1999) (cit. n. 9), pp. xiv–xix, and chap. 7. Ernst Mayr, "Speciation phenomena in birds," *American Naturalist* 74 (1940): 249–78. Mayr to R. E. Moreau, 13 May 1947, EM-HU box 3.

13. Mayr, *Systematics and the Origin* (cit. n. 9), pp. 173–85. Mayr, "Speciation phenomena" (cit. n. 12), pp. 266–74. Mayr to Theodosius Dobzhansky, 13 May 1937, EM-APS.

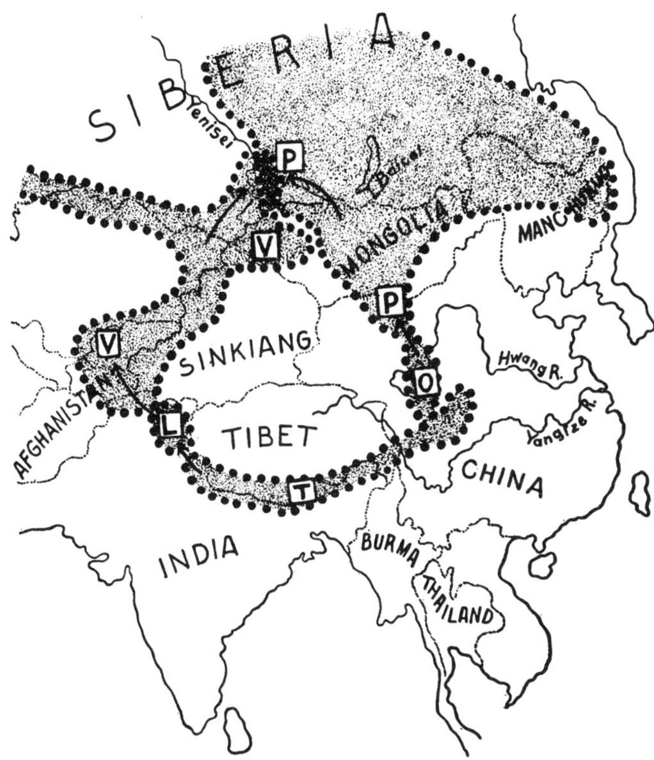

8.1 A ring of intergrading subspecies of *Phylloscopus trochiloides* (represented by the letters *V, L, T, O,* and *P*) that meet and live as separate species (in the hatched area, top center). Such patterns are evidence of the role of isolation in speciation. Ernst Mayr, *Systematics and the Origin of Species from the Viewpoint of a Zoologist* (New York: Columbia University Press, 1942), on p. 183.

The island archipelagoes of the South Pacific afforded such evidence in an abundance and variety that no other place in the world could equal. As Mayr put it much later, "the material of the Whitney South Sea expedition[s] showed with almost textbook clarity the pathway of geographic speciation." It was sheer luck that brought Mayr to the South Seas islands. Stresemann had initially tried but failed to place him on expeditions to Cameroon and Peru.[14] Had he succeeded, Mayr's belief in isolation might have had a quite different fate. Continental species occupy large continuous ranges, and divergent forms tend to merge back into parental species or go extinct, oblit-

14. Mayr, "Introduction, 1999" (cit. n. 9), quote on p. xvii. Haffer, *Ornithologen-Briefe* (cit. n. 10), p. 68.

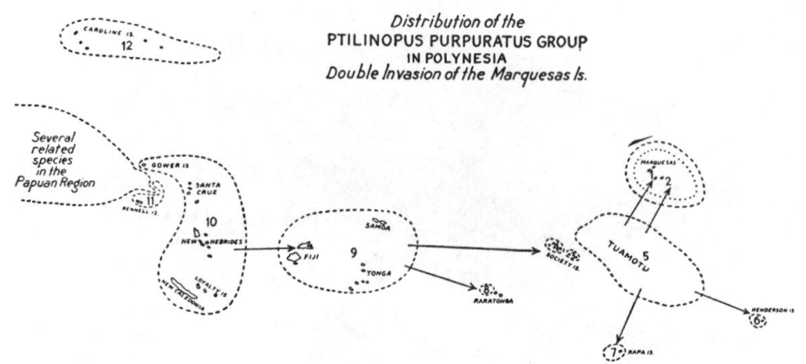

8.2 Distribution of the genus *Ptilinopus*, showing the double invasion of the Marquesas Islands, evidence of the causal role of isolation in the origin of species. (Numbers indicate various species.) Ernst Mayr, "Speciation phenomena in birds," *American Naturalist* 74 (1940): 249–78, on p. 271.

erating evidence of speciation. That was the reason, Mayr surmised, why zoologists who did not work on island fauna found it hard to believe that isolation was essential to speciation.[15] In the South Seas archipelagoes it was easy. The geography of speciation was also a cultural geography of belief.

Mayr's practice was thus a practice of place and could perhaps only have been perfected in the islands of the South Seas, where gaps and discontinuities are not occasional anomalies but the topographic rule. Scan a world map: no other region is so thoroughly made up of islands large and small. It is an allopatric landscape, so to speak, and it is not surprising that an allopatric taxonomic practice would take shape there. Its topography made it easy to reconstruct its biogeographical history. The South Sea islands were not continental margins broken up by a rising sea level; they were always islands and were populated afresh entirely by immigrants from the nearby land masses of New Guinea and Southeast Asia. Knowing migration and settlement history made it relatively easy to reconstruct the random migrations across oceanic gaps that created and destroyed evolving species, and to read the evidence of the role of isolation. The closer an island was to a mainland, the more likely it was to receive accidental migrants, for example, and the less likely that one wave would evolve isolating mechanisms before another wave arrived to swamp new variants by backcrossing. From the topography of successful colonizations—distance, size of target island, ecological matches or mismatches, random extinctions—Mayr could see the

15. Ernst Mayr to Richard Goldschmidt, 13 Dec. 1939, EM-HU box 2. Mayr, "Speciation phenomena" (cit. n. 12), p. 276. Mayr, *Systematics and the Origin* (cit. n. 9), pp. 161–62.

evolutionary meanings in the bewildering geographical variations of island birds.[16] That would have been far more difficult in continental areas, with their deep and complex histories of climate cycles and migrations.

Mayr's practice is exemplified in a set of tables in which he calculated the percentages of island species caught at different stages in their divergence from continental stocks. One set of data dealt with the birds of Biak Island, just off the eastern coast of New Guinea, and Rennell Island, a small atoll near the Solomon Islands. Mayr sorted the resident species of the islands into categories of increasing difference from their presumed parental species on the New Guinea mainland. Many were unchanged or slight variants, but the majority were endemic to the islands and had presumably evolved there in isolation. Mayr further sorted these endemic species into taxonomic categories indicative of increasing phylogenetic distance (subspecies, species, genus); the resulting tabulation showed that the process of speciation in this geologically young region had mainly resulted in divergence to the subspecies level. The closer to the mainland the less complete was the isolation and the less the divergence of island from mainland forms.[17] (See table 8.1.)

Table 8.1. Percentage of birds falling into categories of increasing difference from the presumed parental species.

	Biak (N = 68)	Rennell (N = 34)
Km from mainland	60	145
No differentiation	29%	21%
Slight differentiation	16	21
Endemic subspecies	43	44
Endemic species	13	12
Endemic genera	0	3

Source: Based on Ernst Mayr, "Speciation phenomena in birds," *American Naturalist* 74 (1940): 249–78, on p. 267.

16. Ernst Mayr, "The origin and the history of the bird fauna of Polynesia," *Proceedings of the 6th Pacific Science Congress* (Berkeley: University of California Press, 1940–41), vol. 4, 197–216, on pp. 212–16. Mayr, "Speciation phenomena" (cit. n. 12); Mayr, *Systematics and the Origin of Species* (cit. n. 9).

17. Mayr, "Speciation phenomena" (cit. n. 12), pp. 267; Mayr, *Systematics and the Origin* (cit. n. 9), p. 166.

He made a similar sorting of the resident birds of the Solomon Island group, but with a novel and important twist: he divided forms not into standard taxonomic categories but functional categories representing intermediate stages of species formation. Comparable figures for New Guinea, with partly continuous ranges, and Manchuria, a continental area of continuous ranges, dramatically highlighted the effects of island isolation.[18] (See table 8.2.) Such comprehensive, quantitative, and comparative evidence of species in the making had a weight that even the most striking of "Nature's experiments" could not. It was not proof by laboratory standards, but close to it; and it was achieved entirely by traditional biogeographic and taxonomic practices amplified by counting and tabulating.

Mayr's tabulations were the visible tip of a veritable iceberg of fieldwork, the essential distillation of over two decades of arduous collecting, sorting, and analyzing of tens of thousands of specimens. Systematic zoologists, and especially ornithologists, had been collecting large series of specimens since the late nineteenth century, and the American Museum's Whitney South Seas expeditions were the apotheosis of modern mass collecting. For almost twenty years the boldest and most experienced professional collectors were dispatched with instructions to get specimens of every last species from

Table 8.2. Percentage of birds representing intermediate stages of species formation.

	Manchuria	New Guinea	Solomons
Continuous, uniform	14%	7%	2%
Continuous, slight subspecies	55	41	22
Subtotal: no or slight change	**69**	**48**	**24**
Subtotal: continuous, definite subspecies	**28**	**29**	**24**
Variable, isolated subspecies as different as good species	1	11	34
New species restricted ranges	2	12	18
Subtotal: species or near species	**3**	**23**	**52**

Source: Based on Ernst Mayr, *Systematics and the Origin of Species From the Point of View of a Zoologist* (New York: Columbia University Press, 1942), on pp. 160–61.

18. Mayr, "Speciation phenomena" (cit. n. 12), pp. 276; Mayr, *Systematics and the Origin* (cit. n. 9), p. 161.

every island and islet in the South Seas. The result was a resource for quantitative analysis of variation that was probably unique.[19] As the curator in charge of the Whitney bird collections, Mayr spent the 1930s working up the specimens flooding in from the field, measuring, sorting, mapping, and seeking some pattern in the extraordinary variety of island types.

Mayr's official job was to assign each form to a species or subspecies category and catalog it; but his real ambition was to use this unique collection to prove the theory of allopatric speciation. To realize this ambition he was obliged to change the rules of taxonomic practice. The traditional ideal of taxonomy is to sort every specimen unambiguously into one category or another with a minimum left over (imagine growing piles spread out on large tables). There were always some left over, of course, oddities that did not fit anywhere and had to be explained away as hybrids or sports, or just ignored, like experiments that do not work. But among the Whitney birds there were an unusual number of such embarrassments.

Distinguishing species and subspecies in island forms is always a problem, because the usual test of "good" species—reproductive isolation—does not apply. One simply could not know whether similar forms from separated islands would interbreed if they lived together. Nor does the test for good subspecies (interbreeding in a zone of contact) apply to island forms, since islands have no such zones, just clean gaps. These problems were exacerbated by the extraordinary geographical variation of South Sea forms. Each island seemed to have its particular variants—a taxonomist's nightmare.

However, a taxonomist's nightmare is an evolutionary biologist's happy dream. If we sort not to pigeonhole but to understand the process of speciation, then unsortable specimens can be reconceptualized as transitional forms or "semispecies" caught in the act of differentiating; not symptoms of faulty taxonomic categories but vital clues to evolutionary process. Forms intermediate between local variety and subspecies were early stages; those between subspecies and species, later stages. And with birds from an archipelago of numerous islands, comparable stages in different species could be lumped together, quantified, and correlated with the presence of geographical isolating barriers. That was how Mayr constructed his crucial tables. For evolutionary taxonomists like Mayr, the more unsortable transitional forms the better. And island archipelagoes, with their abundance of forms unclas-

19. Mayr, "Speciation phenomena in birds" (cit. n. 12), pp. 249–50. Mayr, *Systematics and the Origin* (cit. n. 9), chap. 1. On the Whitney expeditions see *Annual Report of the American Museum of Natural History,* 1920 to 1939.

sifiable by traditional methods, were special opportunities.[20] Populational conceptions of species and mass-collecting practices in the field went hand in hand, the one dependent on the other, coevolving.

The point is that Mayr's practice was thoroughly of the field. Little of what he did day to day would have been unfamiliar to a traditional taxonomist, except for the counting and tabulating and his concern with minute differences of form that would normally have disappeared into standard taxonomic categories. As Mayr has tirelessly asserted, his case for allopatric speciation involved no more than the skills of traditional field-workers and taxonomists, quantified and comprehensively applied. Thus amplified, fieldwork became a new kind of evolutionary practice.

Amplification gave Mayr's practice a lablike feel. He may have experienced his work as a kind of experiment. "Since speciation is far too slow a process to be observed directly," he recalled, "I borrowed a technique from the cytologists of the 1870s and 1880s. What they did was to place in sequence scores, if not hundreds, of microscope slides, each representing one particular moment in the process of mitosis or meiosis. Thus, they converted a series of still pictures into a movie. And that is precisely what I did."[21] Quite so. The comprehensive empirical data of the Whitney collections were like the data produced by repeated and varied experiments in a lab. Their sheer abundance and variety made them resemble a vast biogeographical experiment repeated over and over in varied permutations and combinations. For every variable—size of breeding population, ecological factors, frequency of immigration—there were likely to be islands in which one variable was different and others constant. In island archipelagoes every degree of incomplete speciation or "semispecies" could be found and analyzed. Though entirely of the field, such evidence is as good as any experiment could produce, and entirely independent of laboratory tools and protocols—a border practice.

Mayr had deep respect for the methods and values of laboratory science. As Streseman's disciple he felt the appeal of experimental physiology and anatomy. In the mid-1930s he made repeated efforts to do experiments on bird behavior but was too preoccupied by heavy museum duties to take them very far. In the late 1940s he carried out experiments with Dobzhansky on mate selection in *Drosophila,* publishing several papers, and flirted with other kinds of experiments on reproductive isolation. He was a scientific modernist, no heels-dug-in defender of old ways. But he also recog-

20. Mayr, *Systematics and the Origin* (cit. n. 9), chaps. 5–7, esp. pp. 151–53, 164–73. Mayr, "Speciation phenomena" (cit. n. 12), pp. 259–61, 273–76.

21. Mayr, "Introduction, 1999" (cit. n. 9), quote on p. xxiii.

nized the limits of laboratory methods and values for work on speciation. He had little use for the theoretical population genetics of Sewall Wright and R. A. Fisher, though historians assumed an influence. As he later said, he had not studied their work well enough to use it; his inspiration came from the rich literature of taxonomy and biogeography, which he knew inside out. And though Mayr admired, even idolized Dobzhansky, with whom he began to converse and correspond in 1936, the evidence of their early relation suggests that Dobzhansky learned as much from Mayr about natural populations and their dynamics as Mayr learned from Dobzhansky's analysis of their genetic structure. Population genetics only confirmed what Mayr had learned from studying natural populations by field methods.[22] He worked as a field naturalist but to laboratory standards of inference and proof—the essence of border practice.

Hybrid Introgression: Edgar Anderson

The same pattern can be seen in the work of another, less well-known, evolutionary biologist—Edgar Anderson, who shared with Mayr the honor of delivering the Jesup Lectures on Evolution at Columbia in 1941. Had illness not prevented Anderson from turning lectures into chapters, *Systematics and the Origin of Species* would have been coauthored and written from the points of view of both a botanist and a zoologist. (And historians might not have taken so long to recognize Anderson's central role in the evolutionary synthesis.)[23]

Anderson and Mayr had not met or corresponded before they began their brief but intense collaboration on the Jesup lectures.[24] Brought together by Dobzhansky and Leslie C. Dunn, they discovered that they had arrived independently at a remarkably similar way of conceptualizing species and speciation. Both saw local populations, not species or subspecies, as the basic units of evolution, and both focused on transitional forms and the geographical circumstances in which speciation could occur. Anderson

22. Mayr, "Introduction, 1999" (cit. n. 9), pp. xiv–xix. Mayr to F. B. Kirkman, 14 Dec. 1938, EM-HU box 2. Mayr to Theodosius Dobzhansky, 25 Nov. 1935, EM-HU box 1. Dobzhansky to Mayr, 18 April 1937; Mayr to Dobzhansky, 13 May, 25 July 1937, 28 Feb. 1939; all in EM-APS.

23. Mayr, "Introduction, 1999" (cit. n. 9), pp. xvii–xviii. Kim Kleinman, "His own synthesis: Corn, Edgar Anderson, and evolutionary theory in the 1940s," *Journal of the History of Biology* 32 (1999): 293–320.

24. Ernst Mayr to Edgar Anderson, 19, 27 Dec. 1940, 6, 18, 21 Jan., 7, 18 Feb. 1941; Anderson to Mayr, 9, 16, 23, 25 Jan., 13 Feb. 1941; all in EM-HU box 2.

was bowled over by the coincidence: "When I read your last letter and . . . went through much of this first chapter," he wrote Mayr, "I had a curious sensation which I have never had before. It was as though I were reading something which I had written and forgotten about, or as though there were another me, working elsewhere and independently on the same problem." It was uncanny. No botanist, Mayr later recalled, had done more than Anderson to turn populational thinking into operational field practice.[25]

Anderson is best known for his conception of "introgressive hybridization." Briefly, the idea is that plant species occasionally produce interspecific hybrids where local conditions are just right. These hybrid swarms then backcross into the parental populations, carrying into each species genetic material from the other, which enlarges the normal range of variation and becomes the raw material of speciation by selection and gene shuffling.[26] It is not the first-generation hybrids themselves that become new species; they seldom survive for long, being less well adapted to their local environments than either parent. It is their introgression into parental stocks that opens the way to variants superior to the parent species. Introgression preserves genetic variability that would otherwise be lost as hybrids are crowded out by established species. Introgression explains how species that remain unchanged over vast reaches of time can in certain places become something new.

Place is crucial to Anderson's conception of introgression and to his field practice. Introgression can occur only in special places where species with different ecological requirements happen to live together, or where natural or human disturbance deprives parental species of their adaptive advantages and enables hybrids to survive long enough to introgress. Introgressed populations, being genetically highly variable, have equally varied ecological requirements, so they survive only in environments that afford multiple microniches. As Anderson put it, the habitat itself has to be "hybridized"—as if places had the vital qualities of their living inhabitants.[27] Anderson loved turning conventional categories inside out; he was famous (or notorious) for it. His mind, one might say, was a hybrid habitat where odd and fruitful combinations could survive long enough to be translated into workable practices.

25. Edgar Anderson to Ernst Mayr, 25 Jan. (quote), 13 Feb. 1941, EM-HU box 2. Ernst Mayr, introduction to section "Botany," in *Evolutionary Synthesis,* ed. Mayr and Provine (cit. n. 2), pp. 137–38, on p. 138.

26. Edgar Anderson and Leslie Hubricht, "Hybridization in *Tradescantia,* III: The evidence for introgressive hybridization," *American Journal of Botany* 25 (1938): 396–402. Anderson, *Introgressive Hybridization* (New York: Wiley, 1949). Anderson, "Introgressive hybridization," *Biological Reviews* 28 (1953): 280–307. Anderson and George Ledyard Stebbins, Jr., "Hybridization as an evolutionary stimulus," *Evolution* 8 (1954): 378–88.

27. Edgar Anderson, "Hybridization of the habitat," *Evolution* 2 (1948): 1–9.

Because it depends so crucially and unpredictably on specific places, introgression cannot be studied in an herbarium or garden but only in nature, in the places were it actually happens. Herbarium collections are generally too small and randomly collected to reveal telltale patterns of variation. Gardens make hybrid crossing and backcrossing easy, but they are too homogenized to simulate the "recombinant habitats" in which introgression occurs naturally. Experimental gardeners can make hybrids but not hybrid environments. What was needed, Anderson understood, was a field method that made visible the dynamics of variation in natural populations in situ.[28]

Seeing introgression in natural populations is not easy. The differences between introgressed and normal populations are nearly invisible even to a field botanist's experienced eye, and biometric techniques must be used to make them visible. And, because introgression is a statistical property, it can be made visible only by measuring variation in a species' entire geographical range, not just in local hybrid swarms. To catch species in the act of change requires a field practice that is both intensive and extensive; it requires total knowledge of a species—its taxonomy, biogeography, biometrics, and ecology. It is essentially a traditional taxonomic practice, amplified by biometric technique and methodical survey of entire regions, plus occasional checks by lab or garden experiment.

One key element of Anderson's practice was mass collecting. Botanists at the time had not yet acquired the habit of collecting extremely large series of specimens, as zoologists had several decades earlier. (Plants vary their form so greatly in response to local climate that collecting all variants seemed pointless). Anderson pioneered botanical mass collecting in the 1920s (along with Norman Fassett and W. B. Turrill). It is arduous, time-consuming work to collect large numbers of plants across a species' entire range and especially in the ecological border zones where species overlap. (Taxonomists collect in centers of ranges, where they are more likely to find typical forms and be less confused by aberrations.) It helps a little that for biometric work one has only to collect parts to be measured and not whole plants, but gathering data for a study of introgression is still far more demanding than ordinary taxonomic fieldwork: essentially like it, but more exacting and comprehensive.[29]

28. Anderson, "Hybridization in American *Tradescantias*" (cit. n. 1), pp. 511–12. Anderson and Hubricht, "Hybridization in *Tradescantia,* III" (cit. n. 26), p. 369. Carl L. Hubbs to George S. Myers, 27 Dec. 1932, CLH box 24 f. 73.

29. Edgar Anderson, "The technique and use of mass collection in plant taxonomy," *Annals of the Missouri Botanical Garden* 28 (1941): 287–92. Norman Fassett, "Mass collections: *Rubus odoratus* and *R. parviflorus,*" *Annals of the Missouri Botanical Garden* 28 (1941): 299–368.

Quantitative measurement was the second key element of Anderson's field practice. For large-scale biometric work, speed and simplicity are essential. So Anderson devised a technique of weighting biometric characters according to the degree of resemblance to one parent or the other, then combining these values in a composite index that distinguished different populations. (Individual characters vary continuously; composite characters make gaps visible.) The technique was statistically crude but fast, easy to use, and consistent. Anderson also devised a technique of graphical representation from which different populations could be apprehended at a glance. As he noted, this visual practice was a compromise between taxonomists' visible morphological distinctions and biometricians' mathematical abstractions.[30] It was a biometric amplification of ordinary taxonomic practice—in short, a border practice.

The graphimetric practices devised by Anderson and others—Carl Hubbs invented a similar "character index" for local populations of fishes—are a curious blend of the arcane and ordinary. "The point," Carl Hubbs explained, "is to bring into a single index figure the composite of characters, thus statistically accomplishing what perhaps we do subjectively when we look at a person and appreciate his face." No single feature alone distinguishes one individual from another, but combining several eliminates overlaps and makes it possible to distinguish local populations as unambiguously as we distinguish faces, quickly and confidently.[31] Hubbs was famous for his ability to recognize "with lightning rapidity taxonomic differences far too subtle for most of the rest of us." Biometrics enabled anyone to do the same. Anderson made a similar point when he observed that graphic indexes combine the best of mathematics and morphology: "Like mathematics, it is accurate and objective. Like morphology, it leaves something to the trained eye."[32] Clearly Anderson and Hubbs thought it a good thing to leave something to the experienced eye of the field observer. and not to stray from the realities of natural populations into biometric abstractions. Their practices were introgressive: traditional field practice extended beyond its normal limits by an infusion of mathematical objectivity and precision.

30. Edgar Anderson to Carl L. Hubbs, 19 May 1942, CLH box 35 f. 13. Anderson, *Introgressive Hybridization* (cit. n. 26), pp. 92–99.

31. Carl L. Hubbs to Carl Epling, 26 April 1939, CLH box 35 f. 12.

32. Kenneth S. Norris, "To Carl Leavitt Hubbs, a modern pioneer naturalist on the occasion of his eightieth year," *Copeia* (1974): 581–94, quote on p. 582. Edgar Anderson, "The problem of species in the northern blue flags, *Iris versicolor* L. and *Iris virginica* L.," *Annals of the Missouri Botanical Garden* 15 (1928): 241–332, on pp. 241–46, 285 (quote). Hubbs to Anderson, 14 May 1942; Anderson to Hubbs, 19 May 1942; both in CLH box 35 f. 13.

8.3 Edgar Anderson and his wife, Dorothy, in the field, circa 1920s. Missouri Botanical Garden Archives.

Unlike Ernst Mayr, Anderson was not a taxonomist by training or avocation. He was a lab-trained cytogeneticist with a broad and quirky mind and a taste for natural history. He acquired his interest in variation and evolution during his early years at the Missouri Botanical Garden, where he taught genetics and accompanied his taxonomist colleagues on trips into the surrounding Ozark Mountains. This research institution, with its unusual mix of laboratory and field botanists, was the disturbed cultural ground—the "hybridized habitat"—where laboratory and field practices could introgress and form new kinds of border science.[33] Anderson and Mayr came to the lab-field border from opposite directions but with the same openness to combining laboratory and field values.

Hybrids were far from Anderson's mind at first. Around 1926 he decided he could solve the species problem by a large-scale, intensive study of two related species of blue flags. A comprehensive and minute study of their variation would, he thought, reveal how individual variations merged into racial, racial into varietal, and varietal into species differences. For five

33. John J. Finan, "Edgar Anderson 1897–1969," *Annals of the Missouri Botanical Garden* 59 (1972): 325–45. Edgar Anderson, "What we do not know about *Zea mays*," *Transactions of the Kansas Academy of Science* 71 (1968): 373–78, on p. 373. Kleinman, "His own synthesis" (cit. n. 23).

years Anderson crisscrossed the United States and Canada, spotting roadside colonies of the plants and collecting them en masse. But his labors brought him no closer to his goal: though his two species of flag varied widely, each remained distinctly itself, and it remained quite unclear how that absolute gap between them might be breached.[34]

Hybridization was one way of getting genes across the species gap, at least in theory. Plants commonly form interspecies or even intergenus hybrids, and Anderson had seen colonies of hybrid flags in his travels, recognizing them from their similarity to hybrids artificially produced by fanciers in experimental gardens. However, hybrids were taxonomically tainted. Botanists had too often used "hybrid" as a residual category to get rid of the piles of specimens that fit no pigeon-hole, or proclaimed as good species forms that turned out to be merely hybrids. As Carl Hubbs put it, "a hybrid explanation is a convenient but dangerous pigeonhole for any aberrant specimen." Local amateurs were particular sinners, but even experienced botanists were occasionally fooled. Hybrids were disreputable, associated with mistakes, self-deception, even fraud. Hubbs advised that hybrids be invoked only as a last resort and when there was ecological and experimental evidence to support the field data.[35]

No wonder, then, that Anderson was a little shocked when a colleague, Robert Woodson, declared that a genus he was revising consisted of four species and fourteen hybrids. After some heated arguments they decided to resolve the issue experimentally, by producing hybrids and seeing if Woodson could correctly sort out a shuffled pile of parental, hybrid, and backcrossed specimens. He did, easily, and Anderson realized that hybrids could be the missing factor in the transformation of species. The data, he quipped, were "unusually trustworthy, even for a scientific experiment."[36] In a way Woodson himself and his taxonomist's experienced eye were the real objects of the experiment, which demonstrated that taxonomists' "subjective"

34. Anderson, "The problem of species" (cit. n. 32). Edgar Anderson, "The species problem in *Iris*," *Annals of the Missouri Botanical Garden* 23 (1936): 457–509.

35. Carl L. Hubbs, Laura C. Hubbs, and Raymond E. Johnson, "Hybridization in nature between species of catostomid fishes," *Contributions from the Laboratory of Vertebrate Biology of the University of Michigan* 22 (1943): 1–76, quote on pp. 11–12. Hubbs to David H. Thompson, 5 April 1932; Hubbs to J. O. Snyder, 9 Nov. 1931; both in CLH box 73 f. 68. Karl M. Wiegand, "A taxonomist's experience with hybrids in the wild," *Science* 81 (1935): 161–66. Charles B. Heiser, Jr., "Natural hybridization with particular reference to introgression," *Botanical Review* 15 (1949): 645–87.

36. Edgar Anderson, "An experimental study of hybridization in the genus *Apocynum*," *Annals of the Missouri Botanical Garden* 23 (1936): 159–68, quote on pp. 159–60. Anderson, "Concordant versus discordant variations in relation to introgression," *Evolution* 5 (1951): 133–41, on pp. 133–34, 141. Anderson, "Introgressive hybridization" (cit. n. 26), pp. 282–83.

distinctions, though invisible to inexperienced observers, were in fact as objective as any lab result. In time Anderson also learned to spot introgressed populations on sight, by their subtly aberrant patterns of variation.

Anderson's field craft was also a practice of place, like most border practices. Knowing that introgression occurs in certain kinds of places—at boundaries of species ranges, in disturbed ground—he could look with an eye prepared to see the subtle variations that betrayed introgression. Place became in effect a distinguishing character of introgressed populations, as morphological features are of species. The most remarkable such place that Anderson found was in the Mississippi Delta near New Orleans, a locale famous among iris fanciers for the unusual forms that suddenly appeared on the horticultural market in the 1920s—species hybrids, it turned out. The Delta is the meeting place of species of iris that normally are well separated by different ecological preferences but in the shifting environment of a large river delta live as near neighbors, one on the raised alluvial banks of delta streams in partial shade and fresh water, another in swamps and bayous in open sun and brackish water. Hybrids could occasionally form where nature arranged a more intimate meeting, as when an abandoned delta stream bed intersected a bayou, but these were rare events.

Opportunities for hybridization vastly increased in the 1920s when large tracts of Delta landscape were cleared of trees and ditched for cattle raising, creating hybrid habitats where neighboring species could mingle and introgress. Soon local residents began to notice striking iris, appearing spontaneously in astonishing profusion, as if by magic.[37] And through the fancy flower market, these hybrids came to the attention of local naturalists. It was one of these evolutionary hot spots that attracted Edgar Anderson in 1936: three adjoining Cajun farms on the banks of an abandoned delta stream where it crossed a recently dredged bayou. There a spectacular display met his eye: "Hybrids . . . of terra-cotta, wine, purple, and blue flooded out of the swale until it had almost the appearance of an intentionally created iris garden." Locals reported that the flowers had appeared suddenly in 1926 when drainage ditches were cut and forest cleared.

The layout of Cajun farms made the place a kind of natural evolutionary experiment. Each farmstead consisted of a long narrow strip (the vernacular landscape of early-modern France, brought to the Delta by Acadian refugees). In the middle, one farmer had partly cleared the woods and stocked it with cattle, which kept the vegetation clipped and the wet areas churned to mud—

37. Percy Viosca, Jr., "The irises of southeastern Louisiana: A taxonomic and ecological interpretation," *Bulletin of the American Iris Society* (April 1935): 3–56, on pp. 17, 22–23, 28–46.

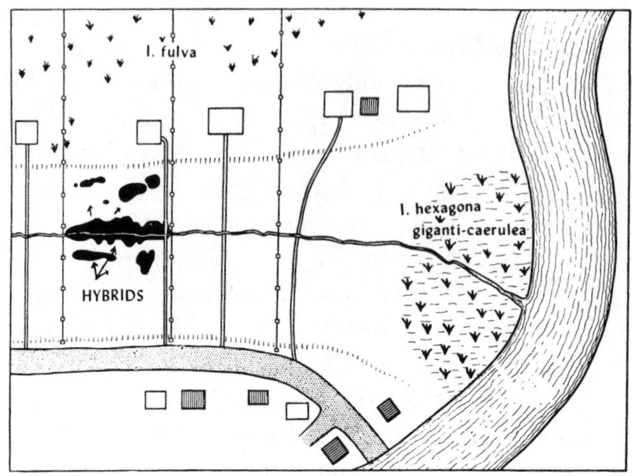

8.4 Three Cajun farms in the Mississippi Delta that demonstrate Edgar Anderson's concept of introgressive hybridization. On the right is a bayou; at right angles runs the abandoned bed of a delta stream. At their intersection two species of iris came together and hybridized. Edgar Anderson, "Man as a maker of new plants and new plant communities," in *Man's Role in Changing the Face of the Earth,* ed. William L. Thomas (Chicago: University of Chicago Press, 1956), 763–77, on p. 771.

ideal iris habitat. One of his neighbors had completely cleared his woods, and the other kept his as a woodlot—neither inviting to iris. The hybrid colonies of the middle farm flooded up to the fence lines and stopped—a striking visual display of the ecology of introgression. It looked like an experimental garden with a plot of experimental hybrids and two controls, and that visual form was powerfully persuasive. As Anderson observed, "the land-use pattern had produced something as demonstrable and convincing as a laboratory experiment."[38]

However, these Cajun farms were not a place of experiment. Anderson did no actual research there himself. He would not have been permitted: the plants were far too valuable. He had to pull strings just to visit. The Delta farms were like the set-piece demonstration experiments that lecturers perform to persuade their audiences of the plausibility of a theory or an experimental practice. It became in Anderson's literary work the proof of intro-

38. Edgar Anderson, "Man as a maker of new plants and new plant communities," in *Man's Role in Changing the Face of the Earth,* ed. William L. Thomas (Chicago: University of Chicago Press, 1956), 763–77, quotes on pp. 771–72. Herbert P. Riley, "A character analysis of colonies of *Iris fulva, Iris hexagona* var. Giganticaerulea and natural hybrids," *American Journal of Botany* 215 (1938): 727–38, on pp. 736–37.

gression, making visible and real a process that proceeds too slowly and obscurely to be witnessed except in such special places.

But special places were not essential to Anderson's field practice. He worked primarily in unremarkable places where introgressed populations were barely visible. The eroding bluffs of the Mississippi River valley where it cut through the ancient Ozark plateau afforded such worksites. Here two species of *Tradescantia* (spiderwort) lived, ordinarily in separate habitats, one on dry sunny cliff tops and the other on shady slopes and river bottoms, so that hybrids were rare. But in the gorges eroding back into the cliffs were small areas of intermediate habitat where the two species overlapped and mingled, and there Anderson knew he could find introgressing populations.

Areas of human disturbance (as with the Cajun farms) could also be counted on to nurture hybrid swarms: railroad cuts and ditches subjected to repeated grading and graveling were favorite places, and rural graveyards

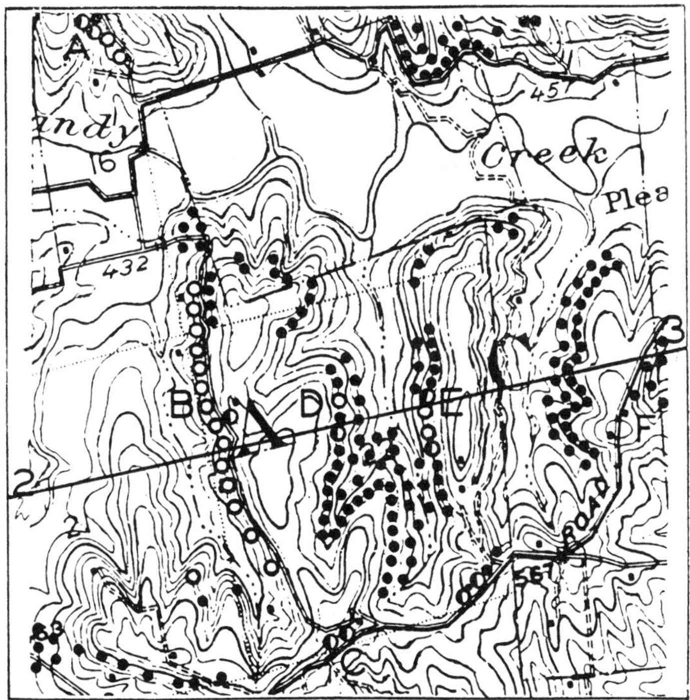

8.5 Distribution of two species of the spiderwort genus *Tradescantia* (closed and open circles), and hybrids (half-open circles), showing how hybrids are a highly local and place-dependent phenomenon. Edgar Anderson, "Hybridization in American *Tradescantias*," *Annals of the Missouri Botanical Garden* 23 (1936): 511–25, on p. 518.

where relicts of species driven out by agriculture lived near new occupants. Wherever there was "a churning and rechurning of the habitat by ditching, pasturing, lumbering, road-building, etc."—there were the "recombination habitats" that introgressed populations require to survive. The ancient Indian roadways cutting across the Arizona desert were another such place. Or an abandoned olive grove in California that was so completely reverted to native oak forest as to be virtually undetectable as second growth, were it not for the swarm of hybrid wild sages that had invaded the open orchard, got established, and lived there ever since. "The orchard became the site for evolutionary catch-as-catch-can," Anderson wrote, where "the new and variable had a decisive advantage." It would be interesting, he thought, to see what would happen next in this mundane place where something special was in progress.[39] In the scraped and scumbled landscapes of North America there were many such places.

Anderson's practice, though quantitative and in some ways lablike, was not a garden practice. It worked wherever introgression happened to occur, which could be just about anywhere. That, I think, was generally true of the border practices of the 1930s and 1940s. They were not limited to places where nature's work could be read as experiments, or places that could be remade or manipulated experimentally. They were as good as experiments but were of the field; not mechanical imitations of experiments but distinctive practices specific to natural places. They were traditional field methods made precise and objective by counting and methodical observation. Anderson and others were quick to see and use cases like the abandoned olive grove or the Delta iris farm to make a point, but their real work was everywhere and anywhere in nature.

Ecosystem Ecology: Raymond Lindeman

Amplified natural-history practices also became a characteristic feature of ecology in the 1930s, though they have been obscured by ecologists' desire for a progressive history: that is, one that proceeds toward laboratory modes. Practicing ecologists today see the 1940s as the time when their discipline became truly scientific, casting off concepts and practices of natural history

39. Anderson, "Hybridization in American *Tradescantias*" (cit. n. 1), pp. 517–19, 523–24. Anderson and Hubricht, "Hybridization in *Tradescantia,* III" (cit. n. 26), pp. 396–400. Anderson, "Hybridization of the habitat" (cit. n. 27), p. 6. Anderson, "Man as a maker" (cit. n. 38), pp. 766–70.

for those of quantitative and experimental science. This is not entirely an imagined past, as a browse through midcentury ecological journals will reveal. The earlier concern with classification declines, and descriptive and historical papers give way to resolutely quantitative studies wrapped in the rhetoric of mathematics and experiment. Practitioners of gradient, population, and ecosystem ecology, the three main branches of ecology in the postwar decades, were especially inclined to statistical and theoretical abstraction.[40] However, these new varieties of ecological practice all originated in the field, in work that was distinctly of the natural-history type. They became more lablike in the 1950s and 1960s and disavowed their historical roots, but in fact they were initially border practices, traditional fieldwork amplified by numerical and sampling techniques to the point where they were as good as experiment though of the field.

Ecosystem ecologists became the most puritanically lablike of these border practitioners. They reduced their interest in nature to the flow of nutrients and energy through natural biotic systems—a kind of environmental "metabolism." Their typical product and emblem is an abstract flow diagram of energy inputs and outputs between biotic or environmental components, with actual organisms and species lumped invisibly inside physicochemical black boxes. The point is evident: only energy relations matter; biological particulars are incidental, and history and biogeography are irrelevant, out of the picture. Although its ancestry is contested, ecosystem ecology is generally agreed to have sprung from the theoretical essay published in 1942 by Raymond Lindeman, a brilliant and ambitious young ecologist who died of a congenital illness at the age of twenty-seven just before his masterwork appeared. His "trophic-dynamic" essay was a blueprint for a mode of practice that takes chunks of biotic and physical environment as units and treats them in much the same way that biochemists treat experimental organisms. Lindeman perfected his trophic-dynamic conception as a postdoctoral fellow at Yale with the geochemical limnologist George Evelyn Hutchinson, but the roots of his idea were in the field, in traditional field practice.[41]

40. Frank B. Golley, *A History of the Ecosystem Concept in Ecology* (New Haven: Yale University Press, 1993). Joel B. Hagen, *An Entangled Bank: The Origins of Ecosystem Ecology* (New Brunswick: Rutgers University Press, 1992). Robert P. McIntosh, *The Background of Ecology: Concept and Theory* (New York: Cambridge University Press, 1985).

41. Raymond L. Lindeman, "The trophic-dynamic aspect of ecology," *Ecology* 23 (1942): 399–418. Robert E. Cook, "Raymond Lindeman and the trophic-dynamic concept in ecology," *Science* 198 (1977): 22–26. Hagen, *Entangled Bank* (cit. n. 40), pp. 87–99. Golley, *History of the Ecosystem Concept* (cit. n. 40), pp. 48–56.

Lindeman's trophic-dynamic ecology was essentially an elaboration of ideas of trophic levels conceived by Charles Elton fifteen years earlier. Elton argued that food chains are the constitutive, organizing structure of ecological communities, and that the passage of matter through food chains is the process that make biotic communities more than mere congeries of species living in the same places. Elton lumped the animal species of communities into a few functional units or levels: herbivores at the bottom of the chain, then primary predators, and higher levels of predators up to a maximum of four or five. He also introduced a novel quantitative potential into his scheme—the famous "pyramid of numbers." Since the creatures of each trophic level harvest only a small proportion of the level just below, the number of predators at each level must be substantially less than the numbers of prey. At some point the mass of prey is too small to support another level of consumer—hence the limit on the number of trophic levels to about five.[42]

Elton did for ecology what Ernst Mayr did for taxonomy: gave it a new set of functional units and categories based in the dynamics of populations. And he did it, as Mayr did, entirely in the field (on Bear Island, in the Arctic Ocean) without the help of instruments, quantitative methods, or mathematical models. He introduced a theoretical structure into ecology that was derived from field observation, not imported from sciences higher on the disciplinary pecking order. The key for Elton, as for Mayr, was a simple act of classification: lumping species into functional categories based on genetic or trophic relations. It was Elton's most pregnant invention, because it enabled ecologists to study fundamental processes without abandoning traditional field practices of observation and survey. Elton's pyramid of numbers invited ecologists to develop quantitative methods, though he developed none. His insistence on the crucial importance of food relations invited others to measure flows and conversions of matter and energy, though Elton himself remained a traditional field observer.

Lindeman transformed Elton's trophic pyramid into a program for counting and measuring trophic dynamics in the field. Historians have either ignored Lindeman's Eltonian side or highlighted the differences between Elton's natural history and the physiological elements of Lindeman's practice, but I find the similarities more striking than the differences.[43] Lindeman himself was conscious of the resemblances, describing his work to one col-

42. Charles Elton, *Animal Ecology* (London: Sidgwick & Jackson, 1927). Elton, *The Ecology of Animals* (London: Methuen, 1933). Hagen, *Entangled Bank* (cit. n. 40), pp. 51–62, 96–99. McIntosh, *Background of Ecology* (cit. n. 40), pp. 88–93.

43. Hagen, *Entangled Bank* (cit. n. 40), pp. 90–94.

league as "somewhat in the Eltonian tradition," and to another confessing his heavy reliance on the ideas of Elton, Hutchinson, and August Thienemann.[44] Ecologists who knew his work also recognized his connection with Elton. The entomologist Alec Hodson, one of the most perceptive of Lindeman's colleagues, agreed with him that "many rather cloudy ideas concerning food chains, Eltonian Pyramids and succession may be clarified by an analysis of energy exchange."[45]

Lindeman's field practice, like Mayr's and Anderson's, was an intensified natural history. It was mostly a matter of collecting, sorting, counting, weighing and tabulating large numbers of all the species of animals and plants inhabiting a place—Cedar Creek Bog, a eutrophic pond in Minnesota's boreal forest—over a period of years. Numbers and weights were converted into biomass and population densities (per square meter of pond surface), and biomass in turn converted to energies. Lindeman then lumped species with similar food relations into trophic levels in two interconnected food chains: one of plankton, plankton predators, and swimming predators; the other of pond weeds, plant browsers, and benthic and swimming predators. Through these trophic levels matter and energy flowed, creating an interconnected and interdependent pond community that changed dramatically in its roster of species from season to season and year to year, but remained trophically the same, bound together in the vital business of eating and being eaten. By further manipulation of the data, plus some rough assumptions about losses to the decomposer microbes of the bottom ooze (which he did not measure), Lindeman could compute the annual energy production of each trophic level and the losses of energy to respiration and decomposers at each stage of consumption, and thus the efficiencies of transfer between levels.

These were rough calculations—some little better than mere guesses or "shots in the dark," Charles Elton complained—but they were the first attempt to give Elton's trophic-dynamic scheme a quantitative shape based on actual field data. As Lindeman observed, "rough approximations of quantitative trophic values are better than none."[46] And it was accomplished by

44. Raymond L. Lindeman to J. Richard Carpenter, 18 Feb. 1942; Lindeman to Victor E. Shelford, 13 Jan. 1942; Shelford to Lindeman, 2 March 1941; all in RLL box 1 ff. 9 and 41. Lindeman read August Thienemann's work in 1939 but did not grasp its significance until he reread it in 1941 at Hutchinson's suggestion. See Lindeman to G. Evelyn Hutchinson, 7 April 1941, GEH box 2; and Hutchinson to Lindeman, 2 Apr 1941, RLL box 1 f. 23.

45. Alexander C. Hodson to Lindeman, 17 Nov. 1941, RLL box 1 f. 21.

46. Charles Elton to G. Evelyn Hutchinson, 21 April 1943, cited in Hagen, *Entangled Bank* (cit. n. 40), on p. 96. Raymond L. Lindeman. "Seasonal food-cycle dynamics in a senescent lake," *American Midland Naturalist* 26 (1941): 636–73, quote on p. 667.

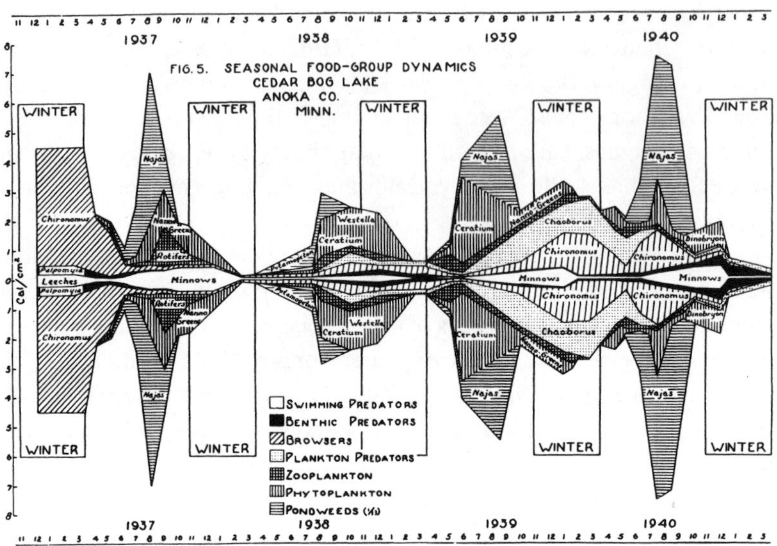

8.6 Seasonal changes in the fauna of trophic levels in Cedar Creek Bog, Minnesota. Note the extreme variability in species from season to season and year to year. Raymond L. Lindeman, "Seasonal food-cycle dynamics in a senescent lake," *American Midland Naturalist* 26 (1941): 636–73, on p. 662.

nothing more than intensive and systematic gathering, counting, and weighing. It was only the subsequent manipulations of the data that would have struck any field naturalist as alien and lablike.

Lindeman was no biochemist but a field naturalist through and through. Readers who know only his theoretical "trophic-dynamic" paper will find his accounts of fieldwork at Cedar Creek Bog surprisingly descriptive and attentive to particulars of place, with vivid observations of bog creatures and their life histories and habits, the number and timing of reproductive cycles per season, food preferences, and predator-prey relations. He clearly loved the place. Despite the abstract quality of the final result—a table of thermodynamic data—and his frequent use of physiological terms like "intermediary metabolism," Lindeman's work resembles Elton's scientific natural history far more than the physiological style of later ecosystem ecologists, or of limnologists like Birge and Hutchinson, whose work he greatly admired. Lindeman deployed no elaborate limnological apparatus: he collected with a standard dredge modified for shallow water, which took in the whole three feet from surface to bottom ooze in a single sweep. He measured respiratory gases but seemed little interested in using the data. His field reports are a vivid account of a place and its inhabitants.[47]

Later ecosystem ecologists set the natural-history elements of Lindeman's work aside as old-fashioned and unnecessary, but it is clear that Lindeman felt no incongruity between theoretical, trophic-dynamic ends and natural-history means. His was a border practice, and he was a border practitioner, inspired by laboratory values and standards but not infatuated. He valued biochemical knowledge and skill but also prided himself on his knowledge of taxonomy and kept it up to date. He welcomed the arrival at Minnesota of a graduate student with a background in organic and physical chemistry as well as field experience with Joseph Grinnell. Another he judged "a bit on the theoretical-argumentative side as yet" but with "a fine chemical background" and too good to waste on "purely distribution problems" or counting stream fauna. Lindeman had a taste for the biomathematical modeling of theorists like Royal Chapman, Robert Park, Vito Volterra and G. G. Gause, but was quick to abandon abstract models when they did not yield practical results in the field. He studied metabolic physiology to understand how its principles applied to field situations, but never meant to go into a laboratory and do it himself. He declined Hutchinson's suggestion of an experimental simulation of winter anaerobiosis, preferring to work on food cycles in the field.[48] He never assumed that progressing in his career meant leaving field for lab, as an earlier generation did.

Place was also an essential part of Lindeman's practice, as it was generally for border practitioners. Cedar Creek Bog, where Lindeman did all his important work, is a small and unremarkable eutrophic pond in the last stage of its life cycle and just 150 years from being dry land. (It only came to the attention of local ecologists as the result of an aerial survey in 1931, and it shrank measurably in the five years that Lindeman worked there.) Just 14,480 square meters in area and no more than three feet deep, it is one of a large family of glacial pothole lakes in the Anoka Sand Plain, a "pitted outwash" created during the final retreat of the Wisconsin ice sheet when a temporary advance pushed a thin lobe of ice over an older area of kettled

47. Lindeman. "Seasonal food-cycle dynamics" (cit. n. 46). Hutchinson, addendum to Lindeman, "Trophic-dynamic aspect" (cit. n. 41), pp. 417–18. Lindeman to Carl L. Hubbs, 20 Nov. 1939, RLL box 1 f. 21.

48. Raymond L. Lindeman to Walter Moore, 29 Oct. 1940, RLL box 1 f. 32. Lindeman to Alexander Hodson, 30 Nov. 1941; Hodson to Lindeman, 17 Nov. 1941; both in RLL box 1 f. 21. Golley, *History of the Ecosystem Concept* (cit. n. 40), pp. 53–54. Lindeman to G. Evelyn Hutchinson, 11 Nov. 1940, GEH box 2.

8.7 Cedar Creek Bog, Minnesota, midsummer 1937: the eutrophic glacial pond where Raymond Lindeman invented ecosystem ecology. Lindeman, "Seasonal food-cycle dynamics," on p. 639.

drift. As the ice retreated, large blocks of ice in the kettle holes were buried by outwash and formed lakes, as they melted.[49]

Cedar Creek Bog was a place no self-respecting limnologist would have looked at twice. It froze solid every winter, killing off its top trophic level. Connected to nearby streams only in wet years (via a woodland rivulet), in dry years it was a dry field of wild rice. An odd place for an aspiring limnologist, to be sure, but an ideal one for quantitative natural history. It was small enough to be intensively studied by a single person. Its uniformly shallow water and homogeneous oozy bottom made it easy to sample from a flat-bottom boat. No violent waves mussed biotic levels and disrupted quantitative counts. Most important, Cedar Creek Bog had an impoverished inventory of resident animals and plants, which greatly eased the task of defining trophic levels. (The depauperate fauna of Bear Island had given Elton a similar advantage.) The ecosystem of Cedar Creek Bog could be taken in at a glance, so to speak, and trophic techniques and concepts could be worked out there more easily than in a proper lake or meadow. And coincidentally, the place had already been staked out for long-term observation and mea-

49. Raymond L. Lindeman, "The developmental history of Cedar Creek Bog, Minnesota," *American Midland Naturalist* 25 (1941): 101–21, on pp. 101–11.

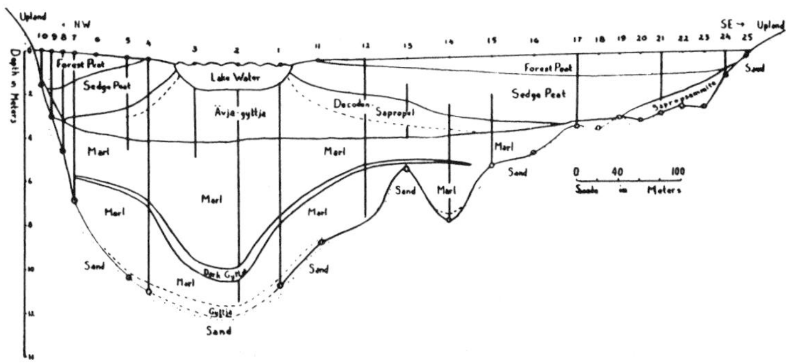

8.8 Transect of Cedar Creek Bog. From the composition of these layers of organic sediment Raymond Lindeman attempted to infer the metabolic dynamics of succession. Raymond L. Lindeman, "The developmental history of Cedar Creek Bog, Minnesota," *American Midland Naturalist* 25 (1941): 101–12, on p. 107.

surement by biologists interested in its quaking bogs.[50] This unexceptional, almost invisible patch of Minnesota's North Woods thus became a site of intensive fieldwork—staked-out, charted, counted, and measured—and the birthplace of an expansive new field practice.

With its thirty-three feet of layered mud, silt, and organic debris, Cedar Creek Bog was also ideally suited for studying trophic history. (Larger lakes like Mendota, though older in years, were physiographically much younger: virtually without history and useless for studying succession.) To judge from Lindeman's published work, one would think he was not much interested in succession, but archival material reveals that the subject had always been foremost in his mind—he just felt he could not write about it until he had proper field data.[51] His keen interest in history is yet another link with the naturalist tradition.

Lindeman's idea, briefly, was that succession in aquatic ecosystems was characterized by dramatic changes in biomass productivity. Initially he surmised that the productivity of a pond ecosystem rose in a simple logistic curve, but he abandoned that idea when a forest ecologist at Yale, Harold J. Lutz, informed him that climax forests were in fact less productive than

50. Lindeman, "Seasonal food-cycle dynamics" (cit. n. 46), pp. 636–39, 651, 657–63. Murray F. Buell and Helen Foot Buell, "Surface level fluctuations in Cedar Creek Bog," *Ecology* 22 (1941): 317–21.

51. Raymond L. Lindeman to Murray F. Buell, 20 Aug., 22 Oct. 1940, RLL box 1 f. 6. Lindeman to J. Richard Carpenter, 26 Feb. 1942, RLL box 1 f. 9. Lindeman to Samuel Eddy, 12 Jan. 1942, RLL box 1 f. 16.

earlier successional stages. He then conceived a more complicated scheme in which productivity increased sharply in the earliest stages of succession, declined to a minimum in the final stage of eutrophication, then rose stepwise through a sequence of terrestrial stages to climax forest. The scheme is boldly depicted in his "trophic-dynamic" paper as a (famous) wiggly curve.[52] It was pure speculation, like Elton's pyramid of numbers, but definite enough in its predictions to invite empirical testing. In the final year of his short life Lindeman worked obsessively to devise field techniques for getting that evidence, once again in the field at Cedar Creek Bog.

Again Lindeman relied on amplifying traditional field practices, now to squeeze trophic data from the meager bits and pieces that remained of historical communities. His source of evidence was a series of over three hundred bottom cores from Cedar Creek Bog, which recorded in its layers of mud and organic debris the history of its changing inhabitants. From this record Lindeman hoped to reconstruct past communities of plants and animals and their metabolic productivities and trophic relations. From the ratio of organic matter to inorganic silt he could approximate rates of photosynthetic activity (silting was assumed to be constant). Undecomposed plant matter he took as a measure of energy lost in the transfer between plants and primary consumers. From undecomposed fragments of different animal species, Lindeman estimated the biomass of the several levels of consuming organisms and, by simple subtraction, the matter and energy passed from one trophic level to the next. And from bits of preserved moss and wood and taxonomic analysis of pollen and plankton skeletons, Lindeman hoped to reconstruct the biotic structure of successional stages. Chemical analysis of dissolved nutrients and chlorophyll promised to yield a rough measure of the potential of biotic communities for biomass production.[53]

Some laboratory work was involved, but mainly it was a matter of simple chemical analysis plus sifting, sorting, and weighing the different kinds of organic remains in cores and interpreting their trophic significance—a tour de force. Lindeman never published this final work, but his correspondence reveals that he spent his final months, day and night, picking through mucky

52. Lindeman. "Trophic-dynamic aspect" (cit. 41), p. 413. Lindeman to J. Richard Carpenter, 26 Feb. 1942, RLL box 1 f. 9. Lindeman to Robert Park, 13 Dec. 1941, RLL box 1 f. 15.

53. Lindeman, "Trophic-dynamic aspect" (cit. n. 41), pp. 411–15. Lindeman to G. Evelyn Hutchinson, 11 Nov. 1940, GEH box 2. Lindeman to Hutchinson, 17 Feb. 1941, with application for a fellowship, RLL box 1 f. 23. Lindeman, "Proposed research," 17 Feb. 1942, with application to the National Research Council, RLL box 1 f. 34.

cores, learning quantitative spectrographic methods, mastering the arcana of rhizopod pollen taxonomy—in short, doing scientific natural history.[54]

The contrast could hardly be starker between the tidy equations and curves of Lindeman's "trophic-dynamic" essay and the messy, nitty-gritty field practice on which he based his theoretical edifice. The theory became famous and inspired a generation of ecosystem ecologists, who hastened to forget the old-fashioned practices on which it was based. But if Lindeman's theory was of the laboratory, his practice was of the field: a border practice, natural history made quantitative and systematic. He conceptualized community and succession physiologically but was in no way dependent on anything in a physiology lab. No instruments, no arcane statistics: it was just a matter of painstaking handwork, sorting, counting, weighing, tabulating. Lindeman addressed theoretical questions and valued quantitative evidence, as experimenters also did, but he worked in the field with field methods. He never supposed that progress meant leaving places like Cedar Creek Bog behind.

Gradients and Continua: Robert Whittaker

Lindeman's trophic-dynamic ecology was well within the tradition of Clements and Cowles. He took for granted that communities were natural units and succession a real process, and that ecology was or would become a physiology of the field; he simply invented a new way of realizing that old ideal. But other ecologists were ready for a more radical change, and if there was a defining feature of postwar ecology, it was the general disenchantment with concepts of communities and development. In the decades after 1950 conceptions of nature as a mosaic of organic communities were replaced by ideas that privileged individuals and continua. Belief in the orderly development of succession and the ultimate stability of nature gave way to randomness and local contingency and a vision of a world in perpetual flux.

Discouraged by the plethora of competing and incompatible schemes of associations and successions, many ecologists in the 1930s and 1940s became agnostic about all theories and systems. They began to think that no system

54. Raymond L. Lindeman to Edward S. Deevey, 11 April, 10 Oct. 1941, RLL box 1 f. 12. Lindeman to Samuel Eddy, 30 Nov. 1941, RLL box 1 f. 16. Lindeman to Frank Hooper, 28 Sept. 1941, RLL box 1 f. 22. Lindeman to Walter Moore, 2 Nov. 1941, RLL box 1 f. 32. Lindeman to Edward Thatcher, 11, 13, 30 Oct., 2 Dec. 1941, RLL box 2 f. 44. Hutchinson to Edward S. Deevey, 20 May 1942, GEH box 1.

at all would be better for their collective reputation than the spectacle of endless contention over rival systems. William Cooper thought it was time to "unsystem" ecological knowledge, and many agreed. Younger naturalists like Alfred Emerson and Orlando Park were relieved when the older generation—Shelford, Clements, Weaver, Cooper—stayed away from a symposium in 1938. It "allowed greater freedom of discussion and . . . a healthy agnosticism . . . concerning all ecological concepts and methods." The English ecologist John Phillips was exasperated with "the formal, academic clap-trap that is called Ecology" and thought much of what was taught as ecology should be entirely scrapped.[55] Some took the radical step of asserting that associations were not natural units but merely contingent assemblages of species produced by local circumstances and histories. The most famous and outspoken of the radical skeptics was Henry Gleason, whom the new generation adopted as an honored ancestor and prophet of the new order (or disorder).[56] The task for younger ecologists was to invent methods for putting Gleason's ideas into practice in the field.

One of the most ambitious and radical members of this new generation was Robert Whittaker, and the field practices that he devised best express the new principles of continua and contingency.[57] His techniques of "gradient analysis" were to the new ecology of continua what the quadrat was to the ecology of community units: a set of practices that enabled ecologists to imagine nature as discrete units or continua and to create representations that made it difficult to see it otherwise. Practices embody and enforce theories and systems of classification. Stake out tidy quadrats and you will see nature as definite, bounded, classifiable communities. Walk gradient transects and you will see nature as jumbled species and continua.

Like Raymond Lindeman, Whittaker was a graduate student when he carried out his formative work. He aspired to make ecology an inductive, lablike science but without adopting the material culture and methods of the laboratory. Gradient analysis was consciously a border practice—traditional fieldwork made objective and rigorous by the use of quantitative

55. Hagen, *Entangled Bank* (cit. n. 40), pp. 28–32. Robert P. McIntosh, "The continuum concept of vegetation," *Botanical Review* 33 (1967): 130–87. William S. Cooper, "The fundamentals of vegetational change," *Ecology* 7 (1926): 391–413, on p. 396. Warder C. Allee to John Phillips, 18 Nov. 1938 (quote); Phillips to Allee, 3 Oct. 1938; both in WCA box 21 f. 2.

56. Robert P. McIntosh, "H. A. Gleason, 'individualistic ecologist,' 1882–1975: His contribution to ecological theory," *Bulletin of the Torrey Botanical Club* 102 (1975): 253–73. Nicolson, "Henry Allan Gleason" (cit. n. 3). Henry A. Gleason, "The individualistic concept of the plant association," *American Midland Naturalist* 21 (1939): 92–110.

57. W. E. Westman and R. K. Peet, "Robert H. Whittaker (1920–1980): The man and his work," *Vegetatio* 48 (1982): 97–122, on pp. 106–7, 113–14.

data-handling techniques systematically applied on a grand scale. Like the other border practices we have examined, gradient ecology was invented in the field and, for all its abstract and tabular end products, remained a field practice.

Whittaker invented the gradient method in 1947 in the Great Smoky Mountains of North Carolina and Tennessee, where he and three other graduate students at the University of Illinois pursued their dissertation research under the direction of Charles Kendeigh. He was just twenty-seven, bright, ambitious, and determined to do something big and important. The Illinois ecologists were divided along theoretical lines, Kendeigh and Victor Shelford promoting the Clementsian views, Arthur Vestal taking the side of his friend and mentor Henry Gleason. Seeing authorities disagree, Whittaker determined to put the issue to an empirical test, and Kendeigh arranged a summer expedition to the Great Smoky Mountain National Park, whose mature deciduous forests seemed ideal for the purpose.[58]

Whittaker's plan was simple: he proposed to map the distribution of individual species by counting them along transects at fixed intervals up a mountain slope or across a mountain valley, without first grouping them into discrete associations. He would make no effort to sample sites in the centers of communities or to avoid boundary zones, as community ecologists normally did, to get "typical" samples. (That would only ensure that the result expected would be produced.) If biotic communities were real entities, then the distribution of species should display sharp discontinuities at the boundaries between communities, and species associated in communities should have identical patterns of distribution.

Probably Whittaker expected to find that associations were real things, but with complex border zones of intercommunity hybrids.[59] What he actually found, to his surprise, was that each species was distributed in a binomial Gaussian curve with a peak of density at an optimal habitat and tailing off smoothly on both ends. He saw no abrupt discontinuities suggestive of well-defined units, no overlapping of the distribution of different species: the distributional curve of each species was unique and "individualistic." Gleason was right after all, Whittaker concluded: "associations" were nothing more than congeries of species whose distribution peaks happened to coincide, artifacts of ecologists' desire to deal with definite and classifiable

58. Ibid. Robert H. Whittaker, "A criticism of the plant association and climatic climax concepts," *Northwest Science* 25 (1951): 17–31, on pp. 17–19. Whittaker, "Vegetation of the Great Smoky Mountains," *Ecological Monographs* 26 (1956): 1–80, on pp. 30–32. Hagen, *Entangled Bank* (cit. n. 40), pp. 31–32.

59. Westman and Peet, "Robert H. Whittaker" (cit. n. 57), pp. 99–101.

units and of their habit of choosing "typical" sites for study. As Whittaker saw it, biotic communities were a human order imposed on a continuous nature; like the colors of the spectrum they had no objective existence. Systems of classifying communities were useful for practical purposes but had no theoretical significance, Whittaker concluded. Vegetation was a "loosely ordered complexity," not "clear and orderly associations, successions, and phylogeny . . . [but] a veritable shimmer of populations in space and time." It was a shifting kaleidoscope, "a pattern of colored lights cast upon a screen . . . a blended, intricate and fluid pattern."[60]

Whittaker's field practice was simplicity itself, though exceedingly labor intensive. Transects were laid out up a mountain slope or across a mountain valley with a dozen or so sampling stations from bottom to ridge. At each station trees were identified by species and counted in quadrats (one-tenth hectare later became standard, and one hundred to three hundred "stems"). It was the old field transect, familiar to any field ecologist or, for that matter, to any practical timber "cruiser" reckoning board feet. The only novelty was the intensity with which Whittaker applied it. Three hundred samples became the minimum for a field project, taken every one or two hundred feet along the gradient, fifty to sixty per thousand feet of elevation, twenty-five thousand stems per sample. These transects along gradients of elevation or moisture (from moist valley bottoms to dry ridge tops) were supplemented by semirandom samples, done by observers who wandered at random along the numerous trails that crisscrossed the Great Smoky Mountain National Park and paused at each new slope or exposure to count a sample.[61]

Such massive sampling created mountains of data, and methods of data processing were an essential part of gradient practice. (Like taxonomists, gradient ecologists collected extensively in the field but did most of their work indoors.) Whittaker could assume that the distribution of species mapped the distribution of environmental factors (rather than the presence of other species). The problem was deciding what factors to plot against species frequencies to reveal the underlying causal relations. Whittaker used several

60. Robert H. Whittaker, "A study of summer foliage insect communities in the Great Smoky Mountains," *Ecological Monographs* 22 (1952): 1–44, on pp. 8, 10–11, 14, 29–31. Whittaker, "Criticism of the plant association" (cit. n. 58), pp. 19–29. Whittaker, "Recent evolution of ecological concepts in relation to the eastern forests of North America," *American Journal of Botany* 44 (1957): 197–206, on pp. 202–3, p. 203 (first quote). Whittaker, "Vegetation of the Great Smoky" (cit. n. 58), pp. 1–2, 8–9, 21–24, 43 (second quote).

61. Whittaker, "Criticism of the plant association" (cit. n. 58), pp. 18–19. Whittaker, "Vegetation of the Great Smoky" (cit. n. 58), pp. 4–7.

graphical methods. Plotting species frequencies against elevation was the easiest but theoretically the least meaningful, since elevation was a compound variable of temperature, moisture, aspect, soils, and topography. Another method was to plot species frequencies along an environmental or topographical axis: valley coves, canyons, flats, draws, ravines, sheltered slopes, open slopes, and ridges and peaks. This was essentially a moisture gradient, but with factors of place included as well, and it was useful for some purposes. But the simplest kind of graph was the moisture gradient, constructed by plotting frequencies against an axis of mesic, submesic, subxeric, and xeric types of vegetation. There was an uncomfortable circularity in this method, Whittaker realized, because the wetness or dryness of sites was not measured directly but inferred from the known ecological preferences of the dominant species present. However, of all these methods it was the closest to being a single-factor plot.

Finally, Whittaker constructed two-dimensional maps by defining arbitrary types of forest (not "communities," he insisted) and plotting them on a grid of elevation and moisture. These mosaics corresponded remarkably well to what an observer might see standing on a peak in the autumn and looking at the colored mosaic of vegetation on a distant mountain slope. None of these graphical methods revealed much about the biophysical causes of plant distribution, but that was not really what mattered to gradient ecologists. Their representations of nature were more mimetic than analytical and lablike. What impressed Whittaker was that nature's order could be reproduced graphically from data gathered without a priori theoretical as-

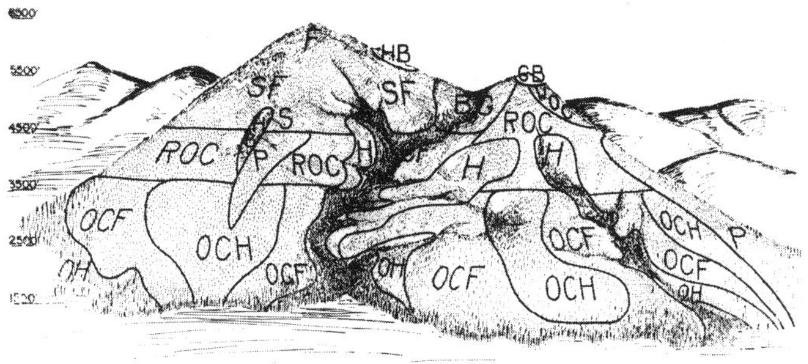

8.9 Map of vegetation types on a mountain slope of the Great Smoky Mountains of Tennessee, created by the methods of gradient ecology. Letters denote dominant tree species. Robert H. Whittaker, "Vegetation of the Great Smoky Mountains," *Ecological Monographs* 26 (1956): 1–80, on p. 59.

sumptions about natural communities. To Whittaker it was this mimetic quality that made his method objective and scientific, and superior to the qualitative judgments of community ecologists.[62]

Whittaker had little interest in instruments or physical measurement. He did set out a few hygrothermographs in his first weeks in the Great Smokies, but when bears found and mauled the instruments he shelved them for good. Besides, the vast scale on which he gathered data made regular measuring quite impracticable, and he felt no need to repeat what Shreve and others had already established.[63] What gradient ecologists did was walk and count, walk and count.

For all his radical protestations, Whittaker was personally very like Clements, and gradients were quite like quadrats. Whittaker was intense, combative, and socially awkward. He tended to take extreme and dogmatic theoretical positions and was intolerant of opposition. He aspired to nothing less than the total reform of ecology as a discipline, on his terms.[64] Even their favored techniques were alike. Gradients, like quadrats, were simple, all-purpose techniques of sampling and numeration, and they were widely adopted by ecologists without the theoretical doctrines that Whittaker attached to them—as Clements's methods had been. Gradient ecology was essentially descriptive, rather like community ecology without the typology and nomenclature. In both cases field methods profoundly shaped the way practitioners saw nature, as quasiorganisms or as chaotic continua.

Place was another essential feature of Whittaker's practice: choosing the right place and knowing it intimately. If you see nature as continua, then it is best to work in places where topography and vegetation are varied but continuous. If nice smooth gradients are what you want, then avoid places with patches of physiographic change and succession. The ancient and well-eroded Southern Appalachians, of which the Great Smoky Mountains are a part, are a relatively undisturbed place. It is a landscape of continuous slopes and relatively stable climax or second-growth forests, with few areas of bare-ground succession. Abundant rainfall has left it heavily dissected, with narrow valleys deeply cut back into the mountain massif (ninety percent of the area has a slope of ten percent or more), and mature soils. The moist climate also prevents regular burning, so there are few areas of pri-

62. Whittaker, "Criticism of the plant association" (cit. n. 58), pp. 28–30. Whittaker, "Vegetation of the Great Smoky" (cit. n. 58), pp. 5–7. Robert H. Whittaker, "Gradient analysis of vegetation," *Biological Reviews* 42 (1967): 207–64, on pp. 207–208, 212–25.

63. Whittaker, "Vegetation of the Great Smoky" (cit. n. 58), pp. 4–7. Whittaker, "Study of summer foliage" (cit. n. 60), p. 5. Whittaker, "Criticism of the plant association" (cit. n. 58), pp. 18–19.

64. Westman and Peet, "Robert H. Whittaker" (cit. n. 57), pp. 111–12.

mary succession to mess up data samples. As Whittaker himself observed, one could support either a Clementsian or a continuum view by picking the right place. For Clements, highly disturbed places, like the fire-prone forests of young mountains of the West, were the right place. For continua and gradients the place to be was in the forests of old mountains.[65]

The history of human land use in the Southern Appalachians also made them a good place for gradient ecology. There was virtually no virgin forest left by 1947, but there were few areas of complete disturbance. Mountain people lived mainly in the narrow bottoms and used the slopes and mountain-top "balds" lightly, for seasonal grazing, selective timbering, hunting, and recreation. These nonintensive uses left the area with large areas of uninhabited forest well supplied with good trails and easily accessible for transects and random walks. The botanical complexity of the region was another advantage (a consequence of its history as a refuge for northern forms driven south by ice sheets), assuring a variety of different species to count. Topographically and ecologically it would be hard to imagine a better place to invent an ecology of continua. No special, lablike places were required, no preserved records of nature's "experiments." Virtually every mountain slope or valley offered an opportunity to walk and count a transect.[66]

Whittaker's gradient method displays all the features of a border practice. It consisted of traditional field methods, scaled up and quantified. It dealt with populations not individuals, required no experiments, deployed no instruments in the field, and did not project a laboratory future for its practitioners. Granting that the causes of plant distribution would only be revealed by research in genetics and physiology, Whittaker did not think that experiments with a few individual organisms in labs would tell much about the relation of populations to environments (a dig at Shelford?). Populations had to be studied directly in nature.

Nor was Whittaker much impressed by fancy mathematical treatments and modeling; they might reveal hidden causal relations but quickly reached the point of diminishing returns. Advances in understanding population structure and classification had more often come, he thought, from "simpler and more direct methods, carried out with skill and perceptiveness." He knew that his methods contained an irreducible element of subjectivity, and that they required skill and judgment that came only with experience. But

65. On the relation between place and belief see Paul B. Sears, "Some notes on the ecology of ecologists," *Scientific Monthly* 83 (1956): 22–27.

66. Whittaker, "Study of summer foliage" (cit. n. 60), pp. 1–2. Whittaker, "Vegetation of the Great Smoky" (cit. n. 58), pp. 2–4. Whittaker, "Criticism of the plant association" (cit. n. 58), pp. 18–19.

he accepted that: the complex problems that ecologists had to deal with were more like those of sociology than physiology, and in his view quantitative but simple field methods were the most appropriate and effective.[67] Whittaker was obsessed with making ecology "rigorous" and scientific, but he was never tempted to achieve that end by turning ecology into a branch of applied physiology. He adopted experimental standards of evidence and inference but not the whole package of laboratory culture. He gave up classification but without embracing experiment as the one best way. The practice he perfected was of the field, not a colonial but a border practice, and it exemplifies the changes that had occurred in the lab-field border zone since Cowles, Clements, Shelford, and Shreve embarked on their careers.

Conclusion

Field biologists at midcentury partly realized the ideals of a scientific natural history that had inspired their turn-of-the-century predecessors. New naturalists like Whitman, Ganong, Cowles, Sumner, Crampton, or Adams could not have anticipated the particular ideas and practices devised by Mayr, Anderson, Lindeman, Whittaker, and others. But the practices of the border biologists of the 1930s and 1940s were in principle, at least, just what the earlier generation had called for: practices that combined the best features of traditional natural history and experimental lab biology. They were of the field but methodical and quantitative; as sound and credible as any laboratory experiment, and more appropriate to nature's complex, ever-changing objects, places, and conditions. The theories embraced by the new border biologists were strikingly different from those that the first new naturalists had taken as gospel. (Who in 1910 could have anticipated that population dynamics and natural selection would dominate theories of speciation, or that chaos and contingency would replace ideas of biotic association?) But the practices that produced these grand new insights into nature's work were recognizably the sort that the first new naturalists had envisioned as ideal for a scientific natural history.

However, there is a striking contrast between the practices described in this chapter and those with which the first new naturalists attempted to realize their ideals. As we saw in earlier chapters, they imported laboratory in-

67. Whittaker, "Vegetation of the Great Smoky" (cit. n. 58), pp. 25–30. Whittaker, "Gradient analysis" (cit. n. 62), pp. 254–56, quote on p. 254. Westman and Peet, "Robert H. Whittaker" (cit. n. 57), p. 102.

struments and methods directly into the field in ways that were often not congruent with the realities of natural objects and places. It was a natural thing to do for biologists who were generally lab trained and operated in a scientific world still enthralled by the laboratory revolution, but who had virtually no experience of making laboratory methods work in the field. That the results were generally disappointing is not surprising. Border biologists, in contrast, began with familiar field methods—survey, mapping, transect, observation—and brought them up to laboratory standards of evidence and inference by simple expedients of counting and systematic data gathering. Such practices adopted the general ideals of laboratory science but avoided instruments and practices that had been devised to work specifically in a laboratory environment. This modest strategy proved highly effective in ways that could hardly have been foreseen. The surprises of the 1930s and 1940s, unlike those of the 1910s and 1920s, tended to be pleasant ones.

The new natural history circa 1900 was an ideological movement, a top-down reform by lab-trained biologists. It was impelled by experience of the limitations of purely laboratory methods, not by experience of proven field methods. Scientific naturalists projected an ideal or imagined community based on field practices that had yet to be invented. Their vision of a shared future reflected the inexperience of a generation who expected that laboratory practices would work in the field but who had not actually made them work there. In the event, their hopes were false.

In contrast, the border biology of the mid-twentieth century was based firmly on several decades of experience in the field. Border biologists were not colonizers from labs; they were mainly field biologists who worked in nature but were familiar and sympathetic with laboratory ideals and methods. Border biology was programmatically less self-conscious than the new natural history, but it was more soundly based in field practices that actually worked and had potential for wide application. It rested not on fond hopes but on real achievements. The practices described in this chapter and others not discussed, like Paul Errington's population ecology, proved to be the foundations of whole disciplines that at the turn of the twenty-first century are still going strong.

The difficulties experienced by the first new naturalists have a positive side, too. The failed first attempts to create scientific field practices disabused biologists of the belief that natural history could be made scientific by simply importing laboratory paraphernalia or protocols. This belief was a powerful impediment to the creation of effective field practices, at least as powerful as the practical difficulties of dealing with natural populations and complex environments. It was an understandable assumption at a time when a dominant and confident lab culture encountered "big nature" for the first time after half a century of astonishing achievement in dealing with "little nature" indoors (the terms are Eugene Odum's).[68] Nonetheless, unrealistic expectations that laboratory methods would work as well in the field served at first mainly to diminish the appeal of more humble but doable ways of working "big nature." It took a generation or more for field biologists to learn through hard experience to have confidence in their homegrown practices, to borrow critically and selectively from laboratory culture, and to value their own identity as border practitioners.

The experience of doing fieldwork in a world dominated by laboratory values is what gives border biologists' practices their coherence. Though highly disparate in their subject matter, these practices share a common strategy of approaching the complexities of big nature with the precision associated with investigations of little nature. These are, as outlined earlier, a preference for populations as natural units, a modestly quantitative and comprehensive approach to complexity, and an inventive use of natural places for analytical ends. The distinctive style of border practice thus reflects both the experience of immodest cultural borrowing from laboratory culture, and the more appropriate and effective combination of lab ideals and field realities that came out of that experience around midcentury.

The ambitions and disappointments of the first border crossers—Sumner, Shelford, Crampton, Shreve, Adams, Tower, Cowles, Clements, and others discussed in previous chapters—became a shared past, figures in the stories, heroic and cautionary, of a new generation who became permanent residents of a cultural border region.

68. Eugene P. Odum, "Energy flows in ecosystems: a historical review," *American Zoologist* 8 (1968): 11–18, on p. 16.

CHAPTER 9

Border Biology: A Transect

In the 1940s and 1950s a group of border disciplines were taking shape along the lab-field frontier. What in 1900 was a defensive boundary demarcating the different worlds of field and laboratory biology had by midcentury become a broad zone of scientific subcultures that were neither pure lab nor pure field but mixtures of the two. Between the worlds of lab and field science a third had been inserted.

The border practices discussed in the previous two chapters were crucial to this new cultural geography: not just symptomatic of fundamental change but constitutive of it. Mixed practices defined the new provinces of biology, gave them their identities, distinguished them from the older provinces that lay on either side of the border zone. Practices altered the experience of fieldwork. Traffic between laboratory and field no longer necessarily involved passage across a cultural frontier, or even physical movement from field to laboratory or vice versa. It was possible to work in the field without cultural borrowing and without thinking that fieldwork was a preliminary to a future in the lab—without a sense of incongruity between laboratory ideals and the realties of fieldwork. Status relations between lab and field were less one-sided: never equal, but distinctions between dominant and marginal, donor and borrower were less clear-cut than they were in 1900, and cultural exchange was less of a one-way street than before. Field biologists were less ready to concede the superiority of laboratory methods and more willing to assert the equal value of their own practices—of the field but like the lab in analytical force.

The first new naturalists envisioned a scientific landscape in which field and laboratory biologists could work cooperatively, but it was an imagined cultural space not a lived, experienced one. Fifty years later field biologists inhabited a stable and expanding border region and possessed effective mixed practices and a distinctive border identity. Tensions between lab and field values had hardly vanished (I will give a few anecdotes later on), but they had relaxed, and an infrastructure of formal and informal institutions was emerging in which lab and field biologists could interact: reading groups and seminars, journals, societies, cooperative projects, and networks of friendship.[1] It was not the imagined landscape of the new naturalists, but a working cultural landscape of diverse border practices.

As I suggested earlier, the analogy with imperial frontiers helps us visualize the cultural geography of border biology. The two kinds of border regions took shape in similar ways. Along the Roman frontier in "barbarian" Europe, a controlled military boundary was transformed by continual interactions across it into a region of border societies that were neither Roman nor barbarian but Frankish, German, Burgundian, and Gothic. The accumulated experience of daily frontier traffic—economic, diplomatic, military—reordered social alignments and interests to create new provincial societies. Likewise, cultural interactions across the lab-field frontier—exchange of "material" and specimens, borrowing of material culture and protocols, the physical to-and-fro of fieldwork—gave rise in the space of a few decades to novel forms of scientific culture that were different from both laboratory and field.

Physiologists and naturalists as Romans and barbarians? Ecosystem ecologists and evolutionary taxonomists as Franks and Allemans? It sounds farfetched, but despite the obvious differences of geographical scale and time, I think the processes of cultural-geographic change are enough alike for us to think about them in the same way. Cultures of all sorts evolve new

1. See, e.g., Joseph A. Cain, "Common problems and cooperative solutions: Organizational activities in evolutionary studies, 1936–1947," *Isis* 84 (1993): 1–25; Cain, "Ernst Mayr as community architect: Launching the Society for the Study of Evolution and the journal *Evolution*," *Biology and Philosophy* 9 (1994): 387–427; Vassiliki Betty Smocovitis, "Organizing evolution: Founding the Society for the Study of Evolution (1939–1950)," *Journal for the History of Biology* 27 (1994): 241–309; and Frank B. Golley, *A History of the Ecosystem Concept in Ecology* (New Haven: Yale University Press, 1993), chap. 5.

forms and create new space along boundaries by the accumulated experience of daily interactions. Why should we assume that science works differently from other forms of culture? A cultural-geographic approach reveals the fundamental similarities among phenomena that superficially seem quite different.

The Border Zone: A Bird's-Eye View

Border biology circa 1950 was not homogeneous but a mosaic of practices that combined laboratory and field elements in diverse combinations. The various practices described in this book do not constitute a developmental series: one did not succeed the other, but all accumulated in layers, and in the postwar decades all were carried on, if not with equal fervor. To describe postwar border biology would be an enormous and complex task—another whole book. So I will conclude this book with a quick survey of the various kinds of practice that were available to border biologists at midcentury—a kind of cultural transect. Practices are a rough-and-ready measure of cultural terrain but a useful preliminary to full-scale mapping. After all, production practices are the central features of cultures of all kinds, the source of livelihoods and identities, the focus of beliefs and customs, the object of hopes and fears. So a transect that takes practices as indicators of cultural location is not simply a convenient shortcut but a good foundation for future empirical work.

My cultural transect will not resemble Robert Whittaker's gradients, with their large samples closely spaced; it is more a bird's-eye transect, the sort of preliminary overview that a naturalist might do of a mountain forest, scouting out good sites for intensive study, standing on a peak in autumn and seeing how the patches of different colors on distant slopes suggest a complex ecological mosaic. Even such a casual survey reveals a broad and complex cultural topography, ranging from pure field to pure lab with many shades of mixtures in between. (Not all are covered in this book, in which I have intentionally stayed on the field side.)

Starting at the laboratory edge of the zone, where it adjoins physiology and genetics, we see an ill-defined belt overlapping with the laboratory disciplines, in which lab-bred experimenters apply standard methods to wild plants and animals, making forays into the field to collect "material" but operating mainly indoors—a kind of scientific poaching. Some try out potentially new "standard" animals, seeking an advantage along the margins of a crowded and competitive world. Some aspire to make laboratory practices more relevant to natural phenomena, but they generally do so with no view of altering the essential purposes and standards of laboratory culture. Here the dominant laboratory culture conducts casual colonizing, extending its reach where opportunity presents itself but without thought of taking over areas of fieldwork. Early-twentieth-century journals of genetics and experimental zoology are full of reports in this mode.

Close to this area of casual border crossing is another, in which we find field biologists applying genetic and physiological methods to field material with the intention of making their disciplines more lablike and scientific. Much of "autecology," that is the ecology of individual organisms, was of this sort in the period we are surveying; likewise a good deal of animal behavior or ethology. Good examples of this kind are Charles Kendeigh's or Eugene Odum's researches in the 1930s and 1940s on the physiology of bird behavior. Experiments were done mainly with individual organisms rather than with populations, and they were generally conducted in controlled laboratory or garden conditions, rather than in nature. It was physiology, but performed on nonstandard material to understand not cells or tissues but activities like reproduction and migration that are fundamental to the lives of animals in nature. Autecologists and ethologists did not intend their disciplines to merge with physiology (though some thought they might eventually). The idea was to create experimental disciplines of field biology equal in standing and authority to physiology.

Somewhat further from the purely laboratory disciplines we encounter a zone of practices that are carried on with laboratory techniques but in the field, on natural populations or ecosystems, and sometimes in places resembling natural laboratories. Theodosius Dobzhansky's population genetics is one such practice, as is the experimental taxonomy of Jens Clausen and his group—and in a slightly different mode the ecosystem ecology invented in the 1950s by Eugene and Howard Thomas Odum. These practices are in a way mirror images of the border practices discussed in the previous chapter: not field methods made lablike by amplification, but laboratory methods made natural, diverted—hijacked one might say—to the study of natural populations and assemblages.

Take Dobzhansky's population genetics, for example: it dealt with a wild and highly variable species of fly, *Drosophila pseudoobscura,* not the standard domesticated *D. melanogaster.* It required extensive field trips to collect local populations and varieties, which Dobzhansky enjoyed enormously (he began his career as a field naturalist and taxonomist). Yet the work was essentially classical genetics—mapping marker alleles and translocations, measuring salivary chromosomes, crossing and sorting flies—performed indoors in standard genetic labs. Its subject, however, was not the structure of chromosomes and the mechanics of genetic transmission, but the genetic structure and dynamics of natural populations and their functioning as units of evolutionary change.

In population genetics, chromosomal mapping was transformed from an end in itself to the means to new, evolutionary ends. The products of Dobzhansky's practice were not improved genetic maps but maps of distinct local populations and graphs of their phylogenies and gradual changes. It was a kind of genetic biogeography, doing by means of genetic techniques essentially what biogeographers like Ernst Mayr did with taxonomic and biogeographic methods. The convergence was unexpected. Dobzhansky was astonished at the genetic variability even in very local populations; Mayr was not, nor would any taxonomist be who has studied the morphological variations of large series of specimens collected over the whole range of a species or genus.[2]

In inventing population genetics Dobzhansky went well beyond the experimental biologists' usual custom of going into the field for organisms to domesticate and assimilate into the controlled environment of the lab. In Dobzhansky's laboratory, variability and locality were not complications to be eliminated but the very objects of study. It is perhaps not surprising that some of Dobzhansky's laboratory colleagues were scandalized when he took up with his wild flies. As Mayr diverted taxonomic methods to evolutionary ends by subverting taxonomic categories, so Dobzhansky diverted genetics to evolutionary ends by applying its techniques to natural and variable populations. Population genetics became a large area of intermediate color on the map of border biology.

Dobzhansky did do the odd field experiment. For example, he released batches of genetically marked lab flies into a natural population to see how fast the flies and their mutant allele spread into the native population. (Expecting random dispersion, he selected "experimental fields" that were topographically homogeneous—the ideal of laboratory measurement is plain enough.) These experiments were first tried in 1941–42 at a tourist resort on

2. William B. Provine, *Sewall Wright and Evolutionary Biology* (Chicago: University of Chicago Press, 1986), chaps. 10–11. Robert E. Kohler, *Lords of the Fly* (Chicago: University of Chicago Press, 1994), chap. 8. Mark B. Adams, ed. *The Evolution of Theodosius Dobzhansky* (Princeton University Press, 1994). Richard C. Lewontin et al., eds. *Dobzhansky's Genetics of Natural Populations I–XLIII* (New York: Columbia University Press, 1981). Theodosius Dobzhansky to Ernst Mayr, 21 Feb. 1939, EM-APS. Mayr to Carl Epling, 9 Jan. 1948, EM-HU box 5.

Mount San Jacinto, California, and repeated with improvements at Jens Clausen's field station at Mather, in the Sierras, in 1945–46.[3] But the vast majority of Dobzhansky's work on natural populations was straight genetics: collecting in the field (preferably in places with mountain scenery and attractive camp sites) followed by extensive genetic analysis in the lab. There was some justice in Jens Clausen's remark on the occasion of Dobzhansky's stay at Mather in 1945 and 1946 that to lab geneticists Dobzhansky might seem radical, but to working field biologists he was barely emerging from the "milk-bottle" stage of genetics.[4] His was a distinctly laboratory practice, though its objects of study were natural populations.

Moving along toward the field side of our transect we find Jens Clausen's experimental taxonomy. Although its name might seem a contradiction in terms (is not taxonomy the quintessential nonexperimental science?), it was in fact a remarkably congenial blend of experimental methods and taxonomic ends. Simple transplant experiments made visible the local populations, or "ecotypes," of which species are composed, and the physiological adaptations crucial to species' distribution and evolution. In a sense Clausen created new diagnostic characters—responses of plants to standard gardens—that exist only in experiments, not in nature. He also hijacked the method of interspecies crosses, a standard genetic technique as we saw in chapter 5, to map the phylogenetic distance between related species. Clausen turned a character that was a mere nuisance to lab geneticists—interspecies sterility—to a novel taxonomic and evolutionary purpose.

Dobzhansky, Clausen, and Mayr shared an interest in local populations and evolutionary process but were differently situated in the cultural geography of the border zone. Mayr relied exclusively on taxonomic and biogeographic methods, as we have seen; he was a man of the field. Dobzhansky and Clausen combined field and laboratory work but in different ways. Dobzhansky was the most lablike of the three, making regular collecting excursions but working mainly in a lab. Clausen was in between, spending more time in the field doing ecological survey and working in the transplant gardens at Mather and Timberline. But if the experimental taxonomists could slight Dobzhansky as a greenhorn and "milk-bottle" geneticist, they were themselves open to a similar critique by field botanists who worked with plant species that could not be bred in experimental gardens and thus

3. Theodosius Dobzhansky and Sewall Wright, "Genetics of natural populations, I: Dispersion rates in *Drosophila pseudoobscura*," *Genetics* 28 (1943): 304–40. Dobzhansky and Wright, "Genetics of natural populations, XV: Rate of diffusion of a mutant gene through a population of *Drosophila pseudoobscura*," *Genetics* 32 (1947): 303–24.

4. Herman A. Spoehr to W. M. Gilbert, 8 Oct. 1945, CIW f. Dobzhansky.

had to be studied in the field. Population geneticists and experimental taxonomists pursued varieties of border practice, neither pure lab nor pure field but a blend of both; two adjacent patches of mottled color in the cultural landscape.

Another hybrid practice, closer to the middle of the border zone but combining extreme elements of both lab and field, is the ecosystem ecology conceived by Eugene and Thomas Odum in the mid-1950s. Thanks to its practical uses in environmental management and to government funding of large-scale projects, ecosystem ecology came to exemplify ecology in the Cold War period. As defined by Eugene Odum in his famous 1953 textbook, ecosystem ecology was essentially Raymond Lindeman's trophic ecology with the biochemistry greatly amplified and most of the natural history stripped away. (The scientistic tone of ecosytem ecology, and its field methods, owe much to Thomas Odum, a biophysicist and evangelical promoter of physics and chemistry in life science.)

In the Odums' version of trophic ecology, biotic systems were reduced to the single function of production and metabolism. Ecosystem ecologists were interested in measuring the efficiency of conversion of radiant energy into biomass and of transfers of energy between trophic levels.[5] They treated ecosystems very much as experimental biochemists do individual animals, measuring metabolic inputs and outputs. It was perhaps the closest anyone ever came to realizing the old ideal of ecology as a physiology of the field, though not in a way that Henry Cowles or William Ganong could have foreseen.

Technically, measuring the metabolism of an ecosystem was not too different from measuring a tissue slice in vitro or a living animal in a respirometer: chemical analysis of oxygen, carbon dioxide, and perhaps a few essential nutrients, a biomass balance of inputs and outputs—period. But it was one thing to enclose a mouse or bit of tissue and another to bound a whole ecosystem: its complex exchanges with its atmospheric and terrestrial environment are simply too vast. One could not dig it out or put it under glass, so the trick was to find a place that was naturally contained and so arranged that measurements of nutrient concentrations could yield total energy flows.

One such place was found on the tiny coral atolls that dot the South Pa-

5. Frank B. Golley, *A History of the Ecosystem Concept in Ecology* (Yale University Press, 1993), chap. 4, esp. pp. 62–66. Joel B. Hagen, *An Entangled Bank: The Origins of Ecosystem Ecology* (New Brunswick: Rutgers University Press, 1992), chap. 7. Eugene P. Odum, "The strategy of ecosystem development," *Science* 164 (1969): 262–70. Odum, *Fundamentals of Ecology* (Philadelphia: Saunders, 1953). Howard T. Odum, "Ecological potential and analogue circuits for the ecosystem," *American Scientist* 48 (1960): 1–8. Eugene P. Odum, "Energy flow in ecosystems: A historical review," *American Zoologist* 8 (1968): 11–18.

cific, rings of low islets connected by shallow reefs, like beads on a string. On the windward side of these islet chains the prevailing easterlies blow a steady, one-way flow of water over the breaker zone and down the gently sloping reef into the inner lagoon, "like a broad mountain stream rippling over a rocky bed," as Howard Odum put it. Because the flow was steady, ecologists could measure its rate, as well as the changes in nutrients and respiration gases resulting from the photosynthesis and respiration of the reef's biota between the surf edge and the inner lagoon. (The daytime respiration and photosynthesis could be calculated by measuring respiration at night, when photosynthesis had stopped.) From this data, plus measurements of solar flux, the reef's production of biomass could be computed. The basic techniques of sampling and data manipulation of the ecosystem method were worked out in a few months at Rongelap and Eniwetok in the Marshall Islands.[6]

But the place that puts ecosystem practice most clearly on our cultural-geographic map is Silver Springs, Florida, one of the numerous large warm springs that are a peculiar feature of the peninsula. Some of these springs are huge: underground rivers that flow through gravel strata, come to the surface in "boils," and flow in "runs" to the sea. Silver Spring pours out 300 million gallons a day of water free of all organic matter, at a steady rate and a year-round constant temperature of 73 degrees. The boil and run support a distinctive biotic community of bottom weeds, photosynthetic algae, and fish (as well as fishermen, swimmers, and tourists).

To lab-trained ecologists like Thomas Odum, Silver Spring was in effect a natural instrument, a giant outdoor chemostat constant in its nutrients, temperature, and flow, in which the metabolism of a resident community could be in effect bounded and quantitatively measured—and much more easily than on a South Seas reef. By simply measuring dissolved oxygen and carbon dioxide at points along the run, and the intensity of solar energy at the water's surface, Odum could calculate the productivity, respiration, and efficiency of the spring's biotic community. Because the rate of flow is constant and the spring water free of organic matter, distance down the run corresponds to an interval of time following the coming together of water, sunlight, and plants at the boil, as if in a chemical experiment. It was, as Odum put it, a "chemical kinetic apparatus" in which experiments could be per-

6. Odum, "Energy flow in ecosystems" (cit. n. 5). Howard T. Odum and Eugene Odum, "Trophic structure and productivity of a windward coral reef community on Eniwetok Atoll," *Ecological Monographs* 25 (1955): 291–320, on pp. 293–95, 295 (quote), 310–17. Marston C. Sargent and Thomas S. Austin, "Organic productivity of an atoll," *Transactions of the American Geophysical Union* 30 (1949): 245–49.

9.1 Silver Springs, Florida, looking downstream from the main "boil," with tourist boats. Howard T. Odum, "Trophic structure and productivity of Silver Springs, Florida," *Ecological Monographs* 27 (1957): 55–112, on p. 58.

formed with a known starting point and controlled duration in an essentially invariant environment (though corrections did have to be made for the variable shading by trees and clouds).

Not only the place but the experience of fieldwork at Silver Spring was lablike. Because conditions were constant, data were cumulative and reproducible. Data that was not gathered one day could be got the next, a laboratory luxury seldom enjoyed by field ecologists. And one researcher's data could be critically checked by others in the same conditions as in a lab—a "most terrible and healthy [thing] for the poor ecologists," Tom Odum quipped. And comparisons could be made of the "metabolism" of springs having different characteristics, thus turning all of central Florida into a region-sized outdoor laboratory.[7] (Later, measurements of the productivity and metab-

7. G. E. Ferguson, D. W. Lingham, S. K. Love, and R. O. Vernon, "Springs of Florida," *Florida Geological Survey Bulletin* 31 (1947), on pp. 5–6, 37. Howard T. Odum, "Trophic structure and productivity of Silver Springs, Florida," *Ecological Monographs* 27 (1957): 55–112, on pp. 55–56, 58 (quote), 87–88. Howard T. Odum, "Primary production in eleven Florida springs and a marine turtlegrass community," *Limnology and Oceanography* 2 (1957): 85–97, on p. 85.

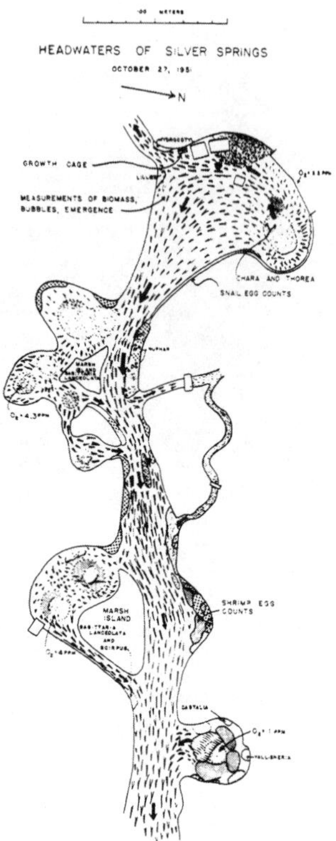

9.2 Silver Springs, Florida, showing research sites where Howard T. Odum measured the material and energy flows of an aquatic ecosystem. Odum, "Trophic structure and productivity," on p. 57.

olism of ecosystems were extended to less exotic places, like the small watersheds of gentle mountain slopes.)[8]

Ecosystems as respiring, quasi-experimental organisms? Natural places as experimental instruments? How are we to conceptualize such exotic cultural terrain? As the vanguard of a laboratory culture expanding more aggressively into the field? Or as relicts of a general retreat from a primitive and simplistic vision of ecology as an experimental physiology of the field? Neither, I think. Rather it was a practice distinctive of a border culture, no longer understandable in terms of a lab-field polarity but something new

8. Frank Herbert Bormann and Gene E. Likens, "Nutrient cycling," *Science* 155 (1967): 424–29. Likens, Bormann, et al., *Biogeochemistry of a Forested Ecosystem* (New York: Springer, 1977), pp. 1–29.

and distinctively itself. Ecosystem ecology as practiced at Silver Spring was in a way a mirror image of the border practices of Mayr, Whittaker, Lindeman, and Edgar Anderson: not a field practice made as good as laboratory experiment, but a laboratory practice adapted to a complex natural place. It is not natural history brought up to lab standards, but laboratory practice made relevant to natural objects and conditions. On our cultural-geographic transect, population genetics, experimental taxonomy, and ecosystem ecology are belts of mottled hue, border practices analogous on the laboratory side to those on the field side discussed in chapter 8. Together they make up a broad middle region of our border zone.

Proceeding on our cultural transect toward the field side, we encounter still other areas, ever greener as elements of laboratory culture disappear and field culture predominates. Here we find practitioners of the "new systematics" who in the 1930s and 1940s reinvented systematics as an evolutionary science and combined it with genetics in the so-called evolutionary synthesis. These synthesizers—people like Ernst Mayr or Julian Huxley—were field biologists broadly trained in experimental biology and too young to have firsthand memories of the events in the early 1900s that created a wall of mistrust and resentment between taxonomists and "mutationists." The new systematists did not do genetics, but they read and understood it and welcomed intellectual exchange with geneticists. And they practiced taxonomy and biogeography in ways that made it accessible and relevant to the geneticists who shared their interest in the mechanisms of speciation. Alfred Kinsey is a good early example of this type. Although his taxonomic methods were somewhat old-fashioned and his favorite organism ill-suited for experiment, Kinsey's massive biogeographic studies of North American gall wasps were the single best source of data on variation and speciation in natural populations before Dobzhansky and Mayr.[9]

Joseph Grinnell was another such figure: a systematic mammalogist and field-worker with an abiding interest in the mechanism of evolution. He encouraged Francis Sumner's experiments, as we have seen, and even did the odd experiment himself when the occasion presented itself (for example, on the ability of small mammals to cross water barriers).[10] Many naturalists of this sort were active in the 1930s and 1940s: for example, the ichthyologist Carl Hubbs, the entomologist Alfred Emerson, and the herpetologist Karl

9. Alfred C. Kinsey, "The Gall Wasp Genus *Cynips:* A Study in the Origin of Species," *Indiana University Studies* 16 (1929): 1–507. Kinsey, "The Origin of Higher Categories in *Cynips,*" *Indiana University Publications Science Series* 4 (1936): 1–334.

10. Francis B. Sumner to William E. Ritter, 7 April 1914, FBS-SIO box 5 f. 31. Hilda Wood Grinnell, "Joseph Grinnell, 1877–1939," *Condor* 42 (1940): 3–34.

Schmidt. Some, like Hubbs and Emerson, combined field and experimental work (on fish hybrids and termite behavior, respectively). Others, like Kinsey, were dubious that the results of field experiments would repay the labor that would have to go into them, but they knew what was going on in relevant experimental disciplines, and there is no question that their biogeographic views were shaped by that knowledge.

Likewise with field ecologists like Forrest Shreve. Forced to abandon experimental work in the early 1930s, Shreve devoted the last decade of his working career to extensive field surveys of the Sonoran and other deserts. Between 1933 and 1937 he made ten extensive field trips in the Sonoran desert, crisscrossing Sonora, Baja California, and the American Southwest—over ten thousand miles by auto (two autos, in case one broke down) over rough roads and trackless desert, forty to fifty miles a day, counting and mapping species, gathering or measuring climatic data, camping and living off the land. It was to be the first installment of a comparative study of the four desert types of North America. (Shreve also surveyed the Chihuauan desert during 1937–44 but never published the results, and he was forced by illness and wartime gasoline rationing to abandon altogether surveys of the Mohave and Great Basin deserts.)[11]

Shreve's account of the ecology of the Sonoran desert is mostly descriptive and biogeographical: lists of plants and their habitat preferences and ranges, descriptions of the distinctive look or "physiognomy" of this most varied and interesting of North American deserts. But it is not merely descriptive. Shreve meant his survey of Sonoran vegetation to be the empirical foundation for an evolutionary history of deserts and desert plants. (In the absence of any preserved or fossil evidence, knowledge of the ecology and distribution of existing plants was all the evidence there was.) Shreve's account was also shaped by his lifelong interest in physiology and physiological life history. As he observed, knowing the physiological requirements of major desert plants guided his work in the less well known areas of the desert south of the border, and "did much to offset the limitations which are inherent in the expeditionary type of research." He knew where to look for ecological boundaries and why desert species could (confusingly) inhabit places with very different soils and climates. Shreve's theory that the four different types of desert vegetation reflect differences of temperature, aridity and—most important—the seasonality and predictability of rain, likewise

11. Janice Bowers, *A Sense of Place: The Life and Work of Forrest Shreve* (Tucson: University of Arizona Press, 1988), pp. 110–115, 138–48.

9.3 Map of the Sonoran desert showing the routes traveled by Forrest Shreve in his study of its physiography and natural history. Forrest Shreve, *Vegetation of the Sonoran Desert*, Carnegie Institution Publication no. 591 (1951), on p. 2.

derived from a knowledge of physiological ecology. (For example, Sonoran vegetation is so diverse because rain occurs fairly predictably twice a year.)[12]

Like Ernst Mayr's systematics, Shreve's ecology may seem to be a pure natural history, but in fact it is natural history done by a man who thinks like a laboratory biologist and understands the biology underlying the form and distribution of species in nature. Shreve's Sonoran work remains foundational for desert ecologists to this day, and in our transect of field practices we may perhaps take it as the border practice closest to the pure sciences of the field. Beyond that we enter the realm of mapping and historical reconstruction: the realm of field biologists little touched by an imperial laboratory culture—paleontologists and paleoecologists (who reconstruct ancient environments and biomes), biogeographers, and so on—who complete our cultural transect.

Conclusion

It is clear even from this swift bird's-eye view that the lab-field border circa 1950 was a broad and very diverse region of field science. At the center are the border practices of naturalists like Mayr, Anderson, Lindeman, Whittaker, Hubbs, Clausen, Shreve, and Paul Errington, who worked in the field but knew the lab and transformed traditional field methods into practices that were as effective as experiments and more appropriate to complex natural objects and places. These, more than any others, exemplify the border culture that had grown around a defended boundary in the preceding half century. But the border zone was not their domain exclusively; it afforded stable and productive careers to biologists with tastes ranging from pure lab to pure field and combinations in between. Border biology is a layered, mosaic culture.

To be sure, the tension between laboratory and field values did not simply fade away with the appearance of these border practices. Differences of status, defensiveness, and resentment remained familiar features of border life. In the mid-1950s, just a few years after Ernst Mayr thought systematic

12. Forrest Shreve, *Vegetation of the Sonoran Desert,* Carnegie Institution Publication no. 591 (1951). Shreve, "The desert vegetation of North America," *Botanical Review* 8 (1942): 195–246. Shreve, "Ecological aspects of the deserts of California," *Ecology* 6 (1925): 93–103, on pp. 94–95, 102. Shreve, "Statement regarding the aims and work of the Desert Laboratory and the biological problems of the desert," 5 Nov. 1931, on pp. 17–24; Shreve to John C. Merriam, 6, 30 April 1932, 30 Jan. 1933; Shreve, "Projected botanical investigations of the Southwestern desert," 30 Nov. 1932, pp. 1–2, 5–8; all in CIW-DL f. general 1931–72.

zoologists were finally getting the standing they deserved, the spectacular achievements of the molecular biologists once again threw a stunting shadow over the field disciplines. The zoologist Edward O. Wilson has recounted how the avatar of molecular biology, James Watson, could with a single disparaging remark derail the appointment of an ecologist to the Harvard department. "Are they out of their minds?" he wondered aloud in a faculty meeting. "Anyone who would hire an ecologist is out of his mind." That was that. About the same time a committee of physiologists surveyed their discipline and projected a future branch of "field physiology"—as if ecology had never existed. What drove Eugene Odum to reinvent trophic ecology in a laboratory style was the threat that systematic zoology would be dropped from his department's core curricula and activities in a reorganization.[13] No doubt many field biologists could tell a similar story. So long as experiment is deemed to be a superior way of knowing, field biologists will be obliged to assert their presence and justify their claims in a way that laboratory workers never are. Since the mid-nineteenth century, field biologists have lived in a world where lab disciplines have the greater credibility and authority, and they do still.

It is no less true, however, that the lives of field biologists have much improved since the first new naturalists became border crossers. Border disciplines afford field biologists stable and respectable identities and a self-confidence founded on proven practices and real accomplishments. Field biologists remain ready importers of laboratory instruments and techniques (isotopes, for example, or DNA fingerprinting), but they borrow less slavishly since they have the option of border practices. They have learned to select what they want from lab culture and leave unwanted baggage behind. Laboratory imports are less likely to be used as talismans to ward off criticism or as barriers to entry by amateur practitioners. Boundaries need not be policed quite so diligently when most practitioners have academic training and credentials.

Career patterns in the border zone show less of the instability that was so striking in the careers of the first new naturalists. There is less lurching between field and laboratory. Field biologists no longer assume that careers begin in the field and end up in the lab, or that observation is merely a preliminary to experiment. They operate not as frontier crossers but as relatively settled inhabitants of border cultures, and their lives are thus more

13. Edward O. Wilson, *Naturalist* (Washington: Island Press, 1994), on pp. 219–220. Ernst Mayr et al., "Biological materials," [1954], on verso of p. 9, Karl P. Schmidt Papers box 12 f. 4, Field Museum, Chicago, Ill. Golley, *History of the Ecosystem Concept* (cit. n. 5), p. 65–66.

predictable. From about the mid-1930s, border biologists could expect to have stable and productive careers. Never again would a large proportion of the most talented practitioners of a field discipline become inactive in mid-career or depart for greener pastures.

Yet for all that, the foundational differences between laboratory and field culture and practices—material, spatial, cultural—still matter in many field biologists' lives. Borrowing remains largely a one-way street, and though the standing of field practices is firmer now, in the world's eye observation is still not as good a way of knowing as experiment. And field practices, though somewhat insulated from fickle ideological winds, are hardly immune to laboratory fashions. Witness the current claims of molecular taxonomists that DNA should supersede morphological criteria for species—Charles B. Davenport's place modes revisited.

The differences between field and laboratory objectives, places, and practices still shape choices and careers in the border zone, because they are not mere conventions but facts of life. The line between nature and artifice can be blurred but not erased. Natural places cannot be made so lablike that they become unnatural; laboratories cannot be made so natural that they lose the artifice that gives them their power. So, too, with field and laboratory practices. Push quantification or modeling too hard and they become meaningless; take experiments too far afield and they are discredited. Differences between lab and field are a major source of creative innovation. In a patchy cultural landscape, isolation and chance opportunities for mixing—or introgression—make possible novel kinds of practice. The art of border biology will always be to borrow, adapt, and blend—and to know the limits of cultural borrowing. Mixed practices are the bread and butter and the soul of field biology: field practices amplified and lablike, natural history made scientific.

ABBREVIATIONS

AA	Annie Alexander Papers, Bancroft Library, Berkeley, California
AGR	Alexander G. Ruthven Papers, Butler Library, Ann Arbor, Michigan
AGV	Arthur G. Vestal Papers, University of Illinois Archives, Urbana, Illinois
CBD-CHS-1	Charles B. Davenport Papers, Cold Spring Harbor Series 1, American Philosophical Society Library, Philadelphia, Pennsylvania
CBD-CSH-2	Charles B. Davenport Papers, Cold Spring Harbor Series 2
CBD-Gen.	Charles B. Davenport Papers, General Correspondence
CIW	Carnegie Institution of Washington Archives, Washington D.C
CIW-DL	Carnegie Institution of Washington Archives, Desert Laboratory Series
CLH	Carl L. Hubbs Papers, Scripps Institution for Oceanography Archives, La Jolla, California
EM-APS.	Ernst Mayr Papers, American Philosophical Society Library, Philadelphia, Pennsylvania
EM-HU	Ernst Mayr Papers, Harvard University Archives, Cambridge, Massachusetts
EM-JH-HU	Ernst Mayr–Julian Huxley Correspondence, Harvard University Archives

f., ff.	folder, folders
FBS-Fam	Francis B. Sumner Family Papers, Scripps Institution for Oceanography Archives, La Jolla, California
FBS-SIO	Francis B. Sumner Papers, Scripps Institution for Oceanography Archives
FEC	Frederic E. Clements Papers, American Heritage Center, University of Wyoming, Laramie, Wyoming
EGC	Edwin Grant Conklin Papers, Princeton University Archives, Princeton, New Jersey
GEH	George Evelyn Hutchinson Papers, Yale University Archives, New Haven, Connecticut
HFN	Henry F. Nachtrieb Papers, University of Minnesota Archives, Minneapolis, Minnesota
HSJ	Herbert S. Jennings Papers, American Philosophical Society Library, Philadelphia, Pennsylvania
INHS	Illinois Natural History Survey Papers series 43/1/1, University of Illinois Archives, Urbana, Illinois
JG-MVZ	Joseph Grinnell Papers, Museum of Vertebrate Zoology, Berkeley, California
JG-UC	Joseph Grinnell Papers, series C-B955, Bancroft Library, University of California, Berkeley
PU-Bio	Princeton University Biology Department Papers, Harvey Mudd Library, Princeton, New Jersey
RLL	Raymond L. Lindeman Papers, Yale University Archives, New Haven, Connecticut
UC-Pres	University of Chicago Presidents Papers, University of Chicago Special Collections, Chicago, Illinois
UC-Trus	University of Chicago Trustees Papers, University of Chicago Special Collections
UM	University of Minnesota Archives, Minneapolis, Minnesota
UM-Bot	University of Minnesota Botany Department Papers, University of Minnesota Archives
UW-LIM	University of Wisconsin Limnology Papers, University of Wisconsin Archives, Madison, Wisconsin
WCA	Warder C. Allee Papers, University of Chicago Special Collections, Chicago, Illinois
WMW	William M. Wheeler Papers, Harvard University Archives, Cambridge, Massachusetts
WER	William E. Ritter Papers, Bancroft Library, University of California Archives, Berkeley, California

BIBLIOGRAPHY

Allen, Garland E. "The transformation of a science: T. H. Morgan and the emergence of a new American biology." In *The Organization of Knowledge in Modern America, 1860–1920,* edited by Alexandra Oleson and John Voss, pp. 173–210. Baltimore: Johns Hopkins University Press, 1979.

———. "Naturalists and experimentalists: The genotype and the phenotype." *Studies in History of Biology* 3 (1979): 179–209.

Anderson, Benedict. *Imagined Communities: Reflections on the Origin and Spread of Nationalism.* London: Verso, 1983.

Aron, Stephen. *How the West was Lost: The Transformation of Kentucky from Daniel Boone to Henry Clay.* Baltimore: Johns Hopkins University Press, 1996.

Becking, Rudy. "The Zürich-Montpellier school of phytosociology." *Botanical Review* 23 (1957): 411–88.

Benson, Keith R. "Laboratories on the New England shore: The 'somewhat different direction' of American marine biology." *New England Quarterly* 61 (1988): 55–78.

———. "Experimental ecology on the Pacific coast: Victor Shelford and his search for appropriate methods." *History and Philosophy of the Life Sciences* 14 (1992): 73–91.

Bowers, Janice E. *A Sense of Place: The Life and Work of Forrest Shreve.* Tucson: University of Arizona Press, 1988.

Bowler, Peter J. *The Eclipse of Darwinism: Anti-Darwinian Evolution Theories in the Decades around 1900.* Baltimore: Johns Hopkins University Press, 1983.

Browne, Janet. *The Secular Ark: Studies in the History of Biogeography.* New Haven: Yale University Press, 1983.

Burkhardt, Richard W., Jr. "Ethology, natural history, the life sciences, and the problem of place." *Journal of the History of Biology* 32 (1999): 489–508.

Cain, Joseph A. "Common problems and cooperative solutions: Organizational activities in evolutionary studies, 1936–1947." *Isis* 84 (1993): 1–25.

———. "Ernst Mayr as community architect: Launching the Society for the Study of Evolution and the journal *Evolution.*" *Biology and Philosophy* 9 (1994): 387–427.

Cetina, Karin Knorr. "The couch, the cathedral, and the laboratory: On the relationship between experiment and laboratory in science." In *Science as Practice and Culture,* edited by Andrew Pickering, pp. 113–38. Chicago: University of Chicago Press, 1992.

———. *Epistemic Cultures: How the Sciences Make Knowledge.* Cambridge: Harvard University Press, 1999.

Churchill, Frederic B. "William Johannsen and the genotype concept." *Journal of the History of Biology* 7 (1974): 5–30.

Cittadino, Eugene. *Nature as the Laboratory: Darwinian Plant Ecology in the German Empire, 1880–1900.* New York: Cambridge University Press, 1990.

Clarke, Adele E., and Joan H. Fujimura. "What tools? Which jobs? Why right?" In *The Right Tools for the Job: At Work in Twentieth-Century Life Sciences,* edited by Clarke and Fujimura, pp. 3–44. Princeton: Princeton University Press, 1992.

Cook, George M. "Neo-Lamarckian experimentalism in America: Origins and consequences." *Quarterly Review of Biology* 74 (1999): 417–37.

Cook, Robert E. "Raymond Lindeman and the trophic-dynamic concept in ecology." *Science* 198 (1977): 22–26.

Croker, Robert A. *Pioneer Ecologist: The Life and Work of Victor Ernest Shelford, 1877–1968.* Washington: Smithsonian Institution Press, 1991.

Cronon, William. *Nature's Metropolis: Chicago and the Great West.* New York: Norton, 1991.

Daston, Lorraine. "The moral economy of science." *Osiris* 10 (1995): 3–24.

———. "Fear and loathing of the imagination in science." *Daedalus* 127:1 (1998): 73–85.

DeBuys, William. *Salt Dreams: Land and Water in Low-Down California.* Albuquerque: University of New Mexico Press, 2000.

De Chadarevian, Soraya. "Laboratory science versus country-house experiments: The controversy between Julius Sachs and Charles Darwin." *British Journal for the History of Science* 29 (1996): 17–41.

Dettelbach, Michael. "Global physics and aesthetic empire: Humboldt's physical portrait of the tropics." In *Visions of Empire: Voyages, Botany, and Representations of Nature,* edited by David P. Miller and Peter H. Reill, pp. 258–92. Cambridge: Cambridge University Press, 1996.

Edward A. Birge: Teacher and Scientist. Madison: University of Wisconsin Press, 1940.

Engel, J. Ronald. *Sacred Sands: The Struggle for Community in the Indiana Dunes.* Middletown: Wesleyan University Press, 1983.

Farrell, Lindsay. "W. F. R. Weldon, biometry, and population biology." Unpublished paper, no date [1970s].

Findlen, Paula. *Possessing Nature: Museums, Collecting, and Scientific Culture in Early Modern Italy.* Berkeley: University of California Press, 1994.

Forgan, Sophie. "The architecture of display: Museums, universities and objects in nineteenth-century Britain." *History of Science* 32 (1994): 139–62.

Frey, David G. "Wisconsin: The Birge-Juday era." In *Limnology in North America,* edited by Frey, pp. 3–54. Madison: University of Wisconsin Press, 1963.

Galison, Peter. "Computer simulations and the trading zone." In *The Disunity of Science: Boundaries, Contexts, and Power,* edited by Galison and David J. Strump, pp. 118–57. Stanford: Stanford University Press, 1996.

———. *Image and Logic: A Material Culture of Microphysics.* Chicago: University of Chicago Press, 1997.

Galison, Peter, and Emily Thompson, editors. *Architecture and Science.* Cambridge: MIT Press, 1999.

Gieryn, Thomas F. "Biotechnology's private parts (and some public ones)." In *Making Space for Science,* edited by Smith and Agar, pp. 281–312.

———. "Two faces on science: Building identities for molecular biology and biotechnology." In *The Architecture of Science,* edited by Galison and Thompson, pp. 423–55.

———. *Cultural Boundaries of Science: Credibility on the Line.* Chicago: University of Chicago Press, 1999.

Gieryn, Thomas F., and Anne E. Figert. "Scientists protect their cognitive authority: The status degradation ceremony of Sir Cyril Burt." *Sociology of the Sciences Yearbook* 10 (1986): 67–86.

Gooday, Graeme. "The premisses of premises: Spatial issues in the historical construction of laboratory credibility." In *Making Space for Science,* edited by Smith and Agar, pp. 216–45.

———. "'Nature' in the laboratory: Domestication and discipline with the microscope in Victorian life science." *British Journal for the History of Science* 24 (1991): 307–41.

Golley, Frank B. *A History of the Ecosystem Concept in Ecology.* New Haven: Yale University Press, 1993.

Gross, Alfred O. "History and progress of bird photography in America." In *Fifty Years' Progress of American Ornithology 1883–1933,* pp. 159–80. Lancaster, Pa.: American Ornithological Union, 1933.

Haffer, Jürgen. *Ornithologen–Briefe des 20. Jahrhunderts. Ökologie der Vögel* 19 (1997): 1–980.

Hagen, Joel B. "Experimentalists and naturalists in twentieth-century botany: Experimental taxonomy, 1920–1950." *Journal of the History of Biology* 17 (1984): 249–70.

———. *An Entangled Bank: The Origins of Ecosystem Ecology.* New Brunswick: Rutgers University Press, 1992.

———. "Clementsian ecologists: The internal dynamics of a research group." *Osiris* 8 (1992): 178–95.

Hannaway, Owen. "Laboratory design and the aim of science: Andreas Libavius versus Tycho Brahe." *Isis* 77 (1986): 585–610.

Heilbron, John L. "Science in the church." *Science in Context* 5 (1989): 9–28.

Henke, Christopher. "Making a place for science: The field trial." *Social Studies of Science* 30 (2000): 483–511.

Ilerbaig, Juan. "Allied sciences and fundamental problems: C. C. Adams and the search for method in early American ecology." *Journal for the History of Biology* 32 (1999): 439–63.

Jack, Homer A. "The Biological Field Stations of the World: A Comparative and Descriptive Study." Ph.D. dissertation, Cornell University, 1940.

———. "Biological field stations of the world." *Chronica Botanica* 9 (1945): 5–73.

Jackson, Myles. "Illuminating the opacity of achromatic lens production: Joseph Fraunhofer's use of monastic architecture and space as a laboratory." In *Architecture and Science,* edited by Galison and Thompson, pp. 423–55.

———. *Spectrum of Belief: Joseph von Frauenhofer and the Craft of Precision Optics.* Cambridge: MIT Press, 2000.

Kendeigh, Charles S. "History and evaluation of various concepts of plant and animal commmunities in North America." *Ecology* 35 (1954): 152–71.

Kim, Kyung-Man. "On the reception of Johannsen's pure line theory: Toward a sociology of scientific validity." *Social Studies of Science* 21 (1991): 649–79.

Kingsland, Sharon E. "The battling botanist: Daniel Trembly MacDougal, mutation theory, and the rise of experimental evolutionary biology in America, 1900–1912." *Isis* 82 (1991): 479–509.

Kleinman, Kim. "His own synthesis: Corn, Edgar Anderson, and evolutionary theory in the 1940s." *Journal of the History of Biology* 32 (1999): 293–320.

Kofoid, Charles A. *The Biological Stations of Europe.* U.S. Bureau of Education Bulletin 440 (1910).

Kohler, Robert E. *From Medical Chemistry to Biochemistry: The Making of a Biomedical Discipline.* New York: Cambridge University Press, 1982.

———. "The Ph.D. machine: Building on the collegiate base." *Isis* 81 (1990): 638–62.

———. *Lords of the Fly:* Drosophila *Genetics and the Experimental Life.* Chicago: University of Chicago Press, 1994.

Largent, Mark A. "Bionomics: Vernon Kellogg and the defense of Darwinism," *Journal of the History of Biology* 32 (1999): 465–88.

Latour, Bruno. "Visualization and cognition: Thinking with eyes and hands." *Knowledge and Society* 6 (1986): 1–40.

———. *Science in Action.* Harvard University Press, 1987.

Lattimore, Owen. *Inner Asian Frontiers of China.* Washington: American Geographical Society, 1940. Reprint edition, New York: Oxford University Press, 1988.

Lillie, Frank R. *The Woods Hole Marine Biological Laboratory.* Chicago: University of Chicago Press, 1944.

McIntosh, Robert P. "H. A. Gleason, 'individualistic ecologist,' 1882–1975: His contribution to ecological theory." *Bulletin of the Torrey Botanical Club* 102 (1975): 253–73.

———. *The Background of Ecology, Concept, and Theory.* New York: Cambridge University Press, 1985.

MacKenzie, Donald A. *Statistics in Britain 1863–1930: The Social Construction of Scientific Knowledge.* Edinburgh: Edinburgh University Press, 1981.

Magnus, David. "In Defense of Natural History: David Starr Jordan and the Role of Isolation in Evolution." Ph.D. dissertation, Stanford University, 1993.

Maienschein, Jane. "Pattern and process in early studies of Arizona's San Francisco peaks." *BioScience* 44 (1994): 479–85.

Maienschein, Jane, Ronald Rainger, and Keith R. Benson. "Introduction: Were American morphologists in revolt?" *Journal of the History of Biology* 14 (1981): 83–87; also papers by various authors, ibid., pp. 88–176.

Mayr, Ernst. *Systematics and the Origin of Species from the Viewpoint of a Zoologist.* New York: Columbia University Press, 1942. New edition with "Introduction, 1999." Cambridge: Harvard University Press, 1999.

———. "Prologue: Some thoughts on the history of the evolutionary synthesis." In *Evolutionary Synthesis,* edited by Mayr and Provine, pp. 1–48.

———. "The role of systematics in the evolutionary synthesis." In *Evolutionary Synthesis,* edited by Mayr and Provine, pp. 123–36.

———. "How I became a Darwinian." In *Evolutionary Synthesis,* edited by Mayr and Provine, pp. 413–23.

Mayr, Ernst, and William B. Provine, editors. *The Evolutionary Synthesis: Perspectives on the Unification of Biology.* Cambridge: Harvard University Press, 1980.

Merrell, James H. *Into the American Woods: Negotiators on the Pennsylvania Frontier.* New York: Norton, 1999.

Mertz, David B., and David E. McCauley. "The domain of laboratory ecology." *Synthese* 43 (1980): 95–110.

Mitchell, Dora Otis. "A history of nature-study." *Nature-Study Review* 19 (1923): 258–74, 295–321.

Mitman, Gregg. *The State of Nature: Ecology, Community, and American Social Thought, 1900–1950.* Chicago: University of Chicago Press, 1992.

Mortimer, C. H. "An explorer of lakes." In *E. A. Birge: A Memoir,* by G. C. Sellery, pp. 165–211. Madison: University of Wisconsin Press, 1956.

The Naples Zoological Station and the Marine Biological Laboratory: One Hundred Years of Biology. Supplement to *Biological Bulletin* 168 (1985).

Newhall, Beaumont. *The History of Photography.* New York: Museum of Modern Art, 1982. Fifth edition, 1988.

Nicolson, Malcolm. "Henry Allan Gleason and the individualistic hypothesis: The structure of a botanist's career." *Botanical Review* 56 (1990): 91–161.

Olby, Robert. "The dimensions of scientific controversy: The biometric-Mendelian debate." *British Journal for the History of Science* 22 (1989): 299–320.

Outram, Dorinda. "New spaces in natural history." In *Cultures of Natural History,* edited by Nicholas Jardine, James A. Secord, and E. C. Spary, pp. 249–65. Cambridge: Cambridge University Press, 1996.

Owens, Larry. "Pure and sound government: Laboratories, gymnasia, and playing fields in nineteenth-century America." *Isis* 76 (1985): 182–94.

Pauly, Philip J. "The appearance of academic biology in late nineteenth-century America." *Journal of the History of Biology* 17 (1984): 369–97.

———. "Summer resort and scientific discipline: Woods Hole and the structure of American biology, 1882–1925." In *The American Development of Biology,* edited by Ronald Rainger, Keith R. Benson, and Jane Maienschein, pp. 121–50. Philadelphia: University of Pennsylvania Press, 1988.

———. "The development of high school biology: New York City, 1900–1925." *Isis* 82 (1991): 662–88.

Provine, William B. "Francis B. Sumner and the evolutionary synthesis." *Studies in the History of Biology* 3 (1979): 211–40.

———. *The Origins of Theoretical Population Genetics.* Chicago: University of Chicago Press, 1983.

———. *Sewall Wright and Evolutionary Biology.* Chicago: University of Chicago Press, 1986.

Rudwick, Martin. "Charles Darwin in London: The integration of public and private science." *Isis* 73 (1982): 186–206.

———. "Encounters with Adam, or at least, the hyaenas: Nineteenth-century representations of the deep past." In *History, Humanity, and Evolution,* edited by James R. Moore, pp. 231–52. New York: Cambridge University Press, 1989.

Schivelbusch, Wolfgang. *The Railway Journey: The Industrialization of Time and Space in the Nineteenth Century.* Berkeley: University of California Press, 1986.

Scott James C. *Seeing Like a State: How Certain Schemes to Improve the Human Condition Have Failed.* New Haven: Yale University Press, 1998.

Secord, Anne. "Science in the pub: Artisan botanists in early nineteenth-century Lancashire." *History of Science* 32 (1994): 269–315.

Shapin, Steven. "Pump and circumstance: Robert Boyle's literary technology." *Social Studies in Science* 14 (1984): 481–520.

———. "The house of experiment." *Isis* 79 (1988): 373–404.

———. *A Social History of Truth: Civility and Science in Seventeenth-Century England.* Chicago: University of Chicago Press, 1994.

———. "Trust, honesty, and the authority of science." In *Society's Choices: Social and Ethical Decision Making in Biomedicine,* edited by Ruth E. Bulger, Elizabeth M. Bobby, and Harvey V. Fineberg, pp. 388–408. Washington: National Academy Press, 1995.

Smith, Crosbie, and Jon Agar, editors. *Making Space for Science.* London: Macmillan, 1998.

Smocovitis, Vassiliki Betty. "Organizing evolution: Founding the Society for the Study of Evolution (1939–1950)." *Journal for the History of Biology* 27 (1994): 241–309.

Star, Susan Leigh, and James R. Griesemer. "Institutional ecology, 'translations,' and boundary objects: Amateurs and professionals in Berkeley's Museum of Vertebrate Zoology, 1907–39." *Social Studies of Science* 19 (1989): 387–420.

Sterling, Keir B. "Clinton Hart Merriam and the Development of American Mammalogy, 1872–1921." Ph.D. dissertation, Columbia University, 1973.

———. *Last of the Naturalists. The Career of C. Hart Merriam.* New York: Arno, 1977.

Theunissen, Bert. “Closing the door on Hugo de Vries’ Mendelism.” *Annals of Science* 51 (1994): 225–48.

Tobey, Ronald. *Saving the Prairies: The Life Cycle of the Founding School of American Plant Ecology, 1895–1955.* Berkeley: University of California Press, 1981.

Traweek, Sharon. *Beamtimes and Lifetimes: The World of High Energy Physicists.* Harvard University Press, 1988.

Weiner, Jonathan. *The Beak of the Finch: A Story of Evolution in Our Time.* New York: Knopf, 1994.

Westman, W. E., and R. K. Peet. “Robert H. Whittaker (1920–1980): The man and his work.” *Vegetatio* 48 (1982): 97–122.

White, Richard. *The Middle Ground: Indians, Empires, and Republics in the Great Lakes Region, 1650–1815.* New York: Cambridge University Press, 1991.

Whittaker, Charles R. *Frontiers of the Roman Empire: A Social and Economic Study.* Baltimore: Johns Hopkins University Press, 1994.

Winsor, Mary P. *Reading the Shape of Nature: Comparative Zoology at the Agassiz Museum.* Chicago: University of Chicago Press, 1991.

Wise, M. Norton, editor. *The Values of Precision.* Princeton: Princeton University Press, 1995.

INDEX

Abrams, LeRoy, 209
Adams, Charles C., 39, 126, 156, 157, 194, 202, 241; career of, 178, 187–89; on ecology, 26, 84–85, 87, 93, 110–11, 118; on lab-field relations, 4, 54–55, 60; on nature's experiments, 216–17, 218–19, 220, 238–39; on new natural history, 26, 31, 32, 34, 36, 57–58
Allen, Joel Asaph, 72, 73
Allen's law, 64, 225
American Museum of Natural History, 191, 258
Alpine Laboratory, Pike's Peak, Colorado, 109, 115–16, 164–65
Anderson, Edgar, 195, 200, 250, 253, 256; and Ernst Mayr, 265–66; and introgression, 265–74
Angell, James R., 45
Arthur, Joseph C., 33, 75, 77
Atmometer, 118–22. *See also* Instrument(s)
Autecology, 134, 296
Automobile, as instrument, 114

Bailey, Vernon, 242
Balds, mountain, 227
Banta, Arthur M., 140, 174, 199; and experimental evolution, 148–51
Bateson, William, 63, 70
Bay of Funday, marshes, 181, 182, 219
Bear Island, 276
Bellamy, Albert, 192–93
Bergman's rule, 225
Bessey, Charles E., 79, 100–1
Biak Island, 261
Biogenetic rule, 28, 241
Biogeography, 186, 224–25, 239–40, 257; and ecology, 74–76, 77; "laws," 63–64
Biological farm, 44–48

Biology: border, 290, 293–95, 306–8; educational reform in, 35–38; and new natural history, 27, 56
Biometry, 62–74, 90–91, 95, 268; and taxonomy, 63, 71–73, 93
Birge, Edward A., 109–10, 113–14
Blatchley, Willis, 29
Blue Mountains, Jamaica, 128
Bogs, as field sites, 230–32
Bonnier, Gaston, 167
Border, lab-field, 1–6, 269, 293–308; dangers to reputation in, 204–5; dynamics of, 60–62, 73–74, 87–88; and ecology, 75–76, 87–88, 95–96, 293–95, 306–8; experience of, 153–54, 156, 175–77, 193, 210–11, 292; experiment and, 136–37, 139, 164–65, 172–74, 190, 212–14, 217–18, 251; as frontier, 11, 14–19, 21, 23, 62, 95–96, 174, 294; genetics and, 89, 91–92, 96; identity in, 194–95, 199, 211, 307; imagined, 34–35, 21, 76, 97, 291, 294–95; instruments and, 97–99, 132–34; "labscapes" of, 41–55 (*see also* Biological farm; Field station; Marine station; Vivarium); and new natural history, 23–25, 34–35, 41, 56, 58–59, 290–92, 294; origins of, 3–4; as place, 4–6; as zone of interaction, 11–12, 14–22, 131–32, 171–72, 199. *See also* Biology, border; Border practice(s); Borders; Careers, border; Field; Fieldwork; Instrument(s); Laboratory; Quantification
Border practice(s), 21–22, 24–25; amplified field practice as, 252–56, 264–65, 274, 275, 279, 283, 284–85, 289–92; overview of, 295–308; practices of place as, 212, 218, 248–49, 271. *See also* Anderson, Edgar; Biometrics; Ecology, gradient; Lindeman, Raymond L.; Mayr, Ernst; Physiological life history; Practices of place; Transplantation; Whittaker, Robert H.
Borders, 12–14; as frontiers, 15–19; versus boundary, 11
Boundary object, 13
Boundary work, 12–13, 199, 215
"Brush Creek," 78, 81–82, 85, 95–96
Burr Oak Grove, Illinois, 227
Bumbus, Hermon C., 29, 138
Bureau of Fisheries Laboratory, Woods Hole, 140
Bush, Vannevar, 210

Camera. *See* Instrument(s), camera as
Careers: border, 175–89, 190–93, 279, 306, 307–8; in ecology, 177–90; in morphology, 30; movement in, 19, 137, 148–49, 151, 157, 174; recruitment of naturalists to, 37, 41, 85–88.
Carnegie, Andrew, 45, 176
Carnegie Institution of Washington, 46, 152, 165, 201, 206, 210; and William Tower, 145, 147, 148, 203, 204. *See also* Desert Laboratory
Castle, William E., 203
Cedar Creek Bog, Minnesota, 254–55, 277–82
Chapman, Frank M., 72, 124
China, Asian frontier, 15–19, 62, 92, 95–96
Clark, C. Walton, 82
Clausen, Jens C., 165, 168–72, 254, 296, 297
Clements, Edith, 196
Clements, Frederick E., 77, 85, 172, 284; advice to colleagues, from, 185–86, 189–90, 197; career of, 178–79, 207; defining ecology, 74, 75–76, 78, 80, 82, 92, 94, 174; and ecological experiment, 154–55, 288; and ecotones, 225–26; on exact methods, 86–87, 106; and experimental taxonomy, 164–65, 168; field course of, 195–96; and instruments, 108–9, 111, 114, 115–16, 118, 125, 126–27; lodgepole pine study of, 228–29; and nature's experiments, 216, 218, 240, 289; on panorama, 239–40; and phytometer, 122–24; and quadrat method, 100–2, 106, 108
Conklin, Edwin G., 26, 31, 32, 43, 45; and vivaria, 48–50
Cooper, William S., 104, 220–21, 224, 249–50, 284
Coronado Islands, California, 243
Coulter, John M., 27, 28, 37, 39, 79, 85, 233
Coville, Frederic V., 53
Cowles, Henry C., 77, 125, 299; career of, 178, 182–85; defining ecology, 78, 80, 86, 92; and experiment, 136, 154–55; and nature's experiments, 218–19, 250–51;

and physiographic ecology, 157–58, 232–35, 238, 240–41; on quadrat, 106–7
Crampton, Henry E., 139, 140, 174, 199; field research, 151–54
Credibility: of field versus laboratory work, 7–11. *See also* Ecology, lab-field tension in; Field, versus laboratory; Laboratory, as place versus field; Laboratory, separation of field from
Curtis, Winterton C., 37

Dall, William H., 30, 36
Darwin, Charles, 3, 28, 33, 175–76, 249
Davenport, Gertrude, 39
Davenport, Charles B., 31, 39, 91, 191–92, 199, 200; and Arthur Banta, 148–50, 199; and biological farm, 45–48; and biometrics, 64–65, 69, 72–73; and experimental evolution, 138–39; and taxonomists, 72–73; and William Tower, 145, 147, 202–4
Death Valley, California, 257–48
Desert Laboratory, Tucson, Arizona, 53, 93, 128–29, 131, 147; late work and closing, 195, 205–10
Dice, Lee, 202
Dobzhansky, Theodosius, xiii–xiv, 96, 257, 264–65, 296–98
Douglas Lake, Michigan, 53
"Driftwood Lake," 78, 82, 86–87, 96. *See also* "Brush Creek"
Dunes, 183, 224. *See also* Lake Michigan
Dunn, Leslie C., 265
Dwight, Jonathan, 72

Earle, Franklin S., 27
Ecology: American versus European, 76–77; and biogeography, 74–76; and botany, 74–75; camera in, 125–26; careers in, 177–90; "car-window," 184; a classifying science, 107–8; community concept in, 283–86; crisis in, 78–80, 179–81; definitions of, 74–78; ecosystem, 274–83, 299–303; experiment in, 154–63; gradient, 283–90; lab-field tension in, 21, 85–87, 95, 106, 172–74, 179–87, 190, 198, 205–9; natural history style of, 54, 78, 81–82, 83–86, 95, 126–27, 184, 189–90, 276–79, 282–83; naturalists and, 74, 77–88; and new natural history, 25, 26, 56, 58, 74; physiographic, 158, 230–38, 239–41; physiologists and, 88, 92–94 (*see also* Spoehr, Herman A.); as physiology of the field, 74–77, 92–94, 162, 207–9, 283, 299, 307; reaction against theory in, 254, 283–84; recruitment to, 21, 77–88, 189–90; synecology, 134. *See also* Autecology; Instrument(s), in ecology; Panorama; Physiological life history; Quadrat; Succession
Ecosystem ecology. *See* Ecology, ecosystem
Ecotype, 163, 169, 170
Eigenmann, Carl H., 82, 140, 149; and biometry, 65–66; and cave project, 191–92
Elrod, Morton J., 54
Elton, Charles, 255, 276–77, 282
Emerson, Alfred E., 284, 303–4
Eniwetok Atoll, Marshall Islands, 300
Errington, Paul L., 194–95, 253, 255, 291
Estes Park, Colorado, 228–29
Evermann, Barton W., 36–37
Evolution: experimental, 44–45, 138–45, 190–93; mechanisms of, 146, 149–51, 225, 242–43, 246–47, 248. *See also* Biometry; Genetics; Mayr, Ernst; Speciation
Evolutionary synthesis, 303
Experiment, 212–13; 307; as artificially simple, 6–7, 18; as better than field methods, 1, 2, 195, 200, 308; as credible, 7–10, 200, 215; as camouflage, 198–99, 204; and ecology, 85–87, 94, 154–63, 176–77, 179–80, 180–85, 211; field, 99, 117–18, 135–37, 153–56, 163–64, 170–74, 207–9; limitations of, 162–63, 215; in "natural" instruments, 300–1; on populations, 70, 74, 76, 91, 154, 254, 296 (*see also* Populations). *See also* Border practice(s); Evolution, experimental; Experimental taxonomy; Instrument(s), Nature's experiments; Physiological life history; Practices of place; Transplantation
Experimental taxonomy, 163–72, 298

Fassett, Norman C., 267
Field, 88–89, 91, 92–93, 101; as by-product of lab revolution, 3–4, 18, 98–99; and identity, 9–11, 194–99; in the lab-field ecotone, 11–12, 60–62, 95–96, 174, 215,

Field (*continued*)
218; versus laboratory, 1–6, 135–36, 162–63, 174, 200; as place, 6–11. *See also* Border, lab-field; Experiment; Laboratory; Place(s); Practices of place; New natural history
Field practice. *See* Biogeography; Biometry; Border practice(s); Ecology, gradient; Ecology, physiographic; Evolution, experimental; Experiment; Experimental taxonomy; Fieldwork; Instrument(s); Mass collecting; Measurement, exact; Observation; Panorama; Physiological life history; Place(s); Populations; Practices of place; Quadrat; Quantification; Transplantation
Field station, 51–55. *See also* Alpine Laboratory; Biological farm; Desert Laboratory; Marine station; Station for Experimental Evolution
Fieldwork, 18, 26, 112, 130, 254, 301; in ecology, 85–86; in field and marine stations, 43–44; and heroism, 10–11; and identity, 194–95, 211; and new natural history, 32–34; and place, 212–13, 255–56; as play, 194–95; social roles for performing, 179–89, 190; teaching, 37, 39–40, 40–41, 57, 179. *See also* Border practice(s); Practices of place
Fischer, R. A., 265
Flahault, Charles, 219
Forbes, Stephen A., 30, 33, 39, 54, 82, 136, 188
Foerst, J. P., 114
Fuller, George D., 183

Gager, Stuart S., 82
Galapagos Islands, 218
Ganong, William F., 27, 39, 85, 219, 299; career of, 178, 181–82; and crisis in ecology, 77–80, 82, 86; and experiment, 155–56
Gates, Frank C., 178, 179
Genetics, 56, 69, 91, 256, 257, 295–96; and evolution, 88–92, 199–203; in experimental taxonomy, 170–71; of populations, 90, 297–98; and taxonomy, 136. *See also* Mutations
Genotype-phenotype concept, 89–92, 96, 201
Glacier Bay, Alaska, 104, 221, 224
Gleason, Henry A., 53–54, 77, 109, 177; career of, 80, 178, 186–87; as critic of community concept, 154, 284, 285; on observation and experiment, 156, 198; on physiographic ecology, 240–41; and quadrat, 105–8; and relict method, 226–27
Gloger's rule, 64, 225
Goldsmith, Glenn W., 115–16
Gradient ecology, 283–90
Graphical methods, and human eye, 268
Great Smokey Mountains, 285–89
Gregory, William K., 216
Grinnell, Joseph, 143, 144, 196–97, 220, 303; and Francis B. Sumner, 242–43; and nature's experiments, 247–48
Gulick, John T., 151
Gull Lake, Minnesota, 51–52

Hall, Harvey M., 207, 209; and experimental taxonomy, 163–71
Harper, Roland, M., 155, 179, 184, 225; and quadrat, 105–6
Harris, J. Arthur, 70, 90, 139
Harshberger, John W., 77, 83
Harvard University, 36, 49
Hedgcock, George, 154
Hiesey, William M., 165, 170
High schools: and new natural history, 35–41
Hodson, Alexander C., 277
Holmes, Samuel J., 136
Holt, William, 40
Humboldt, Alexander von, 98, 225
Humboldt Bay, California, 144, 244–45
Hutchinson, George Evelyn, 275, 277, 278–79
Huxley, Julian, 303
Huxley, Thomas Henry, 3, 36
Hybrid habitat, 266
Hybrids. *See* Anderson, Edgar
Hygrometer, 111–12
Hubbs, Carl L., 90, 192, 204–5; as border practitioner, 253, 254, 256, 268, 270, 303–4

Identity: in the field, 6–7, 9–11, 194–99, 211
Immutable mobile, 13
Instrument(s), 21, 29, 142–43, 307; automobile as, 114; camera as, 124–27; as carriers of lab culture, 97–99, 132–34; deployment

of, in the field, 127–32; in ecology, 85–87, 94, 109, 111, 120, 122, 186; in the field, 108–18; human eye as, 102, 107; natural places as, 300–2. *See also* Atmometer; Hygrometer; Measurement; Microscope; "Microtome-mania"; Photometer; Phytometer; Thermometer
Introgression, 266, 269–71; and place, 273–74
Islands: South Pacific, 257–61, 263–64. *See also* Bear Island; Biak Island; Coronado Islands; Eniwetok Atoll; Galapagos Islands; Isle Royale; Rennell Island; Santa Rosa Island; Solomon Islands
Isle Royale, 224, 241

Jennings, Herbert S., 23–24, 27, 90
Johannsen, Wilhelm, 89
Johnson, Roswell H., 65, 68, 70, 71, 140; career of, 191, 192
Jordan's law, 64
Johns Hopkins University, 36
Juday, Chancey, 109, 113–14

Kearney, Thomas H., 76, 82, 178, 179
Keck, David D., 165, 170
Kendeigh, Charles S., 296
Kellogg, Vernon L., 66–67, 70, 90, 140
King, Charlotte, 83
Kinsey, Alfred C., 191, 193, 303, 304
Kofoid, Charles A., 33, 34, 39–40, 54

Laboratory, 23, 36, 39–40, 86–87; boat-lab ("Megalops"), 67; as natural, 3–4, 94; as place versus field, 6–12, 23–24, 60–62, 94, 112, 135, 194, 212–13; separation of field from, 88–89, 91–93, 98–99; Yellowstone as a, 214. *See also* Biological farm; Border, lab-field; Experiment(s); Field station; Frontier; Instrument(s); Marine station; New natural history; Placelessness; Precision; Vivarium
Lake Michigan: lakeshore ravines, 236–38; dunes, 157–58, 182–83, 218, 232–35, 238, 250–51; fossil beaches, 235–36
Life history. *See* Physiological life history
Life zones, 225. *See also* Merriam, C. Hart
Limnology, 109, 111. *See also* Lindeman, Raymond L.
Lindeman, Raymond L., 253, 254–55, 275–83; Charles Elton's influence on, 276–77
Linville, Henry R., 40, 195
Livingston, Burton E., 78–79, 93–94, 115, 127, 129, 130; and atmometer, 118–21
Loeb, Jacques, 27
Lutz, Frank E., 65, 69, 70, 74, 139, 140, 191
Lutz, Harold J., 281–82

MacDougal, Daniel T., 53, 88, 93, 206; and Salton Sea, 221–24
MacDowell, E. Carlton, 201
MacMillan, Conway, 24, 77, 125, 216, 239; work on bogs of, 231–32; career of, 177–78, 179
Marine Biological Laboratory, Woods Hole, Massachusetts, 43–44, 46, 68, 140
Marine station, 42–44. *See also* Marine Biological Laboratory; Scripps Institution
Mass collecting, 253, 262–64, 267
Mayer, Alfred G., 69–70, 151, 153
Mayfield's Cave, Indiana, 149, 191
Mayr, Ernst, 253–54, 255, 269, 276, 297, 303, 307; and geographical speciation, 257–65
Measurement, exact, 133–34, 186, 268; in the field, 108–18, 122, 127–32; versus good-enough, 114–16; symbolic value of, 98–99. *See also* Biometry; Ecology, ecosystem; Instrument(s); Physiological life history; Shreve, Forrest
Mendel, Gregor, 204
Meteorology, 98, 110–11, 117, 119, 133
Mexican plateau, 145–48
Merriam, C. Hart, 73, 194; on specialists versus generalists, 36, 37, 73; work of, 118, 129, 143, 225, 242, 243
Microscope: as magic wand, 29–30
Microscopy, 28–30, 36, 40, 195. *See also* Morphology
"Microtome-mania," 29, 57
Mississippi River delta, 271–73
Missouri Botanical Garden: as hybrid habitat, 269
Mohave Desert, 144, 206, 242, 304. *See also* Death Valley
Montgomery, Thomas H., 26, 31, 32, 43
Morgan, Thomas H., 43–44, 88, 91, 138, 199, 256

Morphology, 24–25, 42, 44, 55–56; and experimental evolution, 138–40; as fast-track career, 30; in curricula, 35–37, 40, 57; and new natural history, 27–32; "revolt from," 55–56
Mount Robson, British Columbia, 220
Muir, John, 214, 221
Museum of Vertebrate Zoology, 143
Mutations: and evolution, 88–89, 96, 136

Nachtrieb, Henry F., 31, 32, 67
Naples Zoological Station, 43
Natural history: in curricula, 39–41; literary genres of, 196–98. *See also* Naturalists, New natural history
Naturalists: 2, 4–5, 26–27, 37, 54, 124, 126, 176; and ecology, 77, 82–85, 87, 96, 182, 189, 209–10; and new natural history, 27, 31–33. *See also* New natural history
Nature, flux in, 249–51
Nature's experiments, 22, 258; as a border practice, 214–18; as equivalent to lab experiments, 238, 240–41, 243; in process, 218–24, 247–48; results of work on, 248–51; and succession, 227–29, 230–35. *See also* Practices of place
Nature study, 57
Nelson, Aven, 194
New Guinea, 260–62
New natural history, 24–35, 58–59, 60–62, 135–36, 290–92, 294; consequences of, 58–59; decline of movement, 55–58; as ecology, 54; and educational reform, 35–41, 56–57. *See also* Ecology, Natural history
New York Botanical Garden, 178, 187
New York State College of Forestry, 178, 188
Nice, Margaret Morse, 253
Nichols, George E., 77, 179

Observation, 107–8; superior to experiment, 135–37, 155–56, 186; with the power of experiment, 215, 238; inferior to experiment, 3, 7, 85, 87, 172–74, 308
Odum, Eugene P., 292, 296, 299, 307
Odum, Howard Thomas, 296, 299–301
Ohio River: ravines, 218–19
Osgood, Wilfred H., 242

Packard, Alpheus, 29, 30
Pantin, Carl F. A., 215
Panorama, 238–41
Park, Orlando, 284
Pearl, Raymond, 27, 67, 70, 74
Pearson, Karl, 63, 70
Peromyscus, 141–44, 242–47
Phillips, John, 284
Photometer, 113–14, 115, 117
Physiological life history, 156–61, 164, 170–71
Physiology, 27, 38, 56–57, 86, 134, 257, 295–96; as a field science, 92–94, 205–10, 211. *See also* Ecology, as physiology of the field; Ecology, physiologists and; Physiological life history; Plant physiology; Shreve, Forrest
Phytometer, 122–24
Place(s): in border practice, 255–56; and belief, 288–89; changing, as research sites, 218–24; field and laboratory as, 6–11; humanized, as field sites, 219–20; techniques of reading, 224–30, 238–41; natural, as instruments, 300–2; versus "space," 5–6. *See also* Field; Introgression; Placelessness; Practices of place
Placelessness: of laboratories, 7–11, 213
Place mode, biometric, 67, 72–73, 145, 170, 308
Plant physiology: and ecology, 74–77, 88, 92–94; at Desert Laboratory, 207–10. *See also* Ecology, physiologists and; Ecology, as physiology of the field
Plant sociology, 76, 134
Polygon of variation, 63
Populations: in border practice, 253–54, 256; and experimental evolution, 137–38; and genetics, 88–91; and mass collecting, 264. *See also* Biometry; Ecology, gradient; Experimental taxonomy; Introgression; Mayr, Ernst; Populations, natural
Populations, natural, 70, 76, 154; and biometry, 71, 74, 87, 95; and genetics, 264–65, 296–98; as unit of investigation, 137, 172, 253–54, 264
Pound, Roscoe, 75; and quadrat, 100–2
Powers, Joseph H., 138

Practice, visual. *See* Panorama
Practices of place: allowing reconstruction of (ideal) history, 213, 234–37, 240; equivalent to lab set-ups, 212–13, 228; evolution and, 241–49, 260; involving comparison, 217, 227–28; reading spatial elements, zones, and boundaries as, 212, 226, 230; as relying on the visual, 238–41; succession and, 219–24, 230–335; tools of traditional natural history as, 22, 214, 221–22, 224–25, 251; in unremarkable places, 255–56, 271–74. *See also* Border practice(s); Experiment; Field; Fieldwork; Nature's experiments; Panorama; Place(s)
Prairie groves, 227
Prairie peninsula, 226
Precision. *See* Measurement
Princeton University: vivarium, 48–50
Psychrometer (hygrometer), 111–12

Quadrat, 100–8; versus human eye, 107–8; invention of, 100–2; walking, 105
Quantification, 73, 98–99, 100–1, 132–34: in fieldwork, 98, 101, 132–34, 138. *See also* Biometry; Border practice(s), amplified field practice as; Measurement; Quadrat

Reed, Howard S., 78, 85
Reighard, Jacob E., 26, 31, 32, 37, 67, 139
Relict method, 226–27
Rennell Island, 261
Riley, Charles V., 30, 31
Ritter, William E., 23–24, 29, 30, 141; and Joseph Grinnell, 196–97
Rivers: as image of nature's flux, 249–50
River valleys: as research sites, 219, 238
Roman frontier: compared to laboratory-field border, 15–19, 174, 294–95
Romanes, G. J., 44–45
Ruthven, Alexander G., 178, 193

Sachs, Julius, 3
Salton Sea, California, 221–24
San Francisco Mountains, Arizona, 129, 225, 243–44
Santa Catalina Mountains, Arizona, 128–32, 155
Santa Lucia Mountains, California, 128–29
Santa Rosa Island, Florida, 245–47
Schimper, A. F. W., 74–75
Schmidt, Karl P., 303–4
Scripps Institution for Oceanography, 141, 201
Shantz, Homer L., 178, 179
Shelford, Victor E., 48, 105, 213, 284; early views on lab versus field of, 51, 55, 93; ecological survey, 161–62; lab-field swings of, 138, 171, 191; on limits of experiment, 162–63; and physiological life history, 156–61; and publishing, 179, 197; work on successions of, 235–38
Shreve, Edith, 207
Shreve, Forrest, 155, 179, 195, 211; desert surveys of, 209–10, 304–6; and field measurement, 127–32; and field physiology, 205–10
Sierra Nevada Mountains: field stations in, 165, 166, 169, 173, 298
Silver Springs, Florida, 300–2
Snow, Laetitia, 224
Solomon Islands, 261–62
Solution ponds, 227
Sonoran Desert, 205, 209, 304–6
Spalding, Volney M., 24, 29, 30, 33, 76, 198
Speciation, 218, 241, 303; geographical, 257–63
Spoehr, Herman A., 116, 195, 211; and Desert Laboratory, 206–10
Station for Experimental Evolution, Cold Spring Harbor, New York, 47–48, 68, 191, 192; experimental cave, 149
Stresemann, Erwin, 257, 258, 259
Strong, Reuben M., 65, 68, 70, 71–72, 74
Succession, 218–20; in ecosystem ecology, 281–82; lodgepole pine, 228–30; and physiographic ecology, 230–38, 249–50. *See also* Cowles, Henry A.; Physiological life history
Sumner, Francis B., 67, 140, 205; and geneticists, 199–202; and practices of place, 215–16, 241–47, 248–49; *Peromyscus* project of, 140–45, 201–2, 241–48
Synecology, 134
Systematics: new, 303–4. *See also* Experimental taxonomy; Taxonomy

Tansley, Arthur G., 195
Taxonomy: and biometrics, 96; as border practice, 267–68; in ecology, 82–83, 84, 102–3, 107, 109; evolutionary, 263–64; experimental, 163–72, 298; and genetics, 192–93, 202, 205, 254, 256, 303; in schools, 40, 57–58. *See also* Anderson, Edgar; Mayr, Ernst
Texas: coastal prairie, 189–90
Tharp, Benjamin C., 189–90
Thermometer, 111, 117
Thienemann, August, 277
Thornber, John, 154
Tower, William L., 65, 121, 140, 174, 249; reputation of, 202–4; *Leptinotarsa* project of, 145–48
Trading zone, 13–14
Transeau, Edgar N., 77, 178
Transplantation method, 142, 163–70
Turesson, Goete, 167–68
Turkey Lake, Indiana, 65–67
Turrill, W. B., 267

University of Chicago, 147; biological farm, 46–47
University of Illinois: vivarium, 48, 50–51, 158; field station, 54
University of Pennsylvania: vivarium, 48–49
University of Wisconsin: vivarium, 48

Variables, compound, 116–18. *See also* Atmometer, Phytometer
Variation, determinate, 64, 67, 149, 249; failure to find, 70, 148, 153; testing for, 138, 146, 148. *See also* Biometry
Varigny, Henry de, 45
Vestal, Arthur G., 106–7, 155–56, 197–98; career of, 178, 185–86
Vivaria, 48–51, 158
Vries, Hugo de, 88

Ward, Henry B., 31, 37, 39, 188
Warming, Eugenius, 74–75
Watson, James B., 307
Weaver, John E., 86–87, 130, 284
Weediness: as cultural character, 8–9
Weldon, W. F. R., 63, 70
Wells, Bertram W., 189, 227
Welty, Carl, 194–95
Wheeler, William M., 4–5, 27, 31, 43
Whitman, Charles O., 26, 31, 33; and biological farms, 45–47
Whitney South Seas Expeditions, 258, 259, 262
Whittaker, Robert H., 354; and gradient ecology, 283–90
Wilson, Edmund B., 4, 25–26, 28–29, 31, 32, 307
Wilson, Henry V., 68
Woods Hole: Bureau of Fisheries Laboratory, 140; Marine Biological Laboratory, 43–44, 46, 68, 140
Woodson, Robert E., 270–71
Woodward, Robert S., 203
Wright, Sewall, 265

Yerkes, Robert M., 65, 69, 74